Behold the
Black Caiman

Behold the Black Caiman

A Chronicle of Ayoreo Life

LUCAS BESSIRE

The University of Chicago Press   Chicago and London

LUCAS BESSIRE is assistant professor of anthropology at the
University of Oklahoma. He is the producer and director of the
documentary film *From Honey to Ashes*.

The University of Chicago Press, Chicago 60637
The University of Chicago Press, Ltd., London
© 2014 by Lucas Bessire
All rights reserved. Published 2014.
Printed in the United States of America

23 22 21 20 19 18 17 16 15 14   1 2 3 4 5

ISBN-13: 978-0-226-14089-6 (cloth)
ISBN-13: 978-0-226-17557-7 (paper)
ISBN-13: 978-0-226-17560-7 (e-book)
DOI: 10.7208/chicago/9780226175607.001.0001

Library of Congress Cataloging-in-Publication Data

Bessire, Lucas, author.
   Behold the black caiman : a chronicle of Ayoreo life / Lucas Bessire.
      pages cm
   Includes bibliographical references and index.
   ISBN 978-0-226-14089-6 (cloth : alk. paper) — ISBN 978-0-226-17557-7
(pbk. : alk. paper) — ISBN 978-0-226-17560-7 (e-book)   1. Moro Indians—
Social conditions.   2. Moro Indians—Economic conditions.   I. Title.
   F2679.2.M6B488 2014
   989.2004'98—dc23

                                                                2014008904

To Harald Prins

Contents

List of Illustrations

Prefatory Note

This book chronicles my pursuit of the Black Caiman: a shadowy form birthed by the terror of savagery, by the world-ending violence relentlessly stalking certain unruly Indigenous populations and the New Worlds they have tentatively remade, by the social afterlife of ethnographic categories, by broken collections of blood and tradition and souls, by renewed desires to protect the last primitives and the ontological alterity they purportedly embody, by the spirit-filled fragments of forest and history and myth that form the South American tropics, by the unsettling kinds of Native life now engendered by these accreted dynamics and contradictory knowledges of them, by the ironies of those points where fledgling anthropological and Indigenous projects of becoming may split apart and converge again. Tracking the Black Caiman through the everyday actualities and precarious lives of a small group of people in South America's Gran Chaco known as the Ayoreo Indians is the aim and method of this account—its beginning and its end.

My purpose is to upend commonsense ways by which the humanity of people like Ayoreo has become an apparent object for study and governance and gesture to alternatives that are at once epistemic and political. It is the pursuit of an anthropology that is capable of addressing the turbulence of the contemporary not by a search for the primitive and the future it sustains but by accounting for how such searches blur into the conditions of subjective possibility and in doing so exert real and contradictory force on the ranked limits of life itself. Indeed, the intellectual habits and taken-for-granted origins created around Ayoreo life are under investigation in what follows. Early on, I learned that any efforts to describe Ayoreo-speaking people as subjects "pinned like butterflies in a glass case"[1] result in words that are too tense and spaces that are too large. Instead, I have tried to write close to contents to unsettle ideal forms and unmask the inversions they propagate. This means there are no clear answers: the delirium of ethnographic experience is the central aesthetic and interpretive guide.

This book chronicles the nonlinear emergence of Ayoreo life as a contested object as well as the ways Ayoreo people self-objectify these objectifications to their own unexpected ends, both vital and deadly. It is offered in the slightest hope that a more nuanced understanding of the ways certain populations come to be considered victims, targets, residues, or otherwise subhuman is a necessary and useful contribution for furthering their future causes, whatever forms these may take. This book is intended, among other things, as lament, protest, doomed expiation, and critical analysis of the preemptive foreclosure of human becoming.

This ethnographic project and I have grown up together. It evolved through repeated returns over the course of forty-two months of fieldwork carried out in Bolivia and Paraguay between 2001 and 2013. Its focus has sharply changed since I began traveling to the Chaco as a twenty-one-year-old. From the outset, I was an active participant in representing Ayoreo humanity to fellow outsiders, most notably in two documentary films.[2] Yet I soon came to feel that there was something profane about anthropology as commonly practiced in the Chaco. And Ayoreo-speaking people wouldn't let me forget it. Unsettled by the process of making my second video during the aftermath of a 2004 "contact," I resolved that a collaborative project was the only option for the immersive research that I began in 2006 among Totobiegosode—Ayoreo in northern Paraguay.

This elementary recognition turned out to be quite difficult to realize. Eventually I was permitted to stay with them under certain conditions, not least because I gladly offered myself as an immediate source of food, medicine, money, and most importantly, transport. At the time, the relatively remote communities with which I worked did not have access to a vehicle. The children named my old army ambulance *Jochekai*, the Giant Armadillo, and quickly bedecked its rearview mirror with gifts. My presence in the Chaco was as much political and economic as personal and intellectual. Invited to work as an "advisor" to a recently formed tribal organization, I also became a wedge between the Totobiegosode communities and an NGO that controlled the material sources of their daily existence. Early on, it was apparent that there were stark differences of opinion all around. Taking an active role in these daily negotiations gave me standing in the communities; it also made my relations with contracted experts strained and often conflictive. Totobiegosode leaders enlisted me alternately as ally, agitator, and foil to leverage slight openings in a constrained social space. Over the years, these relations thinned and thickened, dried out and grew up. In the process we all assumed forms impossible to foresee. This book, then, is equally a chronicle of lives shared and a genealogy of why this was only partially so.

The common charge leveled against anthropologists in the Chaco is that they do nothing concrete; they take what they need and leave no trace of their passage. What follows is my attempt to respond to this charge.

A New World

The savages of the Amazonian forest are sensitive and powerless victims, pathetic creatures caught in the toils of the mechanized civilization, and I can resign myself to understanding the fate which is destroying them. But I refuse to be the dupe of a kind of magic which is still more feeble than their own and which brandishes before an eager public albums of colored photographs. . . . Not content with having eliminated savage life, and unaware even of having done so, it feels the need feverishly to appease the nostalgic cannibalism of history with the shadows of those that history has already destroyed. Can it be that I, the elderly predecessor of those scourers of the jungle, am the only one to have brought back nothing but a handful of ashes? Is mine the only voice to bear witness to the impossibility of escapism?   CLAUDE LEVI-STRAUSS

And the truth of the matter is that all this was a very slight thing: just one more page of the monotonous census—with more and more precise dates, places and figures—recording the disappearance of the last Indian tribes. . . . The whole enterprise that began in the fifteenth century is now coming to an end; an entire continent will soon be rid of its first inhabitants, and this part of the globe will truly be able to proclaim itself a "New World."   PIERRE CLASTRES

I'd like to think that the woman I will call Tié is my friend, even though there is no word in her language for the absent kind of friendship I can offer.[1]

She was one of a small band of Ayoreo-speaking people who emerged from the dwindling forests of northern Paraguay in March 2004, fleeing ranchers' bulldozers and fearing for their lives. These seventeen people called themselves Areguede'urasade ("band of Areguede") and formed part of the Totobiegosode ("People of the Place where the Collared Peccaries Ate Their Gardens"),[2] the southernmost village confederation of the Ayoreode ("Human Beings").

1

Along with two other small bands that still roam the dense thickets of the Bolivia/Paraguay borderlands, they were the last of the forest-dwelling Ayoreo.

Startling photographs of these brown-skinned people made headlines around the world that spring. Experts jockeyed to declare this one of the final first contacts with isolated Indians. At first everyone wanted in, including me. When the first tremors of the event reached me, I was a second-year anthropology graduate student at New York University. I thought I understood something of Ayoreo people and the Chaco, as I'd already spent fourteen months living among and collaborating with northern Ayoreo-speaking people as a Fulbright scholar in Bolivia three years before. When I heard they'd come out, I couldn't sleep. I rushed through coursework and film training, and by July I was headed south—as if to bear witness.

Reaching the scene of contact meant first passing thousands of other Ayoreo-speaking people struggling to survive in the stark mission stations and makeshift labor camps that lined dirt roads for miles north of the Mennonite colonies. These "ex-primitives," as I started to think of them, were some of the descendants of the nomadic Ayoreo confederacies that had successfully fought off outsiders for centuries. They too had once gained international fame as uncontacted primitives and the über-savages of the Chaco frontier. In 2004, few people cared. Their radical alterity presumably degraded by a series of contacts with North American Protestant missionaries beginning in 1947, the approximately six thousand Ayoreo-speaking people inhabited some thirty-eight settlements on both sides of the Bolivia/Paraguay border. They were routinely treated as less than human: a labor reserve for sex work or charcoal production, a threat to civic hygiene, a source of myths and dissertations and marketable bodily substances, souls already won by missionaries. "We try to avoid the missions," an NGO employee told me as we drove past the camps during my first trip to the scene of the latest contact. "There are some tensions. You never know what might happen."

Like many others, I initially mistook the Totobiegosode holdouts as ideal antidotes to such Ayoreo marginality. They were iconic in the region as "Stone-Age primitives" and "the last great hope for Ayoreo cultural revitalization."[3] Up close, however, the details of their lives defied any primitivist fantasies I had brought with me. The latest arrivals wore sandals made of tire scrap instead of tapir skin. They tipped their arrows with flattened forks or steak knives and carried water in a plastic grease bucket made in Indiana. Much like Ishi a century before, they had developed a way of life centered around the practical problem of concealment from the alien beings—trucks, cattle, bulldozers—they thought were relentlessly pursuing them. They inhabited the literal margins of industrial agriculture. They were often forced to camp in the fifteen-meter-wide strips of brush left as windbreaks around vast cattle pastures. They went long periods communicating only with whistled sounds; even the children were whisper quiet. If they saw a boot print or heard a chainsaw, they would flee far and fast, leaving everything behind. They were not the bearers of a vestigial purity but pieced together the means of survival from detritus. It soon became clear to me that they were by no means isolated. Rather, they had structured precarious lives around the daily logistics of eluding starvation, capture, and a death foretold.

Tié was the youngest woman of the group, around seventeen years old, all unruly hair and pale skin and almond-shaped eyes. Two wild parrots followed her out of the forest and slept near her tarp-covered shack. The other Totobiegosode said she was one of the *uitaque* dreamers who could see the future. At the time, I lived in a tent on the margins of the camp. On good days she would invite me to sit with her and her husband. She spoke, in her halting voice, about the time they all called *nanique*, "before." She told me one of the most arresting stories I heard in the Chaco, one I'm still trying to understand. "We saw tracks. *Cojñone*, Strangers. Where? It was hot. We ran far. Faaar. Swollen tongues. We cried. Crawling low. A water tank. It was full. One *Cojñoi*. Fat. A red shirt. Blood in the water. Trembling underneath. We ran."

Contact meant the Areguede'urasade settled near a group of their Totobiegosode relatives, who themselves had been hunted down and captured by enemy Ayoreo and North American Protestant missionaries in 1986. Since then, these relatives had lived as subordinates to their captors on the New Tribes Mission of Campo Loro. Assisted by a local NGO called Grupo de Apoyo para los Totobiegosode (GAT), the Totobiegosode support group, these Totobiegosode sought to form their own community some one hundred kilometers northeast of the mission, on legally titled land.[4] Two Totobiegosode men from this mission group—Dejai and Cuteri—were surveying a new village site when the contact occurred. The NGO heralded it as a historically unique event. It involved close relatives who personally knew one another, on ancestral Totobiegosode territory, without the presence of missionaries. It was a moment ripe with possibility, but it was by no means obvious who was in control.

What I witnessed—and what I filmed—was profoundly unsettling. When I arrived in July 2004, the Areguede'urasade lived in a separate camp a mile away from the group of their relatives. By October, they decided to form a single village. I tried to tell Siquei, the leader of the

Areguede'urasade, not to go. He ignored my protest entirely. Whether it was due to my then-rudimentary Ayoreo language skills or because he had other ideas I still do not know.

It was a hot day, and the dust from their passage hung in the air for an hour after they had passed. They laid their heavy bundles on the dusty ground before the hearth of Dejai, the man who had contacted them. Many of them knew Dejai from a shared adolescence in the forest, and the rivalry between he and Siquei dated to those days. Back then, they competed for status in the hunt and for the affections of the vibrant young girl Bujote, who later became Siquei's first wife. In 2004, they were competing for something else.

Dejai wasted no time setting down the rules. "Here, you should not follow the traditional taboos," I remember him saying that afternoon in an iron tone, staring steadily at Siquei. "We only listen to the Word of God."

Although none of us quite knew it at the time, that day marked the end of the Areguede'urasade's existence as an independent group. Tié and the others were quickly divided among their captors and reduced to a servile status. They were forced to haul water, chop wood, and provide forest foods to the others. They were ridiculed for being dirty and backward and ignorant, for their reactions to the clothing, soap, and foods of the *Cojñone* Strangers. Dejai himself cut their long hair with a pair of crude sheers. He told them that their painstakingly crafted belongings were *basigode*, or trash, and then sold them to the many outsiders eager to collect such items. He took their clay smoking pipes and beads and feathers. I remember sitting with Asôre and watching in silence while Dejai slowly broke the *catojnai* gourd that contained the potent beeswax used in the annual ritual of world renewal.

Such tensions were invisible to most outsiders with their preconceived images of contact. For advocacy NGOs, the Areguede'urasade were a key resource for fund-raising. Donations aimed at cultural preservation poured into the settlement. It was an economic windfall for Dejai, a ruthless gatekeeper. At the same time, the Areguede'urasade were another kind of capital for the New Tribes missionary named Bobby. By the time I left Chaidi in November 2004, he was visiting every other day. Tié and the others were taught that contact was a radical rupture that implied conversion to Christianity, the abandonment of most past practices, and a fundamental transformation of the form and contents of moral humanity. According to their relatives, Tié and the others were no longer Areguede'urasade. They had become *Ichadie*, "the New People."

Yet life in the Chaco always exceeded any image of it. When I discovered that many of the images I so eagerly recorded had begun to disintegrate—the tapes ruined by the fine dust—it seemed an apt reflection of reality. Whatever the image, Tié refused it. I returned in June 2005 to find that pressure on the New People had intensified. The elder matriarch Ebedai'date starved herself to death. A fervent millenarian religious movement gripped the village, and they held bible study every day. Siquei was morose and withdrawn and spent most of his days alone in his house. He had developed a severe stutter as if words and their shapes didn't quite fit. And Tié inexplicably went deaf. She said she just couldn't hear. Not the wind, the sermons, or the insults. Some days she would not eat. The Paraguayan nurse gave her a medical exam but could not find anything wrong. So the NGO workers gave her antidepressants. She swallowed the pills like an invalid or a child. The following year, she gave birth to a son. She named him Ore Isai, "the captive."

A Death Foretold

What does it mean to write or read yet another ethnographic account of the seeming destruction of yet another small group of South American Indians? What kind of humanity is possible for anyone once the spectacular upheavals of a twenty-first-century "first contact" supposedly subside?

For many earlier anthropologists, like Claude Lévi-Strauss and Pierre Clastres, the answer was clear. So-called primitive society was imagined to be intrinsically opposed to Western civilization. It was, in Clastres's famous writings, a "society against the state." Clastres argued that "primitive man" refused the political and economic forms required for modern statecraft. For Clastres, the modern state was defined by its centralizing drive to erase and homogenize difference while primitive society was outward-facing and deeply antiauthoritarian. Where the modern state was "the One," primitive society was the unassumable "Multiple" that existed beyond it.[5] This primitive multiplicity, for Clastres, was "the conceptual embodiment of the thesis that another world was possible."[6] Yet Clastres and many of his contemporaries imagined this possibility as an ever retreating horizon. Primitive societies, he was convinced, were doomed to disappear before the violent onslaught of capitalist modernity.

Contact was supposed to initiate such disappearances. Because it meant the inevitable loss of this multiplicity, contact made the South American tropics *tristes*, a zone of mourning and what Renato Rosaldo calls "imperialist nostalgia."[7] In the words of Alfred Métraux, the great ethnographer of the Chaco, "For us to be able to study a primitive society, it must already be starting to disintegrate."[8] Such sentiments oriented ethnography. Clastres argued that what was needed was an anthropology capable of interrupting the surrender to singularity by taking seriously the radical otherness of primitive society and experience. It was a political anthropology based on a search for the primitive.[9]

If this binary schema was correct, then the events of 2004 were devastatingly clear. They followed a well-worn script of social disintegration and the loss of future possibility. To be sure, many ethnographers have made exactly this argument. As Ticio Escobar put it long before he became Paraguay's minister of culture, the Totobiegosode Ayoreo are victims of "ethnocide" or "the violent extermination of culture." This process of contact "converts their members into caricatures of Westerners and sends them to a marginal underworld where they end as beings that have no place in their culture nor the culture of others." Escobar argued that this process transformed contemporary Ayoreo-speaking people into a peculiar form of nonhumanity, a population whose "men wander as shadows of themselves through work camps or colonies . . . and their women, defeated, arrive at the towns to give themselves up as semi-slaves or prostitutes."[10] For Escobar, Clastres, and others, the supposed death of culture also meant a wider social death. It was believed to manifest in affective or psycho-pathologies, including "all of the side effects of losing one's cultural identity: alcoholism, social disorganization, apathy, violence, suicide, prostitution and marginalization."[11] In this reckoning, the values of culture and life were conflated. We do not need to read or write about the contact with the Areguede'urasade; we only need to properly mourn their passing.

At first, it was tempting to see traces of decay everywhere I turned. No matter where I went in the Chaco, I found that Ayoreo were disenfranchised as subhuman matter out of place: cursed, subordinated, neither this nor that. On both sides of the Bolivia/Paraguay border, Ayoreo-speaking people were the poorest and most marginalized of any Indigenous people in a region where camps of dispossessed Natives lined the roads and Indians were still held in conditions described as slavery.[12] They confronted a mosaic of violence: enslavement, massacres, murder, and rape were venerable traditions. Many of the girls exchanged sex for money on the peripheries of cities or towns. The pet parrots in one settlement imitated tubercular coughing. People seemed to alternate between nervous motion and opaque waiting. Many sought escape by whatever means were near at hand: prayers, disco music, shoe glue. The violence was unavoidable. Everyone got tangled up in it; we all mouthed its lines.

Fieldwork meant confronting situations I was not equipped to deal with in rational ways, from blunt-force brutality to incredible grace. There was no coherent whole to master or become fluent in. The closest

things I found to Ayoreo cultural institutions were those described so
confidently in books or pieced together from elders' memories and sold
to visiting anthropologists. Sociality swung wildly between extremes of
collective affiliation and agonistic striving for dominance. Merely surviving required a thick skin; I never quite got used to the fact that conflict
and brinksmanship were defining parts of everyday personhood.

Yet the more time I spent in the Chaco, the more clear it became that
death was only part of the story. What appeared initially as losses were,
on more thorough acquaintance, zones of intense translation, rational
calculus, and partial potentials for Ayoreo-speaking people. If such sensibilities reflected a failure, it was the failure of the New World to deliver on its promise and the failure of others to take seriously the kinds
of possibility these emergent attitudes contained. By the time I left the
Areguede'urasade in 2004, I was increasingly convinced that primitivist
narratives of culture death were not any kind of answer at all. Rather, they
were part of the problem. I wondered: How did the New People come to
inhabit a world in which their disintegration was foretold? How were
classic accounts of contact revised and refracted in twenty-first-century
contexts? At which points did the kinds of self-fashioning by Ayoreo and
by ethnographers fall apart, and at which did they unexpectedly merge?

In Search of the Ex-Primitive

I remember the Chaco as a place of white light and scented smoke and
talc-fine dust, vivid in its heat and beauty and brutality. Dust in my food,
dust in my lungs, but after a while it didn't matter whether I was consuming the Chaco or the Chaco was consuming me. Wide roads of pocked
dirt or cake-batter mud straight for miles. Raw earthworks for Australian water tanks and bare pastures like pieces of Kansas or Manitoba laid
somewhere they don't belong. Low forest a thicket so dense you can't see
more than ten or fifteen feet, even when you're crawling through it. Hot
casaake north winds barrel over treetops, crashing like waves on a moving shore. Boundaries of all kinds trace unexpected forms when one is,
as the Argentines say, in Chaco, as if crossing its frontiers makes personal
boundaries suspect too.

But then again, isn't that the truck of the ethnographer in the field,
the ultimate frontiersman? Only I thought you were supposed to tie up
loose ends and come back full of facts and stories that everyone under-
stands. That's not how it happened. I left, for sure. But something stuck.
When I finally returned from the field in the summer of 2008, it took

months before I could write. I could barely speak and only tell stories from the middle out when I could tell them at all. My recollections were more like jokes that weren't funny or puzzle pieces that didn't fit. Like the one about how tree trunks explode when they're hit with a bulldozer. Or the one about Ore Jno, a renowned warrior who was so old they said he was like a flower that wilted with the sun and when he dozed in the morning his great-great-grandchildren would tug on his ears. He walked bent over and panhandled along the train tracks and after so much surviving he was run over by a white man in the street and died. The truck backed up several times to make sure, and by the time we found him in the road there was nothing to be done.

"You ask too many questions," the blonde man in a mechanic's shop in Filadelfia said. "You know, other people can also end up disappearing." He twitched and sweated, "You people don't know anything. I could tell you a lot of things about the savages, things you may not want to hear. But I won't. I don't want to talk to you about it. No one around here does." Would he at least sell me the spear that had been thrown against a bulldozer he had been driving, so I could return it to Asôre, the man who made and threw it? He looked past me at my Ayoreo companion. "You're not allowed in here."

Or like the one about the night when we were sitting around the fire in front of the leader's house in Arocojnadi, my favorite place, and everything was just fine. People were relaxed and joking and passing around *tereré* and including me in the swirl of chit chat and the stars were like something from a planetarium and the wind kept most of the mosquitoes off and the old man sang that one song from long ago when he watched the back of his young lover walking away from his fire. And later there were no night sweats when I slept. A man who has since died woke me from my tent to tell me stories about what the stars did when they were like people in the Original Time and we sat for hours just us two in the red glow of embers laughing at all those stubborn stars and watching them chase each other around the black beyond forever.

Once out hunting in the afternoon, we came upon a camp in the forest that had been abandoned a short time before—ashes, a bundle of flawed arrow shafts twisting slowly from a branch. I thought I could hear whispers in the leaves and see shadows shifting just beyond. The tension with the ordinary sunlight and ordinary flies was nearly overwhelming. I had the urge to run. We stared for a long time at the food scraps and those strange rectangular sandal tracks, a palpable absence and terrible frozen stillness. Like words on a page.

In the aftermath of the 2004 contact, politically pointed feelings of rage, shame, confusion, and despair were impossible to avoid. Tensions of loss and transformation threatened to overwhelm us all. It was difficult to watch as the New People were taught that contact marked a radical rupture of space, time, and being. They learned that the space of the forest / past, *Erami*, was a zone of ignorance and sin that must be transcended. Their relatives called the world they now inhabited *Cojñone-Gari*, or That-Which-Belongs-to-the-Strangers. It was a world where *Dupade*, the Christian God, dictated the rules of life and agency. Their relatives replicated the forms of mission settlements and their imaginations of the lives of *Cojñone* Strangers. They prohibited many past practices as dangerously profane, capable of inciting illness or death sent by a vengeful God. Some Totobiegosode violated the most potent of prior moral taboos to illustrate that the old spirits had lost their power. The New People were converted by ridicule.

My feeble efforts at intervention failed. I struggled to make sense of stories like Tié's and of what I saw as a harsh subjection to a variety of dynamics aimed at reducing and revivifying her humanity in contradictory ways. Perhaps it was impossible for me to fully grasp the range of judgments, sensibilities, and elisions that such transformation so viscerally entailed. But one thing was clear: these common Ayoreo perceptions reflected neither a pure ontological alterity nor evidence of disintegration and "culture death." The Areguede'urasade could not be described as a society against the state, in their sociopolitical forms or subjective horizons, nor could their emergent sensibilities be easily interpreted as a cosmology against the state. Moreover, the ruptures of contact did not seem to reduce multiplicity at all. Instead, it seemed multiplied in unsettling ways. In the aftermath of contact, what counted as difference was fractured, inverted, poorly imitated, doubled back, knotted up. It was all I could to navigate this unstable terrain, let alone force it into the clear boundaries of causal sense.

When I left the Chaco in November 2004, I was not sure I could ever bear to live among the New People again. My fledgling search for the primitive had come to its abortive—and, in hindsight, inevitable—end. Yet I was haunted by images I could not forget and questions I could not answer. The disquieting sense that I might have gotten everything wrong, that I had misunderstood entirely, compelled me to return in 2006 for long-term fieldwork among the Totobiegosode. Time and again, I wondered what kind of humanity was now possible for the New People, once they transformed from iconic primitives into supposedly degraded ex-primitives whose ties to a legitimating past had become impossible, refused, or suspect. I also wondered what kind of anthropology could take their projects seriously. Were there any footings for political anthropology that could account for the profoundly ambiguous being of these people, caught between spectacle and marginality, human and nonhuman, life and death?

What did it mean to abandon the search for primitive society, and stage an ethnographic pursuit of the fractured and dispossessed subjectivities of the ex-primitive instead? Where could it begin? And where would it end?

A Handful of Ashes

"I do not know my own story," Tié said.

We were huddled around a double handful of embers in the wattle-and-daub shack of her captors, late on a cool night in 2007. Her husband, Cutai, had summoned me from my tent when the others were asleep and the owners of the house were away. By that time, Cutai was my hunting partner, the one who gave me his first kill of the day and who received mine in turn: turtles, armadillos, peccaries. I knew how quietly he could move when necessary. One moment I was asleep, the next he was there murmuring to come with him because there was something his wife wanted to tell me. It was an odd request, and I was curious to hear what she would say.

We sat passing a tin cup of *tereré* and breathing fragrant smoke and waiting for the dogs to quiet down and listening to the night wind rustle through leaves and branches and tarps and garbage. Tié spoke in a throaty whisper, eyes averted, long pauses between each phrase:

I do not know where my story will go
I was born in the place called *Aremia*
I do not know what story to tell
I do not know what I will say, I do not know
I do not know my story
We looked for *doidie* roots
We found them near Cucarani
Little birds, in the afternoon
We painted our bodies
We were *sucio*, dirty, in the afternoon
We painted our bodies with ashes and down
Black and white
They were *sucio*, dirty

Sucio
They sang
They told many stories
They saw far away
The old man was there too
He told stories
I do not know what to say
We ate honey
We killed fish
We were dirty
I do not know my story
I do not know what to say
My thoughts and my memories are gone
They will no longer come to me
I do not know my own story
It is finished

When she ended her recollection, we sat quietly for a long time as the embers faded to ashes and the wind rolled through the treetops. I cannot forget what Tié told me that night, even though I still do not comprehend it. What was lost in such tumbling images if not the coordinates of interpretation, the reliable tenets of causality, the possibility for the continuity of self-assured narrative? And what about the mysterious dirtiness that emanates from beeswax adornments to skin soiled by honey or blood until every afternoon is rendered obscene in ways impossible to understand?

Did that mean the knowledge made through ethnography was similarly compromised? Even as I began immersive fieldwork in 2006, I thought I knew what kind of stories I was after. I mistakenly thought I could prejudge their value. But I could never predict what responses my questions would elicit. The future was up for grabs, but so was the past. Many Ayoreo words had meanings that were fractured, unstable, and contradictory. Most Totobiegosode people turned to silence or allegory or interrupted images to communicate the ruptured connections of everyday life. Most of the time I was too eager to find a singular answer, too ignorant or distracted to understand.

Everywhere I went I heard stories so elemental and terrible that if I really got them, I risked them becoming my own. And whether I admitted it or not I was already a character. Sometimes all the stories would fuse together and one could glimpse an almost incredibly momentous scale of human experience, of epic lives; betrayal, murder, revenge, self-sacrifice.

Some took the shape of shadows and shouts in the red penumbrae of firelight. Most of the important ones were hardly stories at all. Like Tié's imagery, they were disaggregated fragments that lingered alongside heat waves.

"Before, our thoughts became reality, and what was true was our thoughts," Siquei told me once when we were out hunting.

Telling stories was a tricky business. There were some that were okay, funny ones about blunders and coarse slapstick and double entendre wit and purely rational explanation. But some of them were contagious. They stayed with you and worked on you and you didn't know it until months or years later. It could be easy or even tedious, but it could also be like being passed something so fragile that it made you feel too insignificant to bear even as it turned to ash the moment it touched your hands and all you were left with was dirty fingers and the lingering impression of a weight so slight as to be imagined and the pressure of a breath you forgot to exhale.

Sublime and profane, the guilty pleasures of ethnography. About halfway through my last long trip in 2007, I quit recording and my notes dried up. They started to sound like the stories I had heard so often. Sometimes everything—roads, bulldozers, smoke, cattle, honeybees, icy Coca-Cola—seemed to be trying to tell me something. But, like Michael Herr in another jungle, the only part I could ever understand was a riddle about loss and emergence and it went like this: why are you here if you cannot put yourself in my place?[13]

Where the Black Caiman Walks

When I returned in 2006, there were two neighboring Totobiegosode settlements and I split my time between them equally. Whereas my clear sympathies for the New People strained some of my relationships in the place called Chaidi ("the Resting Place"), the people of Arocojnadi took me as one of their own. Yoteuoi, a former *daijnai*, "shaman," asked his wife to adopt me and eventually everyone contributed to my education. It became the place I was most warmly welcomed and I felt most at home.

Arocojnadi translated as The-Place-Where-the-Caiman-Comes-Walking. It was an old name, first given to a nearby campsite long ago. The name referred to the strange appearance of a Caiman—*Arocojnai*—walking over dry land into the encampment. Once common throughout the Chaco, the Caiman was usually associated with permanent rivers and rainy season marshes. It was known to lie hidden under black water, capable of sudden crippling attack. And the migratory Caiman was also said to have an ambivalent spiritual affinity with *Jnaropie*, the spirit world of the dead where all color bleeds to gray and where everything is inverted, reversed, upside down. The appearance of one of these large reptiles stalking over parched ground into a camp was a startling event, ripe with portent and menace. In the particular instance referenced by the name of the original place, the walking Caiman was a presage of violence. Enemies massacred most of the inhabitants shortly afterward.

In the stark Totobiegosode settlements of dust and embers, people spent hours, days, weeks apparently doing very little. Sometimes it could seem as if we were all waiting for some nameless event or visitation. When the hunger and heat were particularly acute, my mind wandered. Through fevered hours I imagined those beings that might suddenly appear, walking ribboned paths into the encircled people. As days stretched to months and years, I grew convinced that the walking Caiman had returned in the new ghostly form of world-ending violence relentlessly stalking the New People and their tenuous humanity. This allegorical Caiman I too conjured was ambiguous and elusive, a composite image of Ayoreo senses of being in the world, ongoing destruction, and the reanimated legacies of colonial and ethnographic projects of all kinds. I wondered how my companions and I understood the force of such uninvited arrivals, especially if they were our own spectral doubles—spirit Indians and anthropologists and missionaries conjured by savagery and terror, lurking in the shadows, bedeviling every attempt to make ourselves a space in the world. Which among them were welcomed, which driven away, and which might wait just beyond the edge of the firelight, sensed but unseen, unnamed and unsummoned?

I began to see the attenuated spaces of post-contact life as an uncanny New World; a vertiginous place of constant inversion where poetics became power only for a preordained few, where intention turned to ash, where disorder was order and violence was becoming. I began to imagine it as a place where the space of death was the space of legitimate life, where immanence arose from the negative, where everything had been recast as shadow and doubt: the Place-Where-the-Black-Caiman-Walks.

Actually Existing Alterity

My journeys to the Chaco began with the same imperative as many of my distinguished predecessors—namely, to "take seriously the conditions of the ontological self-determination of the Other." Yet after all of my wanderings, I found myself on unfamiliar and fraught terrain. Only one thing was clear: the archetypes of primitivist anthropology were no-where to be found. Instead of psychosocial types like the Chief, Shaman, Warrior, and Enemy, I found diabolical spirit anthropologists, partially reconstituted souls, bulldozers, madness, addicts. Instead of animism, I found apocalypticism. Instead of jaguars who are humans, I found Indians who were animalized. Instead of wisely multinaturalist primitives crossing human/nonhuman divides at will, I found increasingly sharp and nonnegotiable divides between nature and culture, primitive and human, past and future.

Ayoreo life projects did not fit within most of the analytic categories I initially brought to bear on indigeneity, contact, tradition, or Indig-enous rights. I soon discovered that Ayoreo-speaking people did not offer much common ground to well-meaning activist types who presumed that hope lay only in the reproduction of the past. As I became more proficient in the Ayoreo language, my relationships thickened along with my understandings of the stories told by the New People and other Totobiegosode. I discovered a constant slippage between epistemic crisis and emerging ways of being in the world. I learned that Totobiegosode had made a self-conscious decision to abandon nearly all of the past practices that count as authorized "traditional culture" or ontological alterity in Latin America, such as shamanic rituals, curing techniques, magical songs, myth narratives, and ceremonial aesthetics. Moreover, this decision was not localized but was seen as a necessary precondi-tion for Totobiegosode to join a wider, cross-border network of Ayoreo-speaking people predicated on a shared Christian morality and spread through two-way radio technology. The encounter with modernity, I was told, implied a process called *chinoningase*, or a radical transformation of human form and content. The process was said to particularly target a kind of soul matter called *ayipie*. The *ayipie* is the bodily seat of willpower, memory, rationality, and social sentiments, such as shame. In short, it encompassed precisely the elements of moral humanity. This transfor-mation of humanity was seen as a necessary condition for Totobiegosode survival in *Cojñone-Gari*.

I discovered that such attitudes were a profound moral response to the upheavals of world-ending violence and contemporary political econo-mies. At the time, the sparsely populated Paraguayan Chaco was in the midst of a land rush, and the value of land had increased five hundredfold in the previous ten years. Spurred by the rising price of beef and massive investments by foreign agro-industrial firms, local ranchers turned the slow-growing forest into cattle pastures by bulldozing it bare and burning all matter. Enriched by the ashes, the sandy alkaline earth produced viable grass for a dozen years. On satellite imagery, clear-cut checkerboards exploded around the three Mennonite agro-religious colonies in the center of Paraguay and the soy plantations in southern Bolivia, even as Brazilian agri-capitalists pushed inward from the eastern borders. There were few reliable statistics to quantify the devastation, but it likely outpaced six hundred thousand hectares/year, making it one of the highest local deforestation rates in the world.[14] And this was big business: Paraguay beef exports were a billion-dollar industry in 2010.

Tié and the New People were particularly afraid of the bulldozers used by ranchers to crush the dense forest. The New People hypothesized that the bulldozers were monstrous beings relentlessly following their scent. On several occasions in the 1990s, they attacked bulldozers with arrows and spears but to no avail. The fear of bulldozers was a form of historical consciousness which cited older terrors of genocide. Like all bands of Totobiegosode, the Areguede'urasade were nearly exterminated by Paraguayan soldiers and warriors from hostile Ayoreo groups. Enemy Ayoreo obtained shotguns from missionaries and hunted down the forest bands, which were either killed or brought back to the mission as captives. Such violence dotted the landscape with *dajegeo* places contaminated by human blood, which remained in the soil, capable of inciting future violence. Faced with forest-flattening machines and increasingly surrounded, the former Areguede'urasade began to have visions of group death and a world where no fires would light the darkness.

Many Ayoreo now say these prophecies were fulfilled in the ruptures of contact and conversion. Yet to my great surprise, Totobiegosode elders likened these ruptures to the originary differentiations of human/nonhuman described in the myth narratives now abandoned as dangerously profane. According to these stories, all Beings once shared a proto-humanoid form and limited language. Through a series of transformations, these Beings changed into distinct entities and in the process created the world. When each Original Being transformed, he or she also left behind a moral prohibition and a technique for social becoming, such as how to make fire, find food, paint the body, cure sickness. Proto-humanoids only became moral humans through mastering the power of transformation. Totobiegosode said that contact implied a similarly profound transformation, necessary for transcending savagery and reproducing moral humanity. Such beliefs reclaim self-transformation as the core capacity of being human. Contact and its ruptures, in other words, were not only foundational myths of colonialism but were also meaningful ways for Ayoreo-speaking people to trace new limits around events, causality, and morality in the present.

By 2008, I realized that what at first appeared to be traces of inevitable loss could be considered a consistent form of becoming-through-negation. In the process, Ayoreo-speaking people reworked world-ending violence and colonial terror into the preconditions for the emergence of the modern, moral Ayoreo subject. Bulldozers negated the forest. Conversion negated the past. The transformed *ayipie* negated prior selves, even as the conditions of post-contact life also rendered these new horizons impossible for anyone to realize. New afflictions of body and soul negated both Ayoreo visions of moral humanity and neoliberal negations of it at the same time. I began to think of this process as a formation of negative immanence, through which Ayoreo-speaking people refused their colonial dehumanization, reasserted their capacity to objectify transformation, and crafted off-balance relationships between politically sanctioned culture and moral life. It was a project outside the authorized outside. Instead, Totobiegosode self-objectified the contradictory objectifications of their lives and bent these to their own unexpected ends, both vital and deadly.

This was not a world of stable alterity but one of semantic implosions, interrupted translations, broken lines of flight, vortical flows of rupture, promising horizons. Such nonlinear ontological assemblages were finely attuned to the turbulent conditions of Indigenous life in the Gran Chaco. Asserting the moral value of rupture and transformation was one of the few ways to account for the nonsensical intersections of ecological devastation, soul-collecting missionaries, tradition-fetishizing ethnographers, unscrupulous humanitarian NGOs, and neoliberal economic policies. Ayoreo ways of being were locked in a life or death struggle with modern categories that they also at times embraced or simply ignored. Reducing this complex ferment to a simple dichotomy of incommensurable worlds offered an essential metanarrative of a colonizing power that justified the violent subordination of Ayoreo-speaking people, fractured their senses of the world, and rendered their tentative life projects untenable. Ayoreo notions of the moral human arose as a response to the disjunctures and contradictions of these global politics of Indigenous life.

Yet all too often, such complex Ayoreo subjectivities were intelligible only as the loss of culture or the contamination of cosmology. Outsiders also negated the human possibility implied by such projects. For ethnographers who insisted on reducing the validity of Ayoreo life to continuity with traditional cosmology, a hunter/gatherer moral economy, or a "multinaturalist ontology," contemporary Ayoreo life either contained a kernel of radical alterity hidden under a misleading veneer of change or evidenced the death of a legitimizing essence. It was no coincidence that the "post-multicultural age," where an ethics of difference is imagined to partially reverse the inequalities of neoliberalism, was marked by amplified violence against supposedly "deculturated" Ayoreo-speaking people—and young, urban-dwelling women in particular. Nor was it a coincidence that this violence is characterized by an exceptional degree of brutality: the same patterns of "overkill"—mutilation, intentionally prolonged murders, the use of blunt instruments, etc.—characteristic of " bias-motivated violence" or "hate crimes." Deemed unfit for economic production and already stripped of all potentially valuable commodities—territory, technology, tradition—the culture-less ex-primitive is relegated to her final role as the figure of degradation and death, an avenue for the expenditure of mourning, the receptacle of violent rage and transgressive desire, and the cautionary object required for evoking the cultural revitalization or moral salvation of more powerful others.

Meanwhile, the frustrated desires for pure primitives were once again displaced onto the two small bands of Ayoreo-speaking people that remained in the dwindling Chaco forests. Like the Areguede'urasade before them, they gained international prominence as "voluntarily isolated peoples," and three feuding NGOs formed to protect them from sustained relations with outsiders. They legally resembled elements of nature, and were regularly celebrated as existing in a "single, inseparable unity with their habitat." For such NGOs, the only truly legitimate Ayoreo life was that which is "still fully alive among the uncontacted groups."[15] While actual Ayoreo-speaking people died from starvation, murder, and disease, a transnational NGO economy was mobilized to preserve the haunting fantasy of an Ayoreo life hidden in a besieged wilderness.

Ayoreo-speaking people were thus targeted by several kinds of violence at once: structural violence, governmental violence, biopolitical violence, representational violence, law-making violence, eschatological violence, iatrogenic violence, humanitarian violence, sacrificial violence. Violence was everywhere and nowhere, blunt-force impact and an afterimage so subtle it was difficult to name. Each regime of violence aimed to colonize Ayoreo life in distinct and mutually exclusive ways. This meant that violence interrupted itself. It also implied that in this New World, any Ayoreo life project was always already insufficient, unauthorized, and newly eligible for violent persecution.[16]

On the Trail of the Black Caiman

Today, the primitive has returned as an instrumental figure for pop anthropology, theory, morality, and politics alike. A transnational moral economy is mobilized around the preservation of traditional culture as a "major humanitarian issue," a subjective horizon, and a moral limit of legitimate life.[17] The right to a bounded, stable cultural alterity is codified within national and international law as a grid for citizenship, a "global public good," and a vital principle.

The following pages argue that this development causes real-world effects amenable to ethnographic description. They push against the artificially reduced limits of cultural alterity presumed within these developments. This book offers a different way to account for emergent forms of Ayoreo life than the tired binaries of cultural continuity versus culture death. Doing so also requires going against the grain of an increasingly influential and "ontological turn" in social theory. The ontological paradigm fundamentally updates classic stories about the South American primitive. It advances the premise that a singular "Amerindian cosmology" exists and that this cosmology is opposed precisely to the foundational tenets of "modern, Western, European" philosophy and the binaries of nature/culture upon which this "mononaturalist / multiculturalist" philosophy is based. Instead, or so the story goes, Amerindian cosmology pushes us to imagine a multiplicity of worlds instantiated as "one single culture, multiple natures . . . one epistemology, multiple ontologies." This represents a form of "perspectivism," in which the point of view creates not the object but the subject. This ontology, we are told, is the "New New World."[18]

According to the model of ontological anthropology, Amerindian multinaturalism holds a radical potential because it is presumed to exist external to modernity. Primitive society is valuable not for its political distinction, as Clastres would have it, but for its cosmological alterity. As Eduardo Viveiros de Castro put it, Amerindian ontology represents a "cosmology against the state," based on the "peculiar ontological composition of the mythical world."[19] As such, Amerindian worldviews are imagined to hold a revisionary potential for modernity. Thus, anthropologist Ghassan Hage can argue in all seriousness that the future of anthropological critique lies in the return to the ethos of a "critical primitivist anthropology."[20] Likewise, Michael Hardt and Antonio Negri take up the figure of radical Amerindian alterity as a primary inspiration for imagining an emancipatory "altermodernity" based on the commons.[21]

In such scholarship, critique is located not within the actual subaltern experiences of Indigenous peoples, but in the impending utility of their timeless cosmological difference.[22]

Such models, however, leave little room for the Totobiegosode and their fraught New World. What the following pages suggest instead is that it is also necessary to account for those versions of Indigenous worlding that take up quintessentially modern binaries and their mimetic opposites as reflexive coordinates for self-fashioning. The problem is that applying ontological anthropology as a model of actually existing alterity requires the targeted disavowal of many aspects of ethnographic and Indigenous experience. The instrumental incoherence of indigeneity is reduced to a telos of order imposed on the authorized outside. This, of course, reproduces the same "hermeneutic violence" that Michael Taussig so persuasively located at the core of the colonial nervous system, whereby academic fictions that "flatten contradiction and systematize chaos" sustain terrible violence.[23] Unsettling one binary risks reifying another that is even more crucial for colonizing projects of all kinds: the radical incommensurability of modern and nonmodern worlds. Instead, this book critically examines the conditions under which the emergence of actually existing Ayoreo alterity became possible.

Totobiegosode experience suggests that what is most interesting about Indigenous cosmologies is not their stable difference, but their pervasive refusal to stay put in slots they are assigned within the telos of modern order and its inside/outside limits. The point illustrated in this book is that contradiction and negation do not disrupt lived Amerindian cosmologies. Rather, as Gaston Gordillo shows in another context, Indigenous ontological alterity is constituted through these operations.[24] If there is any opening to altermodernity to be located among those ex-primitives struggling to survive on the margins of lowland South America, it may well lie in the various ways that Indigenous senses of being in the world always already exceed the terms of the radical imaginaries that many outsiders claim they sustain. The challenge we face, then, is not how to more finely parse these sensibilities as redemptive or spurious. Rather, as Michael M. J. Fischer points out, it is how to evoke and consolidate "a new humanistic politics."[25]

The following account asks how such a reanimated humanism could be ethnographically co-envisioned alongside Totobiegosode experiences. To do so, the chapters weave together and develop three linked recognitions. The first is that the category of Native culture matters today not as a flawed epistemology or obsolete heuristic but as a crucial political technology. Increasingly, the sociolegal figure of culture operates as a regime of what Didier Fassin has called "biolegitimacy," or the specific kinds of late liberal governance instantiated through policing the constricting limits of who should live and in the name of what.[26] Indigenous alterity is then held to this constricted ideal form by way of inflexible thresholds and new authorities that arise to govern those boundaries now imbued with the force of self-evident reality.

Second, this dynamic suggests that focusing on Indigenous becomings, rather than the essences of their being, may emphasize their idiosyncratic capacities to endure and transform the worlds they inhabit. Doing so means holding the emancipatory indeterminacy of becoming in intimate tension with the contingencies of interest and rule.[27] Such an approach aims to sharpen attention to both operations, and thus promises to escape from the predetermined analytics that so often haunt ethnographies of Indigenous being.

Finally, a new humanistic politics of Indigenous ethnography requires embracing the peculiar and rotational temporality of knowledge. What makes contemporary anthropology exciting is how its prior ways of knowing are ontologically unruly, looping back into the fabric of the institutions, communities, and subjective horizons we study. Ethnographers venture to the field only to confront obsolete anthropological models reanimated as social fact. Nowhere is this recursive effect more apparent than in the sites and practices of indigeneity. Any new humanistic politics must begin with the kind of reflexive practice that self-consciously attends to this temporality of anthropological knowledge and the force of its ontological afterlives.

What the following account insists on is a shared world of unevenly distributed problems. It is a journey through a world composed of potentialities but also contingencies, of becoming but also violence, wherein immanence is never innocent of itself. In this world, the most potent mode of anthropological critique is resolutely in our present but never confined to it. Such recognitions foreground the decolonizing potential of anthropological knowledge to shuffle back through and with the life projects of Indigenous people. In the process, Tié and the New People may reappear as guides to an anthropology of the contemporary that is critical, urgent, and perhaps even optimistic.

Such are the murky surfaces under which the Black Caiman lurks, and from which he may be flushed. So it is to the first set of faint tracks—his? mine? Ayoreo? no matter—that we may finally turn our attention.

ONE

The Devil and the Fetishization of Tradition

I don't know you, Ayoreo. But I see you shine, smiling in the eyes of Deisy Amarilla. . . . Like all anthropologists, she knew to transmit to us this Je ne sais quoi that transports us to another dimension. Another world. To your world, which used to be ours. . . . My eyes are marveling before you. I admire you for your Innocence, for your wisdom, for your simple lessons, your way of resolving things. Your fatalistic superstitions, your at times perturbing coldness, your loves: sincere, without hypocrisy. Everything brings me closer to you, Ayoreo. Lost race. Happy, however, in these modern times. You knew to tell us your stories, your legends, with openness. From you emanates this smile that I wear. To read about you, to listen to you live, breathe and suffer. POEM IN THE PARA-GUAYAN NEWSPAPER ULTIMA HORA, JULY 11 2009

It is impossible to do real ethnography among the Ayoreo anymore. Only the bones are left. We are grave robbers, digging up bones. ANONYMOUS AN-THROPOLOGIST, 2007

When I first met the New People in their dusty camp at the edge of the forest, I was surprised to discover that Dejai, the leader of the new settlement, had predicted my arrival.

"We knew someone like you would come here," Cutai told me one afternoon as we were taking a short break from hunting honey in the deep shade of the forest.

"Someone like me?" I asked.

"Yes," Cutai murmured. "An *Abujá.*"

"And what is an *Abujá*?" I wondered.

"It is said that they are *Satanas utocaidie*," he replied with a slight smile, looking down. "The helpers of the Devil."

At the time it seemed like an odd remark, but I thought little of it. We both laughed a little (for different reasons of course). Then we hoisted our heavy bags again and moved on through the thick brush.

Only two years later did I begin to question the significance of this brief exchange. It was mid-September 2006, one of the last cold days of the dry season. The low clouds were pushed by a south wind. I had arrived the week before to begin two years of fieldwork. Everything seemed opaque, my relationships tenuous.

I spent my days making preparations to set up camp in the forest and visiting with the fluid group of a dozen Totobiegosode living in a wooden garage behind the brick house in the town of Filadelfia, the Mennonite colony and nerve center of the Paraguayan Chaco where the NGO had its headquarters. At the time, the contrast between the NGO building with its well-fed, live-in staff and the general squalor of the garage seemed to epitomize the ambiguous status of the Totobiegosode.

We were squatting around a handful of burning wood scraps some-one had scavenged from a lumberyard down the road, when a man who appeared to be in his late fifties and whom I had never seen before in-terrupted the quiet flow of conversation, skipping the usual polite for-malities and fixing me with a hard stare.

"Are you an *Abujá*?" He was a small man named Mariano with graying hair and heavy-lidded eyes under a stained knit cap.

"I don't know," I replied. "What is an *Abujá*?"

"What is your work?" He spoke in a strident tone that clipped the Ayoreo and Spanish words.

"I am an anthropologist," I said in a tone that was aiming for polite but came out obsequious. "I am trying to understand the ways [*jmapiedie*] of the Totobiegosode. What their lives are like."

"You are the one who took pictures, who filmed them before, no?"

"Yes. But now I am also working with OPIT [the tribal organization]."

"To bring them money?"

"To support what they want to do."

There was a tense silence while the fire guttered in the wind and the others sat quietly, hands outstretched to the flickering warmth. Mariano stared at me intently and I looked back.

"*Abujádie* are worthless." He turned and spat to one side. "They are only interested in recording the *cucha bajade*, like *ujñarone* curing chants. They want to hear *adode* myths. They try to get people to tell those old bad things. It seems like they must sell them later. They do not respect anything."

Without breaking eye contact, Mariano stood up. Then he abruptly turned and stalked away from the garage, his posture rigid, his steps firm.

At the time, I was unsettled by the confrontation (one among many) and puzzled by what seemed like barely controlled anger directed at the *Abujádie*. I now realize that I could not have received a more appropriate introduction to fieldwork in the Chaco.

Over the following months, I asked the people I came to know to tell me more about the *Abujádie*. I eventually learned that this was an Ayoreo category for "anthropologist."[1] The word literally meant "the Whisperers" or "the Bearded One" and implied a veiled appearance, such as a *Cojñone* Stranger's face or a rain cloud. It was often used as an insult.

According to the stories I heard, *Abujádie* were trickster figures. They were said to possess hidden stacks of money "as long as their beards" and nefarious powers of persuasion. They beguiled people with their charm and smiles. They were known for their ability to "say all the things the people want to hear" but secretly plotted against all that is moral and good. As Jochade, the leader of Arocojnadi, told me in September 2007, "If you see an *Abujá* heading to an Ayoreo community, he is going to have problems. If he tries to go among the Ayoreo, the Ayoreo hate it. They hate him very much. *Ore chijimiji ore idai*, they will cut him out of their settlement."

Everyone agreed that *Abujádie* were difficult to spot. But they all eventually gave themselves away by two defining traits. They shared a propensity to carry cameras. They also shared a profane focus on collecting the set of practices now called *cucha bajade*, the "original" or "first things." It was no coincidence that the *cucha bajade* included all of the very same practices that ethnographers claimed to constitute traditional Ayoreo culture.

I soon discovered that Totobiegosode people (and all other Ayoreo-speaking people in Bolivia and Paraguay) had made a self-conscious decision to abandon these practices in their entirety. During my fieldwork, Totobiegosode elders did not tell *adode* creation myths around nightly fires, nor did they heal one another by sucking out (*ore chigase*) or blowing away (*ore chubuchu*) sickness with *sarode* and *ujñarone* curing chants, nor did they smoke *sidi* tobacco and *canirojnai* roots to enter into shamanic trances, nor did they ritually define kinship through *chugu'iji* and *chatai* performances, nor did they summon the spirits with *perane*, the wordless clan rhythms given to the *Jnupemejnanie*, Those-Whose-Bones-Are-Dust. The rare occasions that someone narrated a myth or a curing chant were catalyzed by the presence and money of a rotating cast of *Cojñone* with a resolute interest in documenting the time now referred to as *Nanique*, or before.

Abujádie may have been physically present, but they only saw traces of the past. One of the many stories Totobiegosode people told about *Abujádie* concerned a taxi that unexpectedly arrived in Arocojnadi at dusk, bearing a tall thin foreigner and a woman. The smiling couple got out, and they were white-skinned, blue-eyed, and beautiful. These people acted very friendly and sat down around the fire. They began to talk and they talked a long time. They smiled at the children. Then they pulled from the taxi a suitcase full of money. Maybe it was dollars or maybe it was euros. It was enough money to buy fifty hectares of ancestral Toto-biegosode territory. These *Cojñone* Strangers said they would give the people this money if they would let them make a film.

They wanted the Totobiegosode to go back to the forest, to take off their clothes and live like they did before. The white couple would go with them and film them. They wanted to take pictures of how people ate, how people slept. They even wanted to take pictures of sex between a man and a woman. When they said this, the people knew that they were *Abujádie*. The people were not tricked by them and became angry. They began to speak harshly. They wanted these *Abujádie* to leave them alone and go away. Chacuide, a very old man who had killed enemies in his youth, was listening. Although he was usually kind and gentle, even he became furious at the words of the *Abujá*. He went to his house and came back waving his spear. Chacuide walked to the man and began screaming at him, "Go away! Go away or I will kill you!" The women be-gan shouting and gathering things to throw at the *Cojñone*; the children started to cry. Faced with this response, the *Abujádie* got in their taxi with all their money and left the village. These two, the people say, were true *Abujádie*.

The Devil Does Fieldwork

Anthropologists have long noted that, for many Indigenous peoples, Devil imagery offers a meaningful and potent way to interpret the partic-ular conditions under which they live. Michael Taussig's classic accounts of the Colombian Putumayo and Gaston Gordillo's compelling writings on the western Toba of the Argentine Chaco are among the finest exam-ples.[2] Both accounts are based on and expand Marx's notion of commod-ity fetishism, which holds that the intrinsic conditions of capitalism cause "the relationship between people to take on the character of a thing and thus acquire a phantom objectivity, an autonomy that seems so strictly rational and all-embracing as to conceal every trace of its fundamental nature: the relation between people."[3] Both argue, in somewhat different ways, that Indigenous peoples use Devil imagery to protest against this phantom objectivity of the commodity-form produced through their alienated labor and how it justifies and sustains their subordinate status relative to wider political and economic processes.[4]

At first glance, the Ayoreo use of Devil imagery appeared strikingly distinct. Ayoreo-speaking people did not join devilry with capitalism per se but with the practice of anthropology. And they did not appear to protest the conditions of wage labor so much as the conditions of fieldwork and the kinds of temporalities that underwrite ethnographic representations of tradition. This distinctiveness raises a number of questions: What, if anything, might an ethnographer of the Ayoreo and a mine owner or plantation overseer have in common? Are we similarly engaged in a labor of allowing the relationships between people to take on the phantom characteristics of an object or thing? What is at stake for Ayoreo and anthropology if this object or thing is traditional culture? And doesn't this Ayoreo critique then require us to accept that it is now necessary to account ethnographically for the palpable social presence of anthropological knowledge and the unequal forces that it conjures and exerts against human life?

As was so often the case, Ayoreo suspicions about outsiders like me contained a kernel of keen insight. By the time I met the New People in 2004, I too was already well on my way to becoming an *Abujá*. It was not my first experience with Ayoreo-speaking people. In 2001–2002, I spent fourteen months among northern Ayoreo in Bolivia, descendants of the Direquednejnaigosode and Jnupedogosode confederacies that had emerged from the forests in the 1940s and 1950s and who had a much longer history of dealing with anthropologists, missionaries, NGOs, and the like. I met these people as a twenty-two-year-old while on a Fulbright scholarship to examine a sustainable forest management project that my host organization, the NGO Apoyo para el Campesino del Oriente Boliviano, was implementing in the Direquednejnaigosode community of Zapocó. I was soon swept into the constant movement of Ayoreo-speaking people from relatively remote settlements like Zapocó to an informal camp of tarp and embers in a notorious shantytown on the outskirts of the city of Santa Cruz de la Sierra and back again.

To prepare for my first trip, I had spent a year reading everything I could find on Ayoreo. What a shock to discover that none of the practices and forms that I had studied so closely appeared to exist in Ayoreo daily life, either as something that people did or even as something they discussed. Could this actually be the case? Like a good *Abujá*, I resolved that it could not, that such appearances were deceiving. There must be some hidden substrate that with dedication, intelligence, and solidarity I could access. I suspected that this difference could be reconstructed from oral histories. This misleading impression was sealed when I discovered a ready set of answers to my probing questions. I soon shifted my full attention to collecting oral histories and spent six months in the spring of 2002 traveling among several Ayoreo communities in eastern Bolivia. I recorded more than five hundred hours of elders' stories—first working with translators and later, as my Ayoreo language skills improved, alone. My justification at the time was that of the salvage ethnographer, imagining that I could document a vanishing difference for future generations.

I was especially eager to record the *ujñarone*: a set of esoteric chants that, according to the books and articles I had read, previously formed the core of shamanic practice. By 2001, these chants had been abandoned across Ayoreoland. Jnupi, my first adopted father and himself a respected elder who accompanied me on some of my journeys, told me that there was only one old man in Bolivia who still knew the chants and might be willing to talk about them.

His name was Simijáné. In 2001 he was around ninety years old. He had been a main informant of every ethnographer who had ever written on Ayoreo "traditional religion." All of Simijáné's assistance had not translated into material wealth. Simijáné was poor even by Ayoreo standards. He survived on the coins he begged from passersby in the city center and from his great-granddaughters' earnings from sex work. His unashamed willingness to talk about *ujñarone* was not merely exceptional but singular. It made him a highly controversial figure among other Ayoreo-speaking people.

I found him hunched unsmiling on the curb of his favorite corner in downtown Santa Cruz, a few steps away from a French boutique. The slight figure looked impossibly alien among the roaring buses with his dirty rags, patched bag, eyes blue with cataracts, and cane made from a broken broom handle. As I approached, he held out his hand and asked for a coin in a surprising bass voice.

I ignored the request but felt awkward about towering over him, so I squatted down on the curb nearby. He seemed to lose all interest in me at that point and sat in silence. Within seconds, a large, middle-aged Ayoreo woman approached in an oversize pink T-shirt and a long, floral-print skirt, holding a toddler. She fixed me with a hard stare and asked me what I wanted. I tried to explain. The woman was his daughter, and her gaze relaxed as soon as she heard that I was an anthropology student.

"So you want him to tell you about the old things," she interrupted.

"Yes."

"Will you pay?"

I flushed. She said all anthropologists had to pay. I said we should talk about it. Both of us knew that meant probably. I was uncomfortable but she was pleased. She told me to come visit them that weekend and gave me directions to the abandoned lot where they were camped. She said their family had been kicked out of the urban camp three weeks earlier and had been beaten with fists and clubs. She didn't say why, but I later heard that Simijáné had been suspected of sorcery. Others in the camp attributed several unexplained deaths to his secretly reciting of chants in his home. They said he was in league with Satan.

When I arrived at Simijáné's camp the following week, I too asked Simijáné if he would be willing to teach me *ujñarone*. I knew there were two kinds of chants: those that could be recited relatively freely and those that were *puyaque*, or taboo. Reciting a *puyaque* chant could reportedly cause great calamity, madness, and death, unless it was indicated by a specific illness. I told him I was not interested in the *puyaque* chants but only in those that were safe. After a short exchange with his daughter, he agreed as long as I would pay him four dollars a day and take him someplace where there were no other Ayoreo-speaking people because even the safe *ujñarone* might cause unintended damage despite the fact that "no one believes in them anymore." I eagerly accepted his terms. Our acquaintance began with a formal instruction in curing chants over two months in 2001–2002 and gradually turned into lengthy storytelling sessions, first in a cheap hotel in the nearby town of Samaipata, then on the street corner, and two years later over a tiny fire in his drafty shack once he was allowed to return to the urban camp.

Whereas most elder Ayoreo refused to speak about the past, Simijáné was the opposite. Like Primo Levi, Simijáné's response to the end of one world was an irrepressible impulse to narrate.[5] For an *Abujá*, what could be more compelling than the prospect of documenting an entire genre of previously sacred practices that were on the cusp of disappearing? Yet, time and again, Simijáné seemed to interrupt himself and any pretensions I had of encountering a vanishing alterity. His stories came out in what seemed to be disorderly tangles, jumping seamlessly from mythic ancestors to his experiences of healing patients to his vivid erotic desires and dream journeys to Hell or Heaven and back again. Precontact cosmology proved remarkably elusive. Every time I thought I could grasp it, it dissolved once again into a stridently surreal narrative.

Simijáné mixed stories of the first clan ancestors with those of Adam and Eve and the Devil. "Eve is our grandmother, Adam is our grandfather. From them come the good people of today. There are also bad people who come from Satan, from the Devil who tempts them. It turns out that men are all bad but when someone has courage he can become a leader and kill enemies but only if he is also lucky."

He told of the first time he received spirit power by drinking green tobacco juice and passing out instead of vomiting and of how he killed an enemy and became a man desired by young women. He interjected pointed advice about the moral imperative to be generous into stories of spirit flights in which he could hear giant Ancestor Beings whispering behind him no matter which way he turned, and he classified Coca-Cola as an *edopasade* possession belonging to five of the seven clans because it was so good to drink.[6] In his stories, it was rarely clear whose perspective Simijáné was voicing—the drowning or the saved or both at once. I struggled to keep up.

Simijáné taught me about the *ujñarone* of Jobe, or Tarantula, in the following way:

Jobe, Tarantula, stayed at the door to his house when the wind blew. Yocaoi [the leader of a present-day Ayoreo community] said to the people, "Could it be that no one knows an *ujñarone* for the wind when it blows too hard?" They were afraid of the strength of the wind that was too strong and always blew dust where they were living. The people where I was living said, "Yes, we know that *ujñarone*." The leader said, "Why don't you do that *ujñarone* for this wind?" So I began to do the secret of Jobe.

You can use the secret of Ango'oto for this wind, but it may kill someone in the village. The secret of Jochin'goi [a small tortoise] is also fine to use for this. I told the Jobe *ujñarone*, and the strong wind went away. When the wind died down, it began to rain.

My daughter told me, "Why don't you do that *ujñarone*, there is too much drought." We heard it even rained in Santa Cruz. Yocaoi said to me, "I had already become afraid of so much wind that was throwing dust in our faces."

So I said I would try. A man cannot say, "I know everything." God told the men [that] one who doesn't know very much should not say that he knows everything. He has to say, "I do not know very much." One should not be proud if he knows a lot of things. He who knows something should not be proud. When he receives something, he shouldn't keep this thing for himself. He has to give it to others as well.

Tarantula and his people, they had grown afraid of the wind. When the wind blows hard, it makes a sound like "kee yee gee yee gee yee gee yee a." The warrior leader of Tarantula's band was Jochin'goi. He said, "I am going to go back to where I was before." He thought about the place where he lived before. He said, "I'm going back to live there again." When someone is tired of too much wind, he can make their *ujñarone*.

Jobe said to the people, "My *ujñarone* is good!"

The people said to Jobe, "Let's see. We'll listen to your *ujñarone* to see if it is good."

The people said to him, "Come on, let's go to that house over there and eat."

Jobe told them, "No, we're not going to go there. We're going to go to the bend in the road. And there we will stay in my house. There is a lot of food to eat there."

Jobe's house was like a grocery store. There was a lot of food to eat there.

Without pause, he then switched to a detailed discussion of the relationships between Jobe and the Uyujnanie, or Strong Winds, as well as

Ajidapaquenejna'gate, a small, stingless kind of honeybee. He then told how Jobe gained power from tricking the Strong Winds and locking them inside his house. Tarantula eventually decided to give his prisoners food and released them, on the condition that they would only blow from time to time. In return, the people of the village gave Jobe many things, and the Strong Winds decided to give Jobe a secret:

When the people saw that it was going to make a strong wind with rain, they can use this *ujñarone*. They can use the *ujñarone* Achiangoi gave to Jobe for this. This *ujñarone* will take apart the storm that is going to come. And it will stop blowing.

I am telling this now because I see that it is cloudy. The people know that a strong storm is bad because the strong wind can damage them. This is the *ujñarone* given to Jobe by Achiangoi.

I am taking myself apart
I am taking myself apart
I am taking myself apart when the sun is covered
I am taking myself apart when the sun is covered
I was the wind that came with strength
I was the wind that came with strength
I am the one that destroys the forest
I am the one that destroys the forest
I am the one that destroys the beauty of the forest
I am the one that destroys the beauty of the forest
I am the owner of happiness
I am the owner of happiness
I am the destroyer of heat
I am the destroyer of heat
I am the destroyer of heat
I sound like se se se *[calming]*
I sound like se se se
Puuuuu *[wind blowing lightly]*
Puuuuu
Puuuuu

This is the secret of Jobe. This secret is used when a strong storm is coming. If this is used, then the storm disappears. The words go up and destroy the storm. When the storm leaves, the weather is cool. The people are very pleased with this *ujñarone*. The people gave gifts to Jobe. Each time that someone had something, they gave part of it to Jobe. Shamans used this secret so that the *Uyujna kiyigijnanie* ("dust devils") would go to another place. The *pujopie* ("soul spirit") of other shamans could be traveling up in these winds.

I said to Yocaoi, "Are you still bothered by the Strong Winds?" He was tired of too much wind. Afterward, it rained and there was not too much strong wind. I used this *ujñarone* to do it. I destroyed all the winds.

Simijáné's power to heal came from the sense that his humanity was flexible, capable of being subordinated to the outside forces that he could channel. Visiting ethnographers likewise expected him to channel Ayoreo alterity writ large. He made particular note of his experiences with the man he called Don Luciano—Lucien Sebag, the noted student of Claude Lévi-Strauss who visited two Ayoreo missions in the early 1960s. Simijáné still blamed himself for Sebag's 1965 death by suicide. He said that Sebag had begged him to recite *puyaque* chants. Simijáné initially refused, but after Sebag promised him a substantial payment, he relented and agreed to let Sebag record a single chant, the chant of *Poji*, or Iguana. This chant was so powerful that it broke Sebag's reel-to-reel tape recorder. In that instant, the words also infected Sebag.

"Before, I did not know how to put my words to one side and Luciano died because the *ujñarone* broke his recorder and went directly to him." Sebag's madness was preceded by Simijáné's own. After recounting the prohibited chant, Simijáné was possessed by the *Pojiode* spirits. They took their vengeance by removing his *ayipie* soul matter. "I do not remember what happened," he said. Others told him that he began to resemble an Iguana. In the midst of his madness, his skin turned yellow. He was overpowered by an irrational fear of his people. After weeks of searching they found him alone in the forest, living in a hole, eating dirt, mumbling nonsensical phrases, covered in excrement. It took a week of curing sessions to restore his health.

It was a telling irony that Simijáné's teachings to me contradicted nearly everything that had been written about Ayoreo traditional spirituality by the same men he had previously instructed.

Tradition as Ethnographic Fetish

Ethnographers have made Ayoreo tradition, and spirituality especially, their primary object of study since shortly after missionaries contacted northern Ayoreo-speaking groups in the late 1940s. Although Ayoreo-speaking people are nearly absent from English-language scholarship, they are the subject of dozens of anthropological theses, articles, and monographs in Spanish, French, German, and Italian. By the mid-1970s, more than fifteen anthropologists had visited and written about various

Ayoreo-speaking groups in Bolivia and Paraguay, notably, Heinz Kelm and his student Bernd Fischermann from Germany, Sebag from France, and Marcelo Bórmida from Argentina. Most of their work focused on Ayoreo spirituality or myth narratives, described in the ethnographic present.[7] Partly because of these accounts, Ayoreo-speaking people are imagined to be a homogenous pueblo or tribe and the most authentically primitive or savage of all Chacoan peoples. As such, contemporary interest in Ayoreo groups is so pronounced that local scholars wryly refer to an industry of "Ayoreología."[8]

By 2001, these ethnographers had developed three general frameworks for interpreting Ayoreo tradition. The crudest example is offered by the work of the "ergon and myth" school based at the Centro Argentino de Etnología Americana at the Universidad de Buenos Aires. First elaborated by Marcelo Bórmida and his students in the 1970s and 1980s, this model continues to inform several recent anthropological accounts of Ayoreo-speaking people.[9] As Gaston Gordillo notes in his trenchant critique, such work is based on the idea that Ayoreo myth narratives are evidence of "permanent structures, independent of space and time" that are not "contaminated" by the "profanity" of European contact or daily cultural life.[10] Together, they form a "timeless," "mythic consciousness," which is antithetical to modern rationality and which assumes a "meta-temporal and transtemporal character by which it is inserted into the present and will be in the future."[11] That is, legitimate Ayoreo life was equated with an Ayoreo culture itself reduced to the permanent structures and causalities of mythic order.

Proponents of this model can thus argue that the themes of death and transformation so widespread in Ayoreo myth stories meant that Ayoreo culture (and, thus, Ayoreo life) was defined as one ruled by "terror and death."[12] For example, Carmen Nuñez argued that there is "a very defined direction in the culture toward a violent, vengeful and bloody world."[13] Anatilde Idoyaga Molina asserted that within a general Ayoreo "ethos of terror and precaution," the "nucleus of the worldview . . . is characterized by violence, multiple homicides, envy and aggression."[14] Celia Mashn-shnek likewise concluded that a "mythical horizon of death and murder invades the cultural life of the Ayoreo."[15] In this schema, tradition was reconstructed and substituted for the value of Ayoreo life.[16]

A similar equation informed the ethnographic construction of Ayoreo people as typical hunter-gatherers of the Chaco.[17] This is most clearly elaborated by Volker von Bremen's work, which argued that Chaco Indians embody a "completely different system of reproduction, values and social order that . . . exercises its influence to this day."[18] Like Bórmida,

Von Bremen described a specific Indian mentality and argued that Indians are only able to interpret modernity by reference to the "mythical time" of tradition.[19]

For Von Bremen, the interactions of Ayoreo-speaking people and capitalist economic systems are predetermined by a hunting-gathering mode of production and by an Ayoreo cosmology that supposedly "makes it feasible to continually reorder the forever-metamorphosing external world without having to alter the essential mythological structure on which it is predicated."[20] According to Von Bremen, these two features form "the principles of a whole way of life," based on "integrating experiences of relatively recent historical time into the myth system, which is itself characterized by a lack of chronological order. As a result, there is no diachronic view of history, and the development of a historical consciousness of progressive sequences is prevented."[21] Von Bremen's argument rests on the premise that all Ayoreo people believe that mythological structures create an external world that cannot be changed by human action and to which humans must adapt. This allowed him to erase both politics and history:

> The specific problems of contemporary Ayoreode do not consist primarily in the danger of not being able to maintain their ethnic identity or their right of self-determination. In their view, it is rather their still existing ignorance about the origin of the new phenomena and their concomitant inability to contact the related ancestors. It is this contact that they regard as an essential precondition for successful living in the present world, since the ancestors always accompany the contemporary phenomena.[22]

Bernd Fischermann offered a more sophisticated formulation of these links between myth, culture, and identity.[23] In his painstaking magnum opus—a continually revised doctoral thesis originally for the University of Bonn—he wrote against "incomprehensible . . . desktop" descriptions of an Ayoreo ethos of terror and death.[24] Instead, he offered an exhaustive catalogue of traditional practices, concepts, and myths. Despite the many ethnohistorical contributions of this work, Fischermann too based his descriptions on an implicit theory that the cosmological order of the world established by myth narratives is eternal and unchanging for Ayoreo. "The results of the mythic happenings are reflected in the actual ordering of the world that for the Ayoreode is valid and immutable."[25] Fischermann argued that myths thus explain "the conformation of the structure of human society and nature in the form in which it is currently encountered."[26] Because this stance presumed that the abandonment of tradition would imply a degradation of Ayoreo life, he had little choice but to argue that contemporary Ayoreo culture is a function of a deterministic core of traditional cosmology that endures beneath a veneer of apparent change: "The traditional culture manifests itself in many aspects of the thought and behavior, and its bases govern [*rigen*] to this day in the Ayoreo culture. The obvious appearances of change only represent tendencies of adaptation to a new environment, but in every case can be identified with the respective traditional structures."[27]

Like von Bremen, Fischermann argued that "new and unknown . . . phenomena of the modern world" are "totally submitted" to the cosmological orders established by myth narratives and time. Fischermann wrote that this "complete subordination" is possible because Ayoreo people assigned all "introduced elements"—including medicines, material goods, and Christianity—"to the time frame and category of mythical ancestors."[28] He thus concluded his sweeping work with the supposedly romantic and empowering claim that contemporary Ayoreo people have an "intact cultural identity," because "modern cultural elements" have proven unable to contaminate Ayoreo tradition.[29] In doing so, Fischermann merely inverted the value attributed to Ayoreo tradition. Like other *Abujádie*, he ultimately reified the reductive equation of tradition and life, and reasserted the exclusive authority of ethnographers to determine the limits of both. Such projects deflected attention from the more interesting question of how tradition became such an eagerly sought object and diabolical fetish in the first place.

The stories that I heard in the field—and those that I did not—suggested two fundamental critiques of these models: first, that the very terms of the questions being pursued were dangerously suspect and pathological; and second, that they were based on a serious misinterpretation of the myth narratives they claimed as their empirical evidence.

On the most superficial level, Simijáné confirmed the general outlines of "Ayoreo cosmology" as described by ethnographers. He told me that the first world was populated by Ancestor Beings known as the *Jnani'bajade* (Original or First Men) or *Cheque'bajedie* (Original or First Women). These beings, according to Simijáné and my other teachers, shared a humanoid form, language, and idiosyncratic personalities and capacities. The fundamental division of society from nature and thus, the human from nonhuman was the result of various processes by which the majority of these Original Beings were transformed into the nonhuman forms apparent in the world today (plants, animals, virtues, etc).

The narratives of these transformations were referred to as *adode* in the Ayoreo language. This word (singular *adi*) referred to manners or customary ways of acting (as well as the joined parts of a body and ejaculated semen). Together, these stories formed a body of sacred knowledge called the *quicujaidie*, the traces of all things. Post-contact narrations of *adode* suggested various reasons for these foundational transformations.[30] Regardless of the reason, at the time of his or her bodily transformation each Original Being "gave" two things to those who maintained a proto-human form. The first was a moral restriction on behavior, a taboo known as *puyaque* or That-Which-Is-Prohibited.

The second was a prescriptive chant that could be used to mediate the illness or malaise caused by transgressing the restricted behavior. These chants were called *ujñarone*, or breathed ones, and along with a certain number of magical songs were known more generally as *sarode*.[31] For Simijáné and my other teachers, *adode*, *puyaque*, and *ujñarone* were inseparable, and they had previously existed for every plant, animal, and abstraction that comprised the time/space of the precontact past, *Erami*.

Simijáné's stories, however, offered a distinct interpretation of the relationship between these practices and the constitution of moral human life. Ethnographers such as Sebag, Bórmida, von Bremen, Dasso, Fischermann, and others described the *adode* as evidence of the integration of culture and nature within Ayoreo tradition.[32] Such descriptions resonate with the late structuralism of perspectivist anthropology, critiqued by Terence Turner and currently an influential way to understand Amazonian cosmologies.[33] The central premise of this scholarship is that all Amerindian peoples share the same timeless cosmology, based on the homogeneous belief that animals subjectively identify themselves as humans, or cultural beings, with whom they share a generic soul.[34] This notion of a spiritual unity and a corporeal diversity represents "a perspectivist theory of transpecific personhood, which is by contrast unicultural and multinatural."[35] As Viveiros de Castro puts it, "One culture, multiple natures—one epistemology, multiple ontologies. Perspectivism implies multinaturalism, for a perspective is not a representation. A perspective is not a representation because representations are a property of the mind or spirit, whereas the point of view is located in the body."[36] In such a schema, not only humans and animals but Indigenous peoples and "moderns" inhabit fundamentally different worlds. Like Bórmida's "mythic consciousness" or Fischermann's "immutable orderings" of the world through myth narratives, this "Amerindian cosmology" is taken as a self-evident and valuable object.

Yet Simijáné's narrations of *adode/puyaque/ujñarone* emphasized the opposite. In his stories, the crucial difference between humans and non-humans lies in the capacity to harness and control bodily transformation itself. This is why such forms were considered the "source of life" in *Erami*, or *yucucuecaringueri garanone u sarode iji monte nanique*. As in the Kayapó models described by Turner, Simijáné said that the First World inhabited by the Ancestor Beings preceded the rise of human society or culture. These Ancestor Beings were not fully human. Rather, they were proto-humans that were fundamentally amoral and unsocialized. In every case, Simijáné stressed that these proto-humans did not possess the basic means for reproducing moral society.

In fact, proto-humans only learned how to become properly moral humans—how to make fire, weave clothing, construct shelter, plant gardens, kill game, wage war, cure illness, navigate the Chaco landscape, and enlist metaphysical alliances—through the lessons and warnings imparted by Ancestor Beings as they transformed. The key elements necessary to reproduce moral human life, in other words, did not already exist from time immemorial but emerged from these foundational transformations of content and form. Thus, Sebag's oft-repeated observation that animals represent "ancient Ayoreo who, for one or the other reason, renounced their humanity," contains a crucial error: even within the terms of post-contact *adode* as recited by Simijáné, humanity itself emerged as a result and condensation of transformative processes.[37] The same error is repeated in other contexts by perspectival anthropologists, who claim that "the original common condition of both humans and animals is not animality but rather humanity. . . . Humans are those who continue as they have always been: animals are ex-humans, not humans ex-animals."[38]

Simijáné, in contrast, emphasized that humanity had always been defined by constant flux. It was never a stable state. As in the Kayapó systems described by Turner, "the whole point of these myths is not how animals became and continue to be identified with humans, thus subverting the contrast between nature and culture, but how animals and humans became fully differentiated from each other, thus giving rise to the contemporary differentiation of nature and culture."[39] Moreover, it is precisely through the capacity to continually manage, contest, and redefine such distinctions that moral humanity was constituted. And it is this plasticity of the human that made healing chants an effective technology.

To cure using *sarode* or *ujñarone* chants, a healer first activated his or her *pujopie* soul matter through the smoke of *sidi* tobacco or *canirojnai* roots. In the following trancelike state, the healer would visit with the particular spirits with whom he or she shared an affinity. These spirits would indicate the source of the illness, sometimes prescribe certain sequences of chants, and allow their spirit/soul matter to be put at the service of the healer. Repetition was a key part of successfully using *ujñarone*. The short lines were repeated between two and five times, the stanzas punctuated by conversation, massaging the patient, spitting on the skin or blowing away offending spirits. The entire *ujñarone*, uttered in arcane, guttural, and rapid-fire diction, was repeated various times, and it was said to be effective only when used in specific groups. Often these *ujñarone* groups were ordered according to kinship ties between Ancestor Beings. Such relations remained apparent in the physical properties of the visible, non-human forms taken by Ancestor Beings after they transformed. Those who were related often had similar transition narratives, spiritual power, and physical features.[40] Nearly all *ujñarone* cures climaxed with a series of sounds that channeled the actions of Ancestor Beings and catalyzed their healing activities within the patient's body. The universal use of the first-person pronoun and these vivid onomatopoeias completed the healer's transposition with spirits he or she embodied through the utterance of the *ujñarone*. This transposition was the core of the healing chant.[41]

The performance of an *ujñarone* chant was successful only if it could realign the relationships between the First World of Ancestor Beings, the linear time of human experience, and the inverted time of death. It required the healer, the patient, and the audience to be transformed by contact with the powerful spirit of an Ancestor Being. The power of such *ujñarone* did not derive from a projection of the past into the present. Rather, the chants were a fluid genre that reconstituted the relationships of past, present, and future through the flexible medium of the healer's body. When Simijáné was teaching me how to use *ujñarone* chants, he directed me to add additional lines and phrases based on my own sense of or relation to the *adode* events. He corrected me only when my references contradicted the primary elements of the story. Simijáné himself never repeated any *ujñarone* in exactly the same way. The point seemed to be less about formal repetition than about the creation of a space in which original transformations could be summoned and further used to transform those in the present—a process that could also alter the past events that were being channeled.

In Simijáné's teachings, the very practices that ethnographers have identified as constituting "traditional Ayoreo ontology" depended not on stable cosmological orders but on a radical break with continuity, on a constantly shifting world, on the capacity to recreate the past, present, and future. These practices took for granted the fluid nature of humanity, the moral distinction between human and nonhuman, and the power of spoken words to move backward as well as forward through time to cause desired effects. This means that, in the early years of the twenty-first century, "traditional Ayoreo ontology" could only be discovered as an ontology of transformation—if, of course, it could be discovered at all.

Where the ethnographer collects and orders, the Ayoreo healer unravels and ruptures. The tension between them is not merely one of disagreement. This friction also creates new figures and forces and sets them loose on the world where they demand responses from us all.

Despite superficial disagreements, most ethnographers of Ayoreo groups were involved in a unified project. This project consisted of transforming the unequal relationships between outsiders and Ayoreo into the object of traditional culture. This object was then transposed with the psyches, history, and value of Ayoreo life. Such ethnographers actively celebrated the phantom objectivity of a traditional being they alone defined *over and above the universal capacities of human becoming, a process* that concealed the fundamental nature of this thing: the vertical relations between people and their unruly capacities to transform. This substitution required an active omission of the conditions and relationships by which anthropological knowledge was possible in the first place. Such a mystifying ethnographic project depended on and actively encouraged the alienation of Ayoreo-speaking people from what they considered to be the defining essence of themselves. Whereas the mine owner trafficked in silver or the plantation overlord in wage labor, the *Abujá* trafficked in tradition. Setting this process over and above Ayoreo projects of vitalizing transformation created and sustained the strange fetish power of tradition, even while the resolute anthropological focus on its extraction conjured and loosed diabolical spirit-anthropologist doubles with whom everyone, including future ethnographers, now must contend.

Such operations were not unfamiliar to the New People and my other teachers. In fact, they were precisely what most Ayoreo identified as the source of shamanic power in *Erami*. Ayoreo people consistently described this power as a kind of magic whereby representing something or someone gave the representer power over it—what anthropologists have long referred to as sympathetic magic. This magic is what gave power to the markings painted on the seven different kinds of shaved wooden posts capable of fending off enemies and sickness, and it is what allowed special verses to lure game or prayers to reach the spirit world. This is what caused the wordless *perane* songs to invoke the potent essence of clan ancestors; it is how wiping out crude rectangular figures carved in the earth in the *ditai* ceremony could purify a warrior contaminated by spilled blood; it is what made group performances like *ore agapi cukoi* drumming on a hollow *cukoi* tree able to call rain; and it is what could transform ordinary men into fearless warriors through *chugu'iji* performances of masculine rage, in which a man would don his potent *cobia* feathers, recite the deeds of brave ancestors, and prepare himself for battle.[42]

What seemed to be missing in the ethnography of the *Abujádie* and in perspectival anthropology was a way to account for the relationship between political power and meaning, or the mimetic faculty underlying the chains of sympathetic imagery that imbued such ancient transformations with potency in everyday life. What reportedly distinguished the human/nonhuman in such discontinued Ayoreo cosmologies "was not merely classification, or even a simple cognitive or perceptual process of objectification, but a reflexive process of meta-objectification, in an abstracted and generalized form: that is, of the process of objectification itself."[43] This capacity for metaobjectification was inseparable from using mimicry to effectively invert negativity and transform the original by shifting from proto-human spirit to animal form and human content and back, using words that seamlessly moved from the time of myth to the colonial present and the ideal future in order to resignify all.

Sympathetic magic was what Ayoreo elders invariably used to explain the power of the *ujñarone*. It was also, as in this description of a successful cure by the Totobiegose man Aasi, what they used to explain why such potent forms were dirty, immoral, and in need of erasure:

We had many beliefs before. We respected and feared everything. It was as if we were formed [*tocade*] from this respect before. It was bad because we were ignorant. We were afraid of these things for no reason. But because we were afraid, they could make problems for us. Once my mother nearly died. We did not know what her sickness was. They came and sat around her. She was weak and she could not get up unless someone helped her. We made her a stick so she could push herself up. A woman named Ugui'date remembered that the sick woman had put color on a *pamoi* sitting band. But it was *puyaque* at that time of year for a woman. A woman should not touch or put color on a *pamoi*. The woman *daijne* took off her skirt and folded it so she could sit on top. She began to "heal by blowing" [*chubuchu*] on my mother. She began to cure her, and she said, "I am going to try to cure her with the *ujñarone* of *pamoi*. That is the way to save this woman." She cured my mother because she knew the correct way to save her. She knew that my mother had touched the *pamoi* that a woman should not touch and that is why she could be saved. When she cured my mother, everyone knew that her sickness had come from the *pamoi*. And my mother was saved.

In the case of the *ujñarone*, this sympathetic magic resided in the power of words to conjure spirit and alter the present. Some people were said to have "heavy words," and whatever they said was likely to come true. Such heavy words could also *churu pajeode*, or "wash away negative emotions," in the present and the future. Others were known as *uto ca'achu*. Someone who was *uto ca'achu* was capable of causing something to happen to another person through his or her spoken predictions. Such an occurrence was called *pugaite*. The *u'e* were another class of people who spoke about good things and caused them to happen. According to these ideas, if someone's prediction came true it was because the utterance itself was animated by the speaker's soul matter, exhaled in his or her breath. The power of a chant depended on a general Ayoreo theory that spoken words can, under certain conditions, travel through time to cause the effects they purportedly describe. As the figure of the *Abujá* attests, ethnographic representations of tradition are premised on the same kind of power. And they may well have the same effects, whether we admit it or not.

The Secret of the Devil

During our first long session, Simijáné seemed most interested in teaching me the basic operations of the *ujñarone* and their relationships to the myth narratives, and in establishing his own authority on the matters in question. He started with around two hundred chants considered generically beneficial but weak, such as those to make children strong and smart, to cause peace and happiness, etc. Soon, he appeared to grow bored with this and began asking me to use the chants after he had demonstrated them. At the time my Ayoreo was abysmal and I invariably butchered the words and rhythms beyond recognition. Simijáné always found this quite hilarious. Each time, he would laugh and repeat, "He doesn't know!"

He never laughed or joked, however, once he began teaching me the more powerful chants and he never asked me to repeat them. These included a set of chants made around Christian figures, such as Adam, Eve, Noah, Mary, and Joseph. Simijáné learned these chants from a visionary woman shaman named Amo'nate, when they were both living on the mission of Tobité in the late 1950s.[44] Simijáné told me he knew the *ujñarone* of God and the Devil:

We cannot imagine the Christian God. We cannot see him, either, because his spirit is so powerful. We cannot imagine, but we know what he is called and we know that he exists. God made all the things that we see in this world. *Ujñarone*, medicine, everything. God gave some of his power to his son, the Christ. The Son of God asked his father for help. He said, "Papa, I want you to help me. Please give me blessings." The father said to him, "Write down these things on a page in your notebook. All the things that you need and then give it to me." The son already had some paper to write down everything. The father told him, "Write down which blessings you want." He wrote down, faith in my father. Also, he wrote love for other people. Then he asked for peace too. Then he asked that he may be good and generous and that he never got angry. Also he wrote that he may be sympathetic to the others. Then he asked to be respectful and responsible. Then he said, "Thank you father. Now I will make my *ujñarone* for the people."

I am the big man
I am the big man
I am the great teacher
I am the great teacher
I am the one who the people obey
I am the one who the people obey
I have killed myself and became the warrior leader
I have killed myself and become the warrior leader
I am the first
I am the first
I bring peace to the world
I bring peace to the world
I bring sympathy to the world
I bring sympathy to the world
I bring peace to the world
I sound like se se se *[calming]*
I sound like se se se

This is the secret of Jesus. You can tell it in the evening, around six o'clock. That is the time to tell it. He is the one who gave us sympathy and peace. God taught his son this *ujñarone*. Then, the Christ tested his disciple. He gave him alcoholic drinks to them. When one is a believer, he doesn't vomit what he drinks. The alcohol doesn't damage a person who believes. But if someone doesn't have faith, he will vomit.[45] Jesus gave them alcohol, but they didn't vomit. He said, "You all want to follow me."

Amo'nate was the one who knew this story. She knew the story of Jesus before we met the missionaries. We knew about Jesus in the forest. We knew about Jesus in the forest, because Amo'nate knew. The Christ learned from the advice of his father. "These are the commandments and I have them guarded in my heart," he said. Jesus brings peace. But God makes war against his other enemies. In the time before, God made a test to the people who didn't believe. He said, "This is my blood, which has run out of my body for you all. Drink it." When they tried it, the people of Satan all vomited. He said, "Look at them. They don't believe in me, that is why they have all vomited."

A wise man named Asi'daquide knew this before. He taught Amo'nate this story. When Asi'daquide died, she kept this story in her mind, like I am doing now after she is gone. Very few people know this story. Jesus didn't want any bad things; that is why he wanted to receive peace, love, and calmness from his father. "I don't want to be among people who are bad. I want to be gentle."

[Interrupted by young man nearby]: "That is not true. You didn't have any of this before."

[Simijáné continued]: "God has given all the *ujñarone* to us. God gave all the *ujñarone* to each thing in this world. Even to the animals, the stars, Jesus, Ayoreo, and God himself. You can use this secret if someone is a thief or if he is bad and angry. The person will leave behind whatever sin or badness with this *ujñarone*.

The Devil also had his *ujñarone*. We knew about Satan a long time ago when we lived in the forest. Satan said, "I am also a warrior leader of the entire world. I am going to see that person you call God. And I am going to dominate him too." But the Lord is more powerful than him. He wanted to be bigger than God. But then he thought about doing battle with God and knew that he would lose because no one can dominate God. When he lost the battle, Satan transformed himself and left his *ujñarone* to the people. It goes like this:

I am the one who causes great pain to those around me
I am the one who causes great pain to those around me
I am their master because I am bad
I am their master because I am bad
I am their master because I am bad
I am the owner of souls
I am the owner of souls
I am the owner of souls
I am bad
I am bad
I am their pain
I am their pain
I am the one who causes great pain to those around me
I sound like huuuuuuu
I sound like huuuuuuu
I sound like huuuuuuu

Simijáné said this *ujñarone* could give the healer power over the Devil. He could out-devil the Devil by first conjuring the Devil's image and then turning this copy against the original at his whim. However, Simijáné said that he himself had never learned how to use that particular *ujñarone* properly. The woman who once knew died long ago. Now, no one could gain power over the Devil except God. He said that no one believed in any of the chants. He thought he was the last one and the chants would lose their power when he died. "The power of the *ujñarone* depends on faith. If I have faith in them, I can cure. If I don't have faith, what use are they? There is nothing there."

With the exception of Simijáné, nearly all Ayoreo people during my field-work said that the *ujñarone* were satanic, as were all of the many practices related to shamanic healing. Even the Totobiegosode people who had been contacted in 1986 agreed. "It was Satan," they told me, "who compelled us to heal by blowing away before." Today only a handful of elders know the *adode* myths, and even fewer know more than one or two *ujñarone*. Many middle-aged Ayoreo in Bolivia do not know what the word *ujñarone* refers to. Those few elders who can still recite *adode* myths for the meager pay offered by visiting *Abujádie* often insist on prefacing every sentence with the word *chiese*, which roughly translates as "it was once said but is now known to be untrue that . . ." The knowledgeable elders refuse to teach the myths and the chants to the few young people who are interested. And it is said that even thinking too much about such things can cause illness and retribution from God, to whom shamanic healing is morally offensive.

If we agree with previous ethnographers that tradition and life or being in the world are indistinguishable for Indigenous peoples and that such spiritual practices constitute tradition, then this decision to abandon the chants leaves us with only one choice: to mourn the death of Ayoreo. The image of Ayoreo is sealed—their lives indistinguishable from a tradition we alone have the authority to classify, delimit, and fetishize. In the process, we deny them the full range of human being, the capacity for human becoming.

But Simijáné's unruly and disordered teachings suggest another possibility. If moral human life emerges from the capacity to define and control the terms of transformation—from the ability to renarrate causality and properly use sympathetic magic—then this collective decision is simply the most recent of many historical transformations of humanity. That is, abandoning the chants is not the end of humanity but a reaffirmation of Ayoreo capacities to transform themselves. It is thus an affirmation not of being but of becoming, one that reclaims a kind of radical agency for ontological self-determination in the face of dispossession and subjection, a stance as radical as that of Simijáné's persistence in using the chants against the wishes of other Ayoreo. Indeed, the abandonment of the *ujñarone* is now required by the very terms of the sympathetic magic that once made them so potent and effective. Continuity implies rupture and vice versa. Only by insisting on the extinction of such shamanic forms can a Christian Ayoreo mainstream be consolidated and its members hope to retain some of the powerful magic they once held so close.

It is precisely this opening to transformation and rupture that the tradition seeking *Abujá* denies Ayoreo-speaking people. The *Abujá* is diabolical because the object of tradition conceals the unequal human relations through which it is produced, a process that means denying Ayoreo the very tools of moral adaptation. Even more than wage labor, the substitution of a limited tradition for life depends on a profound alienation of Ayoreo individuals from themselves, from that which many believe makes them moral humans. At the same time, the *Abujá* claims these very same capacities to define humanity through the power of representation—through sympathetic magic—as something only he may possess.

During my fieldwork, even the words of shamanic practice were fractured and precarious. Many people used *adode* and *quicujaidie* to also refer to biblical narratives, *Jnani'bajade* to refer to Bible characters, and *edopasade* for spiritual gifts bestowed by Jesus. Such slippages illustrated how exceptional Simijáné's perspective was, while also emphasizing that my story of Simijáné's teachings should be impossible to mistake for an encounter with pure difference or an ontological alterity that exists external to the particular relationships between Ayoreo and outsiders. Rather, such slippages require us to ask what fields of force are created and concealed by the fetish of Ayoreo tradition, by the peculiar constellation of people all striving to realize their own competing projects of becoming by staking a claim on prior Ayoreo cosmologies. This is a frontier of difference inseparable from the operations of internal colonialism, in which the always-frustrated desires of ethnographers to gain access to a secret domain of true primitive difference is the key to understanding how the figure of that difference is reproduced and sustained by the same apparatus that consumes it and targets actual Ayoreo lives for extermination in the present.

At this point, the figure of the ethnographer dutifully recording and analyzing images of the *Abujá* who is satanic because he produces copies of tradition out of techniques of healing that were themselves based on the sacred power of a mimicry now rendered profane—relative to the more

fraught mimicries of colonial subjection and the resulting transformations of humanity—threatens to collapse under the weight of mimicry
and make a mockery of its own conceit to describe.[46]

Like me, and my own doomed efforts at escape, Simijáné was never
quite able to fully transform himself. He did not give up his proximity
to the old spirits and he remained on the margins of mainstream Ayoreo
society until his death in December 2011 at a very old age. By the end he
was blind and deaf and his stories dissolved into disjointed fragments.
Yet the things he shared did have a powerful effect. They made it impossible for me to continue as an aspiring *Abujá*. I would like to attribute
this change to the rhythmic words and moving breath of a ninety-year-
old man chanting the secret of the Devil late one night on the curb of a
bustling city street into the tape recorder of a stranger who was so very
young. But I cannot remember which way his words fell and I can no
longer say for sure.

As Mariano made clear that cold day around the fire, most Totobiegosode also demanded this change as the precondition for establishing
any kind of relationship with me. By the time I returned for long-term
fieldwork in 2006, it was impossible for me to ignore the fact that most
Totobiegosode people had a very different sense of the past and of shamanic practices than Simijáné. For them, such things were raw and close
and vital and their magic pressed against us all.

I had little choice but to renounce my quest for tradition. Confronted
with the caricature of the *Abujá* and queried repeatedly about my intentions, I told everyone in those first few weeks of 2006 that I did not want
to know about the *cucha bajade*. I said that I would not ask questions
about myths or curing chants. The story got around. When the Totobiegosode leaders publicly granted me permission to stay among them in
early October, they said it was only because I was not a true *Abujá*.

Over time, this stance became more than a solution to the practical
problems of access. After several months in the Totobiegosode villages,
I too started to feel the force of the *Abujádie*. By the time I left, I thought
I could finally begin to understand why they were the objects of such
distrust and rage.

Ujñari

There was something obscene about the spectacle of well-dressed outsiders in air-conditioned SUVs descending in waves on stark villages,
searching out the oldest people, beseeching them to tell prohibited sto-

ries and then leaving. It was even more disturbing when you realize it's like looking in a mirror.

I'll never forget the first time I met Ujñari. Already old in his late fifties, he was a striking sight. His diminutive body was bowed and misshapen as if time was a physical weight. His hairline was permanently recessed by repeated plucking and burning, to resemble the vulture-spirit who had promised him eternal life in that time everyone called *nanique*, before. His eyes were sunk into the taut face of the starving. Clothed in an over-sized green T-shirt from Gino's Pizza in Toledo, Ohio, tattered jeans sewn together with plant-fiber twine, bare feet the color of old ashes, standing a little apart from the others in the village of Arocojnadi.

Of course, I'd seen him in photographs on fund-raising brochures published shortly after he made "first contact" in 1998, looking serene and strong. His image had circled the globe and it had given cachet to the NGO charged by the state with his well-being. I had already heard the outlines of his story. That his family were Totobiegosode outcasts, that his dying father had told him to marry his sister so that not all signs of them would be erased forever. He had done so, and they had five children together. Until 1998—when he and his family walked out of the forest, trying to talk to a Paraguayan ranch hand who promptly shot at them—these people had been his entire social world. I learned that the woman who was his sister and wife had left him for an aspiring preacher shortly after they came out.

Ujñari wore whatever clothes he was given, women's dresses or just a long shirt, and woke up at dawn to talk to the spirits out loud. His favorite tools were already archaic for Totobiegosode, rarely found even in museums: tapir-skin sandals, a little two-handled tension axe, a pungent hyssop for soaking up honey from tree trunks. He never quite shed the outcast status of incest with the others. Like the Areguede'urasade, he was ridiculed for his failure to comprehend and for his proximity to a dirty, profane life in the forest. But he seemed to tolerate the insults and imperatives without rancor.

Ujñari was one of those old Ayoreo who see through you. I ran into a handful of them, contact survivors who possessed the rarefied gaze of the unreachable, like they had seen beyond the world. Like they could never be surprised again. For all of his strangeness, Ujñari was an extraordinarily gentle man. After he told me that he was starving because the children of powerful families always stole his food, I began to bring him little packets of beans, fruit, aspirin.

Ujñari was a special attraction for those outsiders who knew his story. They treated him differently and offered to pay him by the myth. Usually, Ujñari refused such requests. He had been sick since shortly after contact, when he eventually was diagnosed with tuberculosis. Years later he was still weak and carried his bloody phlegm around in a plastic bag. During my fieldwork the people paid to provide health services to Totobiegosode ignored his wasting lung condition. When he said he had been urinating blood with a high fever, I began taking him to the hospital, to little effect.

Late one night under the fluorescent glare of a bare hospital room during the last trip we made together, he summoned me to his bedside and thanked me in the old way, a rhythmic *chatai* recitation of the deeds of certain clan ancestors. He began to narrate the scars stitched across his body, the puckered gash from an enemy spear in his right arm, and the self-inflicted ones—the burnt crater covering half of his left forearm, the jagged lines he had cut across his torso in mourning for the siblings he had seen murdered.

Then, in a breathy voice barely above a whisper, eyes averted, he began telling me myths—the story of the vulture spirit, the origins of life and death. The stories were offered with a shy smile, like gifts of the most precious kind.

When I left the last time, he was a hunched figure in the rearview mirror watching long after the others had sat down. He died five months later of lung failure on his smoky blankets in a small hut on the edge of the forest. It was death by myth, and I could not forget it.

TWO

The Lost Center
of the World

And the Lord said, What have you done? The voice of your brother's blood cries to me from the ground. And now you are cursed from the earth, which has opened her mouth to receive your brother's blood from your hand. . . . And Cain said to the Lord, . . . Behold, you have driven me out this day from the face of the earth; and from your face I shall be hidden; and I shall be a fugitive and a vagabond in the earth. GENESIS 4: 8—14

What has been lost is the continuity of the past. . . . What you are then left with is still the past, but a fragmented past which has lost its certainty of evaluation. HANNAH ARENDT

The place once called Echoi, or Salt, was located just north of the present-day Bolivia/Paraguay border, in the geographic center of ancestral Ayoreo territories. This place contained two large saline lakes. In most years, the water evaporated from these lakes by the end of the dry season, leaving behind a mile-long expanse of white or pink salt, ready to eat and there for the taking. The salt pans begin twenty-five kilometers west of the flattop mountain called Gososo and some thirty-five kilometers northwest of a vast freshwater lake called Nakaje. The rough triangle formed by these three points, with its fertile soils and abundance of water, game, and minerals, is one of the most beautiful and hospitable areas in the entire Gran Chaco.

Outsiders have long noted annual migrations to this welcoming area by various groups of Ayoreo-speaking people.

In fact, this place figures so prominently within the archival and historical record that it has become indistinguishable from the major movements of Ayoreo history and interpretations of them. For most scholars, however, Echoi figures only as a site of resource exploitation by mutually hostile Ayoreo bands or as the scene of several brutal massacres by enemy Ayoreo as well as *Cojñone*.[1] Fischermann, writing in the ethnographic present about these journeys that were discontinued in the 1960s, thus summarizes a broad consensus when he writes that "war [between Ayoreo groups] occurs when an enemy group penetrates into foreign territory. Bellicose actions are almost regularly produced when, in July and August, the various local groups take the road to the Salinas. . . . The only objective of such trips is to provide themselves with vital salt for the following year. . . . That is why the trip to the Salinas, especially for weaker local groups, is a risky business and could mean the extermination of the entire group. Thus, extreme caution is used on these trips to the Salinas."[2]

When I first met the New People, I presumed that I already knew the rough outline of their recent past. I accepted this history that others had written for them: that precontact Ayoreo life was characterized by extreme violence between the various Ayoreo subgroups, that it derived from traditional ways of managing territory, that it found its maximum expression at the place called Echoi and that this interethnic violence was what ultimately led some northern Ayoreo groups to establish contacts with missionaries in the 1940s. Such narratives were repeated so often they were taken for granted as self-evident, even for younger generations of Ayoreo-speaking people. Many accepted descriptions like that of anthropologist Ulf Lind, who wrote, "The most important possession of the [Ayoreo] tribal group is communal territory . . . according to its importance, territory is defended in the most violent form. . . . Every unauthorized intruder is considered an enemy, and if it is possible, he is killed. One does not enter into the territory of an unknown or enemy group. . . . In a culture as simple as that of the Ayoreo, there could not be a jurisprudence, each person had to defend their own rights."[3]

I only began to suspect that something was missing from this history—something that I had been seeking—because of a song I heard in Arocojnadi in 2006. Every evening, a fire was made in front of the wattle-and-daub shack of Jochade, the founder and leader of the village. The three dozen inhabitants gradually drifted into their customary place in the circle, the men holding makeshift chairs and tin cups for *tereré*, the women spreading blankets on the ground in front of them and the children darting quietly in and out of the circle until they collapsed near their mothers, while behind everyone a ring of a dozen skeletal dogs sat with desperate patience, ready to pounce snarling on any scrap of food tossed into the darkness. This was the hour of jokes and stories and debate and decisions and sometimes, if one of the four older men felt inspired, old-style singing accompanied by a *paka'a* gourd rattle. The art of such singing lay in maintaining a constant swirling of the seeds and stones around the inside of the rattle and punctuating this with delicate shadings of the voice. The lyrics of such songs were always short stories of lost love or old battles or shamanic visions.

On this particular evening, Yoteuoi, my adopted Totobiegosode father, concluded the performance of one song with a casual mention that it had been composed by Ichague'daquide. I was not paying close attention and thought I misheard the name. The only Ichague'daquide I knew about had been a prominent leader of the Direquednejnaigosode Ayoreo, who had died sometime in the 1920s or 1930s and who had spent his entire life approximately five hundred miles north of where we sat. I asked Yoteuoi if this was the same Ichague'daquide, and he affirmed that it was. I then asked if he had learned the song recently, from a cassette sent by someone or on the two-way radio. To cut off my direct questioning, which was slightly rude and threatened to break the peaceful mood, Jochade interrupted with a stern tone. "No. Lucas, listen to me. We know this song because the Direquednejnaigosode gave it to us long ago. They gave it to us directly and said we can sing it. That is why we have it."

I pondered this in silence, as the smooth flow of laughter and chitchat resumed. But in the days and weeks that followed, I began asking other Totobiegosode what they knew about northern Ayoreo groups. Much to my surprise, even the New People knew many of the same stories and events I had heard about in Bolivia from elders like Simijáné, Jnupi, Ore Jno, and others who were supposedly their mortal enemies and who had made contact a half century earlier. Siquei in particular could recite names and details that were astonishingly consistent with the stories I had recorded in Bolivia four years earlier. Yet if different Ayoreo groups had always been involved in violent wars against one another, how could this be the case? How could they know the same stories? I wondered if there might have been another kind of place, a place of peaceful contact that was absent within official accounts but which implied that a different kind of history could be written. The very thought was electrifying to me at the time, not least because it seemed to confirm my own desires to discover a site of resistance. I wondered what kind of future possibilities for Ayoreo and anthropologists such an alternate history might contain. Where could such possibilities be located or at least imagined?

A History of Fragments

Like a dutiful ethnographer, I made a diagram in my notebook of all the places I knew, writing late into the night in the blue glare of a flashlight. Of them all, Echoi seemed the most likely place to begin my pursuit of an alternate history. At the time, I had little to go on except several vague and half-remembered mentions of this place and my eagerness to tell a different narrative about the past. Yet stories about Echoi were not forthcoming from my elder teachers. I had to seek them out.

My efforts to do so were fraught and complicated by what I gradually learned was a striking and profound reluctance to discuss many past events in detail, especially among Totobiegosode-Ayoreo. Usually, my attempts failed. I could never predict what responses a question about the past would elicit, no matter how innocent or general it seemed. Some people would respond casually and at great length, others not at all. One day, a middle-aged woman agreed to tell me her mother's teachings; the next day, she began to shake and sweat at the thought. When I asked one man to tell me about life after contact on the mission, he responded with a lengthy story about hunting peccaries.

Those who agreed to talk to me invariably set limits on what domains of the past they would discuss. When I asked Yoteuoi to tell me about visiting Echoi in his youth, he answered in the following way. "I will tell you the names of those who died. There were seven massacres of our people: Jochadaquide cachodi, Jnacaode cachodi, Chequedie cachodi, Junchaai'nate cachodi, Ichajuide cachodi, Pajine cachodi, Cuteri cachodi." He listed 118 names, then walked inside his house. He did not invite me to follow.

If such stories had a meaning that can be grasped, it was that some histories cannot be whole. At the time of my fieldwork, historical consciousness for Ayoreo-speaking people was an intensely political domain, composed of fragments and pieces that never quite fit together.[4] But its fragmented presence exerted a strange force on us all.

Codé was an elder Totobiegoto woman. She had been captured in 1986, and because she was the mother of another of my adopted mothers, Yijnamia'date, she was my classificatory grandmother. But she seemed to harbor no affection for me. After years of ridicule on the mission of Campo Loro, where she and the others were taken in the days after their capture, Codé specialized in hard doctrine and hard humor. She often sat shirtless around the fire and she reserved especially denigrating comments for visiting *Abujádie* like me. When someone told her that I was saddened by the death of Jnupi—a Direquednejnaigose man who was my first adopted father—she ridiculed my weakness and stupidity at length. "He was bad. It is good that he died," she said, unsmiling, in her high rasping voice. "I hope that all of us old people will die soon so that all those bad old things will die too and be forgotten forever."

While historians offered linear accounts of cause and effect, for elders like Codé the past was something to transcend. It was emphatically nonlinear and rotational, something that made unpredictable demands on the present. I soon discovered that Echoi, in particular, was defined by such profound tensions.[5] Elders across Ayoreoland, from the Direquednejnaigosode and Jnupedogosode in Bolivia to the Guidaigosode and Totobiegosode in Paraguay, concurred that Echoi was formerly known as *Erami Gatocoro*, the center of the world. Many said they had once considered it "the trunk of humanity" (*Ayoreode-ero*), a place where a permanent truce was in effect. Yet, in the same breath, these elders assured me that Echoi has since become irrevocably lost. Returning, they agreed, was now impossible.

It seemed that every attempt I made to approach Echoi was blocked. I began to envision Echoi as a mocking and uncanny place whose power was indistinguishable from three convergent strands: contested Ayoreo senses of the past, systematic efforts to exterminate Ayoreo, and my own thwarted desires for a site of redemption from the stark inequities that were so profoundly apparent in the Chaco. Echoi was uncanny to the same degree that the subject position of the Indian was always partial and violently interrupted. The negativity of Echoi as a space exceeded each of the contradictory limits set around viable Native life. For outsiders who located viable Native life in continuity with the past time of tradition, it was a landscape of loss that prefigured the present as an epoch of crisis requiring intervention. For Ayoreo-speaking people, it was both a site of nostalgic fantasy and immorality that must be transcended. It oppositionally framed modernity as a space only inhabitable for radically transformed humans. This complex terrain was also the severely attenuated zone of Ayoreo political agency, defined by local assertions of autonomy predicated on the value of negation, rupture, and transgression. Yet this negativity also posed a critique of the politics of indigeneity. Such emerging spatial ontologies refused the reduction of human life to the segregated and limited set of practices that counted as authentic Native culture in a neoliberal Latin America, and unsettled the contradictory notions of humanity organized around any coherent narrative of an "Indigenous" or cultural past. Ayoreo-speaking people did not piece together history from the remainders of their lives. Rather, they made the remainders of history into the future space of life itself. As such, Echoi demanded an ethnographic accounting of its shadowy presence, its menace, its powerful allure.[6] It revealed how the past was a domain against which the limits of modern moral life were created and evoked in contradictory ways. I thought I was up to the task.

Mining Memories

It was no coincidence that the same two elders from Bolivia who taught me about *adode* myths and *ujñarone* chants also provided my introduction to this place. Simijáné told me the first story I heard about *Echoi*:

They say that there was a woman named Samá. She went to the south looking for a place to stay. She found a lake, but it was very small, so she continued. Then Samá arrived at a good spot, which was beautiful and rich. She said, "Here I will stay." She tried to put her head into a hole, but it was too large to enter. She had to cut it into many pieces. *Nak, nak*, it sounded when she cut her head up. She had been searching and she found the place of heads. Her head belongs there. Her head rolled backward when her body fell, and it created a depression there that is *Echo babi* (Little Salt). Her bones made *Echo querui* (Big Salt). Her cranium is white and that is why salt is white. Her blood dripped onto a rock. The salt sometimes is red like that rock. She called one Little Salt, and she called the other Big Salt. Later, the Human Beings went there to find salt, but there wasn't any at all. There was a man named Ejeidayabi who made a long trip to the salt, but he only found mud there. The next year, he came back and it was still only mud. He was a clan relative of Samá and had learned her secret. He spoke the secret so that the salt will always be there and will never again run out. And to this day, the salt is still there. This is the secret of salt:

I am the sweet flesh of the earth
I am the sweet flesh of the earth
I am what the people want
I am what the people want
I satisfy the cravings of the people
I satisfy the cravings of the people
I am the sweet flesh of the earth
I am what the people like
I am leaving with her
I am leaving with her
I am looking for a place to live but there is not enough of the world to contain me
I am looking for a place to live but there is not enough of the world to contain me

I am going and I make myself fall
I am going and I make myself fall
I am she who seeks a world to live in that is cold and wet
I am she who seeks a world to live in that is cold and wet
This one fits my body and this one fits my head
This one fits my body and this one fits my head
But I am the one whose head wouldn't fit
But I am the one whose head wouldn't fit
I will send my head away from me
I will send my head away from me

As Simijáné concluded, my young Ayoreo translator turned to me. "Do not believe him," he said. "He is lying."

My adopted Direquednejnaigose father, Jnupi, had visited Echoi several times in his youth and had also met Bajebia'date, his wife, there during one visit in the 1940s. Jnupi provided me with my second set of clues about this place:

Author: Thank you, father, your stories are beautiful. I want to hear your stories about the place that they called Echoi.

Jnupi: There are many stories about that place.

Author: Hmm. It was a good place, right?

Jnupi: Those-Who-Came-Before called it the *Erami Gatocoro*, the Center of the Forest World. They used to say (but are lying) that it came from some old women.

Author: Hmm, that is good! I am looking for the reason why the Totobiegosode and all Ayoreo know the same stories if they were fighting, killing each other before.

Jnupi: They know the same stories because we lived together; we are the same people.

Author: Where did you live together? At Echoi?

Jnupi: That's the way it was. The people met there and each group gave each other things. In those days there was not yet any war between us.

Author: Hmm.

Jnupi: All the people, everyone met each other at Echoi. Direquednejnaigosode, Amomegosode, Jnupedogosode, Guidaigosode, Totobiegosode, Ducodegosode, Garaigosode, everyone. Cochocoigosode were there too. There were many of them, but they camped to one side of Echoi. We were all there together but we did not kill each other. Everything was good. This was not a long time ago, not long at all.

Author: It is good what you are saying.

Jnupi: One group came and another came and there they were together. The others put on their feathers and jaguar-skin headdresses and they did *chugu'iji*. We did the same. They yelled at us, "We are here and we are good inside!" They yelled loudly. We yelled back, "Don't be afraid of us, we are good inside! Here we are and we want to share turtles with you!" The others answered, "We have palm hearts and we want to share them with you." We shared with each other. No one could be angry; they were all happy.

Author: That is good.

Jnupi: That is what we did. We gave each other things at Echoi. We played games. The young women played with the young men. We were strong. It was very good there in that place, before the war. It was our trunk.

Author: Did the people do the ceremony for Asojná, the Nighthawk, at Echoi?

Jnupi: Yes, they did that thing there, off to one side of Big Salt. A place of sandy soil and a spring. Northeast. They said, "Here we eat the world and the world eats us." They called it our bridge. That is what they said. Our bridge. Sometimes we did it together with people from other groups. There are many stories about that. The people liked it at Echoi because they would give each other things. They talked a lot but more than that they would sing. All night long they sang.

This brief exchange not only seemed to confirm my desires, it exceeded them. I became ever more convinced that Echoi was the key to rewriting the past, present, and futures of Ayoreo-speaking peoples. Heartened by this apparent confirmation, I began to collect stories about Echoi whenever I could over the following months and years. These were usually not formal interviews but casual snippets of conversation written down after the fact and only later systematized. What was most striking about these fragments gathered across Bolivia and Paraguay was their consistency. Those Ayoreo elders who had made a journey to this place shared a remarkably similar set of impressions about the past significance of Echoi. A Ñamocodegosi man from the community of Tunucujnai in Paraguay remembered that the road to Echoi was clearly marked:

There were five different roads that our grandfathers traveled to arrive at Echoi. The roads were big and clear. They were the branches to our trunk. The Guidaigosode had one, the Direquednejnaigosode had one. The Totobiegosode and the Garaigosode shared the same road to arrive there. We shared the road because Echoi and its roads had no owner. Anyone can travel the road to Echoi.

I was told that these well-trodden footpaths led from the farthest lands of Ayoreo-speaking groups to Echoi. According to my teachers, people from all the various Ayoreo subgroups gathered there once a year, between July and September, to collect salt and socialize. The annual gatherings at Echoi could include hundreds of people from bands whose territories were several hundred kilometers apart. Each group, I was told, had a customary campsite around the borders of *Echo Querui*, or Big Salt, and together their camps encircled the white expanse. Each group would announce its arrival with self-identifying shouts. As one Ducodegoto elder from Campo Loro put it, "Before, the fires of the Ayoreo went all the way around." Everyone agreed that Echoi belonged to no single band but, rather, "belonged to all Ayoreode."

The place called Echoi was for all of the Human Beings, including the Direquednejn-aigosode and the Totobiegosode. It is a place that does not belong to anyone, and no one can remain there to call it their own. It is a place that all Human Beings can arrive at and pass through without fear.

Furthermore, all the elders agreed that there was a permanent peace in effect within the boundaries of Echoi. They said it was *puyaque*, prohibited or taboo, to wage war or incite conflicts with another group while camped at Echoi. As a Tiegosi man from Campo Loro told me,

Long ago, there was no war separating us. Echoi was like that time. A boy never had to carry his spear when he left the camp. No one ever thought to kill another person, and no one ever thought that another person would be waiting to kill him. Young men would go and visit other people. They weren't afraid of the others before because there was no war there.

During stays that varied from several days to six weeks, people from different groups played games focused on reciprocal giving such as *parachiapidi, carui, cukoi, pimechekua*, and *pu'uguyapidi*. These games were a prelude to more extensive exchanges of gifts between different groups. Exchange at Echoi, according to such memories, mimicked the exchange between *urasade* co-band members in that it was also ideally focused on an exchange of equivalents rather than material accumulation. This was known as *yico oredie*, or "we exchange shadow / images." Thus, one group would give turtles and receive palm hearts. These palm hearts might be given to another group in exchange for turtles, or any one of a number of items perceived to be equivalent, such as one ball of *dajudie* twine, one full *catojnai* honey gourd, one armadillo, or one piece of larger game animals such as peccaries or giant anteaters. These items were common and readily available to any Ayoreo group. Gift giving sometimes even included the exchange of identical items, for example, a turtle for a turtle. I was told that exchange at Echoi did not include any formal or large-scale exchange of highly valued items such as metal axes or materials specific to one zone, such as *curude* red ocher. The possession of such items was reportedly a source of conflict, except in the unlikely case that they remained in a state of nearly continuous motion between owners.

I was told that, at Echoi, young, unmarried people started temporary love affairs and families arranged marriages with people from other groups. "The young men from different groups would play at night together with different girls. They were together for the nights that they stayed at Echoi." Young men would commonly decide to go on visits (*daquiaque, dajetagode,* or *ugumieaque*) to the country of another group, particularly if they were interested in a young woman. This is what happened to Jnupi:

My wife was a Guidaigoto, and her people lived over there between Tres Cruces and Pozo. The people of her mother lived there. My people are Direquednejnaigosode. I met my wife at Echoi and took her back to my people. We have been together since then.

During encounters that would last all night (*a niome yoque*), men and women from different groups would gather to discuss the events of the past year and exchange stories at Echoi. As one Totobiegose from Arocojnadi recalled,

The people loved going to Echoi because they would meet other groups and give each other many gifts. We told our stories, of hunting or if someone killed a jaguar or if someone killed a *Cojñoi*, during those nights at Echoi. They always talked a lot, but more than talking they would sing.

By all accounts, singing figures prominently in elders' memories of Echoi. As I was told, singing "brings out what is inside" (*chijna bajeode a guesi*), referring to an act of publicizing the corporeal seat of emotions. Citing an emotional event in a song is believed to ideally provoke the circulation of *ayipie* soul matter. It is the locus of individual personality, but it can also be shared among two or more people, such as spouses, families, and members of any group focused on a particular task. Those who share one *ayipie* are presumed to possess similar attitudes, particularly about proper morality. Songs were exchanged between groups at Echoi in a highly ordered way. Through such song exchanges, Ayoreo-speaking people established a sentimental register for the face-to-face narration of events. A sweet (*uneja*) song is said to cause people to share the same feeling, while a bitter (*derocoro*) song is seen to be an inadequate vehicle for communicating either the poignancy of the lyrical situation or the potency of the singer. This not only created a forum in which listeners and singers participate in a shared chain of feelings through the event described in the lyrics, but it also foregrounded the social negotiation of the causal relationships between event, place, and feeling.

The salt at Echoi slowly emerged as last season's rains evaporated. By the middle of August, the salt was usually ready for gathering. This process could be expedited by piling salty mud into small wooden frames from which the water evaporated over the course of months. The annual gatherings at Echoi often coincided with the first sightings of the Red Star (*Guedo Carate*) and the first cries of the ritually powerful Nighthawk (*Asojná*), recently returned from her seasonal migrations north. These signaled the time for the main annual ritual of renewal. Ayoreo called the place of this ritual *ore pajapidi*, or the bridge, and believed that it was *yoca'apacadi*, a place where the *ujopie* soul matter of *Asojná* remained. This ritual, which has been described in detail by Heinz Kelm and Bernd Fischermann, reconstituted the relationship between Human Beings and the wider ecology of metaphysical forces. It marked the beginning of the process by which the closed world of the dry season, with all of its strict rules for human behavior, would transform into the open world of the rainy season, a time of abundance and relaxed taboos. It was a time in which adolescents were initiated into adult status, food items were gathered and redistributed, and potent songs were performed.[7]

In addition to Jnupi, seven other people told me they remembered that there was a designated place in Echoi where Ayoreo regularly celebrated this two-day ritual. "In that place we had the fiesta of *Asojná*. We all knew that place. If we heard *Asojná*, we did it there. Our fathers and their fathers told us about that place." Such ritual sites were considered to be places of lingering power and were avoided after the ritual was performed. Moreover, Jnupi and the others were also able to recall years in which the people from more than one *uage* subgroup performed the Asojná ceremony, a rite of annual renewal, together at this place.[8]

The Mists of History

I was astonished by these stories, not least because I had never previously heard anything about this place. Was it possible that I had failed to pay sufficient attention, that my ethnographic sensibilities were not sensitive enough? I asked Yoteuoi in Arocojnadi why this might be, why elders rarely mentioned Echoi even in formal interviews with outsiders.

"Because the *Cojñone*," he told me, "only want to hear about the time when we killed each other." This answer seemed to suggest a profound point also emphasized by Walter Benjamin and other theorists of history. Echoi was an image of the past which threatened to disappear, in part because it was not recognized as a meaningful concern of the present.[9] That is, the figure of the past deemed important reorganized the causal links and meaning of Ayoreo existence within history. History made and unmade places. Yet the past itself was highly contested terrain.

The stories that Ayoreo-speaking elders told about Echoi between 2001 and 2008 seemed destined to disappear, not least because Ayoreo informants self-consciously edited them out of their interactions with outsiders and younger Ayoreo alike. Yet Echoi initially appeared to me in these stories as a place around which a powerful counter-history could be organized—a counter-history that resonated with my aspirational sense that ethnography could identify and expand the limits of possibility in zones where death was otherwise a foregone conclusion. Such a revisionist genealogy seemed to be articulated in Echoi's presence as an abandoned *axis mundi*, or in Ayoreo terms, *yequerodie Echoi nanique ome ayoreode uyoque*: "Echoi was the trunk of all Ayoreo before." This phrase implied that Echoi as a place and a set of practices was the origin or fundamental source for the humanity of Ayoreo-speaking people. In the figure of Echoi I began to cocreate and evoke together with my elder teachers, I thought I had found a serious challenge and alternate line of descent to much of the history outsiders had written for these people.

Ayoreo stories about Echoi seemed to suggest the existence of historical joint-use areas, a finding that challenged the widespread notions that Ayoreo-speaking people inhabited rigidly defined and exclusive territories and that territorial affiliation was the primary source of band composition or individual identity in the decades preceding contact.[10] The existence of peaceful zones also upended the limited idea that historical violence between Ayoreo-speaking groups necessarily arose from territorial disputes or can be explained solely as an internal function of a static and intrinsically violent "traditional culture." In fact, my elder Ayoreo teachers said that precisely the opposite was the case. They emphasized that the profound violence between Ayoreo groups in the first half of the twentieth century was an aberration. That is, they suggested that some other interpretation of violence and abandonment was needed.

Yet try as I might, I could never entirely peer through the mists of history. The stories about Echoi as a place of unity, repeated over and over again, were offered to my tape recorder in exchange for food or a sympathetic ear. They were recounted around smoky fires against the stark backdrops of squatter camps, under tarps, and in scraps of brush and mud huts on overcrowded missions, far from the deep forest they figured. I remember those scenes now as a series of disjointed images of brown skin and muscle cramps and biting flies and acrid smoke and tepid water and earnest voices, but more than anything I remember how the stories of a vanished world were invariably suffused with a dignity impossible to convey or forget.

I've since discovered that many of my recordings are inaudible, drowned out from time to time by blaring disco music and barking dogs and screaming children and roaring motors. Usually the narrators were facing an old age that was precarious at best and often cruel. Many of them were starving and sick. Nearly all have since died. The contrast between the degraded conditions of their everyday lives and the utopian scenes they described was so great that it often strained my capacity to understand even as the same tension reflected what I wished so desperately to encounter and to believe. Oddly enough, the stories became ever more real for me the more predictable they became, the more fully they faded into memory. I thought I could sense the brown bodies and red firelight and hushed voices and epic tales of bravery and betrayal and revenge. I was haunted by the past I could only imagine. It was impossible to separate my own sense of loss from the tones of nostalgia and escapism that saturated these stories of a clouded place of pleasure and peace and eternal youth last visited decades before. As I pondered these stories and their stakes, as I listened to them over and over, Echoi began to take on the shades of a lost paradise, even though it was never entirely clear for whom.

A Lost Paradise

Colonial archives revealed that Echoi has long figured within frustrated Euro-American geographies of desire and redemption. Although many early modern European scholars imagined this area and all American tropics as the benighted domain of savagery, others argued that the Garden of Eden was located remarkably close to the area called Echoi.

As historian Heidi Scott has described, jurist Antonio Pinelo de León popularized this thesis in his mid-seventeenth-century text titled *Paraíso en el nuevo mundo,* or "Paradise in the New World." For Pinelo de León and his later interlocutors, it was a crucial argument for the development of a *criollo* consciousness. Known as "the American thesis," this idea evoked the sense, common at the time, that the unique climate and ecological diversity of the Americas exerted a positive effect on its European inhabitants, in order to justify colonial domination by naturalizing the social distance between Euro-American colonizers and Indigenous peoples. The fact that savages now occupied the former location of the Garden of Eden was thought to reflect the degraded nature of their humanity. God's Grace was found in the bounty of American nature, and *criollos* became its only legitimate cultivators. Accordingly, *criollo* and nationalist scholars continued to argue that the New World was the site of the lost Paradise for the next two hundred years.[11]

At the same time, practical efforts to colonize and thus restore divine order to this region directly affected the everyday lives of Ayoreo-speaking peoples. I pursued Echoi across archives in the United States, Argentina, and Bolivia. And I found that Ayoreo groups and the Salinas were inseparably linked in outsiders' imaginations of them. The Jesuit priest Juan Bautista de Zea provided the first description of Ayoreo-speaking groups in 1711. According to de Zea, they were "people of large stature and good strength." He took special note of the respect they paid to their wives:

They honor their women with the title of Ladies, and truthfully they are so, because they command their husbands, and it is due to their whims that they move from one place to another; they never put a hand to domestic chores, rather, they are served by their husbands, for even the most humble tasks. Although they have chiefs and captains, this does not mean they have either government or religion, and they only have some reverence for the Devil's company. . . . To drink they have some jungles of palm trees, from whose trunks they extract the thick and spongy heart, which when squeezed alleviates the lack of water. In the winter it is very cold there and also freezes, which does not bother the natives, although they go about naked, because their skin is covered with callouses two fingers thick, and because of this they are robust, strong and of great endurance, there are men and women that exceed 100 years of age and die with no other ailment than old age.[12]

Word lists compiled by early eighteenth-century Jesuits (and a linguist named Ignace Chomé, in particular) prove that the language spoken by these people is nearly identical to that of contemporary Ayoreo.[13] Moreover, the Jesuits noted that these Indians shared the same language with several other groups—Curacates, Cucutades, Ugaroños, Sapios, Zamucos, Careras, Zatienos, and Ibirayas. All of these groups, the priests noted, live together "near some Salinas."[14] That is, archival sources not only demonstrate that these Zamuco groups were the ancestors of contemporary

Ayoreo-speaking people but they also establish that Ayoreo have held deep cultural ties and aboriginal title to the area of Echoi for, at the very least, the past three hundred years.

Jesuits, keenly aware that this area was an important geopolitical link between their increasingly besieged missions in eastern Paraguay and those in the Chiquitania area of eastern Bolivia, began an ambitious campaign to contact all Zamuco groups and reduce them to a mission, the southernmost of all Chiquitania.[15] Archival sources indicate that, despite several attempts in which priests were killed and many disputes over its precise location, the mission of San Ignacio de Zamucos was founded in 1723 near Echoi. The first inhabitants were two bands of "Cucutade" who were "fleeing from the Ugaraños" and "ill with the plague." Father Agustín Castañares undertook most of the contact work in the early years of the mission.[16] Abandoning his initial plan to turn Zamuco groups into sedentary agriculturalists, Castañares lived as a seminomadic forager alongside his congregation for the next five years, always in the vicinity of the Salinas. Eighteenth-century maps based on Jesuit sources thus invariably link the salt lakes, Zamuco groups, the mountain labeled Yoibide, and the mission of San Ignacio de Zamucos, albeit in slightly different configurations.[17]

An anonymous Jesuit, perhaps Jaime Aguilar, makes the same link between San Ignacio and the Salinas in an unpublished 1741 account entitled "Relación de San Ignacio de Zamucos." In his account of surveying the mission site two decades earlier, it is clear that salt was an important factor in the placement of the mission and that the salt marketed by San Ignacio came from the same Salinas controlled by the Ayoreo-speaking group called Zatienos:

[For the Fathers] one thing was missing at this location, a thing absolutely necessary, salt. This country had been until that point lacking any salt lakes, but there was some vague suspicion that these could be found in the lands of the Zatheniens. A large number of our Indians [Chiquitanos from San Jose] wanted to find out and clarify this fact. After having traveled through all these woodlands without having discovered any indication that there was salt, one of these Indians went up on a small height to see if he could discover from there what was so dearly wished. He saw in the not-too-far distance a lake of colored water, surrounded by low brush. Because of the hot weather that day, he decided to cross through these low scrubs in order to bathe. While entering the water, he noticed that the lake was covered with some sort of glass [verre], which he took in his hand and found it was half-formed salt. Satisfied, the Indian called his companions, and the missionary, having been informed, took measures to make a road to that place and to put it under the care of the idolizing barbarians.

The mission of San Ignacio, however, proved ill fated and short lived.[18] Jesuit records show that San Ignacio remained the smallest of the Chiquitos missions throughout the late 1730s, with the population listed as 648 in 1742, 666 in 1743, 679 in 1744, and 683 in 1745.[19] Droughts, attacks by hostile Indians, and tensions between various Ayoreo-speaking factions led the Jesuits to question the viability of the mission. Finally, "afraid for their lives," the priests burned the sacred oils, buried the sacred objects, and abandoned San Ignacio de Zamuco on October 24, 1745. This decision was at least partly due to the stated desire of the Ayoreo-speaking neophytes to return to their "primitive life of savages in their old camps."[20] From these Jesuits, the Ayoreo learned their word for the Christian God: *Dupade*, adapted from *Tupã*, the Guaraní name for the sun god.

After the Jesuit expulsion in 1767, references to the various Zamuco groups described by the Jesuits largely vanished from history. It has been shown that after the hurried Jesuit expulsion (in which they commonly burned all the records they could not take with them), colonial administrators, Franciscan priests, and neighboring Indigenous groups substituted other labels and names for Jesuit categories of place and peoples, including the Zamuco groups.[21] Various names applied to them after the Jesuit expulsion include *Yanaigua* (lit. "wild man," a derogatory term for Ayoreo-speaking people from the Chiriguano), *Guarañoca* (a derogatory term from the Chiquitano that became widespread on the eighteenth- to nineteenth-century Bolivian frontier), *Pyta Jovai* (lit. "claw foot," used by Guarani peasants in Paraguay to the present day), and various Spanish names like *Moro* ("moors"), *Empelotudos* ("naked ones"), *Bárbaros* (barbarians), or even, at times, *Guaicuru*. That is, after the expulsion of the Jesuits, all Ayoreo-speaking groups were once again labeled barbarians and the site of their former *reducción* resignified as an untamed wilderness in need of moral and economic colonization.

The Sediments of History

Colonial administrators aimed to exploit the salt of Echoi, regardless of the kind of Indian living nearby. Indeed, under colonial laws Zamuco, Toba, or Guaicuru were largely indistinguishable by the colonials: "wild" or "savage" Indians were not capable of possessing land or property. In 1795, the newly arrived Franciscan priest of Santiago de Chiquitos, a small mission located some one hundred kilometers northeast of San

Ignacio de Zamucos, traveled to the Salt Lakes of Echoi in the company of a number of neophytes—some of whom, we may suppose, were descendants of the several hundred "Zamucos Indians" relocated to Santiago by the priests after the abandonment of San Ignacio in 1745. The Franciscan wrote a letter to the governor reporting his discovery of—and claim to—the great salt deposits in the heart of the land occupied by Ayoreo-speaking people, near what he describes as the ruins of San Ignacio de Zamucos.[22] The strangely shaped mountain he describes is Cerro San Miguel, which is also the landmark noted on Jesuit maps with the Ayoreoized word *Yoibide*:

The Salt Lake is distant from Santiago by some 60 leagues to the south, beginning six leagues from the estancia of San Joaquin. . . . One emerges from the point of a palm savannah, where a mountain is found, very tall, a quadrilateral in shape from south to north, where various fires may always be seen. From the base of this Mountain, looking west from the palmar the Salt Lake can be seen at a distance of 10 leagues. It is about a league and a half long, and a quarter in width. Two arroyos leave from this, one from the south, and another from the north. The salt is very beautiful and clean, colored white and pink. From this Mountain . . . the Salt Lake to the west can be very clearly seen, and to the north all of the Mountains of the province are seen, like those of San Jose with its Salt Lake, as well as those of San Juan and Santiago. Finally, it should be noted that according to the signs and paths there are many people who come here to collect salt, and in part of the mountain there are various cuts in the trees that mark the direction of the road, and that in the direction of Santiago, there is a wide trail opened.

Salt was a centrally important commodity along the Chiquitania colonial frontier. It was traded between missions, used to cure beef shipped to the silver mines at Potosi, and often served as a frontier currency by which missions could purchase goods and pay taxes. The inhabitants of Santiago, for instance, used salt from the Salinas to buy supplies from the neighboring missions of Santo Corazon and San Juan. Chiquitano Indians made regular trips to the Salinas to mine salt, with each salt lake assigned to a different mission town in the last decades of the eighteenth century. A colonial tribunal prohibited Native salt gathering in 1792, stating that the Salinas did not belong to the Indians, and reserving it for commercial exploitation by the Crown. Although this met with severe opposition by the Indians, who "laid around crying" and claimed they had used it since "before the expulsion of the Jesuits," the order apparently stood.[23]

Instead, colonial authorities granted a royal provision to the young German botanist Thaddeus Haenke, then employed by the Spanish government, for a June 1799 expedition to the Salinas of Echoi. He was instructed to inform the colonial tribunal of the "cost and utility" that the public could expect from the "benefits and use of these Salinas," whose salt was widely held to be superior and healthier than that used in the city of Santa Cruz. Haenke's report caused the new government of the Provincia de Chiquitos to issue an order on November 9, 1799, that "all possible means are used to develop the Salinas" found south of Santiago, and to take all necessary measures to ensure "its extraction and benefits, not only to supply that province, but also the province of Moxos." Indeed, the salt of Echoi was reportedly greatly desired in the regional capital of Santa Cruz for the intensity of its flavor and supposed health benefits. Soldiers and administrators began gathering salt from the Salinas in the dry season of 1800.

Approximately ten years later, however, resistance by Ayoreo-speaking people forced the Crown to give up these plans. After the governor of the province sent a military expedition to "conquer" and reduce Ayoreo-speaking bands living near Echoi, these Indians "frequently damaged the exploitation of the Salinas, attacking the Indians of Santiago and San Jose that headed there every year," until they were left "in peace" in the forest.[24]

By 1886, Ayoreo-speaking inhabitants of Echoi were regarded as savages living in a natural wilderness. In his book *Las Misiones Franciscanas entre los infieles de Boliva*, Franciscan missionary José Cardus makes special mention of the "savage tribes" living near the Salinas. Cardus's entry on the "Zamuco" is remarkable for its vivid detail. In his description of these Indians, Cardus mixes frontier anecdotes with information obtained from "an Indian woman of said nation that was taken prisoner." With this anonymous slave as his guide and informant, in 1875 Cardus traveled to "those places that she knew" near the Salinas, surveying village sites, gardens, and discarded tools.

After the expulsion of the Jesuits, Cardus writes, these Indians "returned to their ancient way of life, to their independence, without communication." They lived "in the same place where the Mission was," which according to him, was fifteen leagues to the south of the Salinas of Santiago. Another group of them are nomadic, moving "near the Salinas of Santiago and San Jose and their confines." These Indians, he says, "live a very anguished and miserable life, which is doubtlessly in part due to their customs." He notes the small "ridiculous" gardens, how they would sleep "piled one on top of another like so many pigs," the "well crafted" digging sticks of hard *palo santo* wood, the "very curious" rectangular sandals, and their formidable war clubs.[25]

Cardus reports that these people organized expeditions to steal metal, and he was struck by how they cut up a single knife to make several wood-hafted tools. Like D'Orbigny, Cardus linked Zamuco groups with the area of Echoi and their defense of it to violence against them:

The Zamuco before were somewhat gentle, or at least one could talk to them, and they came out to converse with the Chiquitanos and whites that went to mine salt, and they knew how to ask for some tools or food. The Christians, however, were the first to commit outrages [*tropelias*] against them, even killing some of them: and since that time [1845] . . . they have declared themselves to be enemies of the whites and Christians, taking revenge whenever they can. Because of this, to mine salt at the Salinas of San Jose and Santiago, it is necessary to go in a group of many people, and always with great care.[26]

Archival sources, then, reveal how changing visions of this area and its value were reflected in changing perceptions of Ayoreo humanity. For outsiders, the place known as Echoi was alternately a zone of diabolical savagery, a possible location of the Garden of Eden, a key geopolitical outpost, a site for economic production and resource extraction, and finally, an untouched wilderness. Likewise, representations of Ayoreo-speaking people mirrored these shifts. Images of their humanity flashed quickly from savages to converts and back again, before, in the early twentieth century, they were perceived as "uncontacted" and thus in need, once again, of contact and conversion. Each shift required a new version of history to legitimize and sustain its meaningful projects.

Place Making

One alternate way to apprehend Echoi was as a manifestation of the complex ways that Ayoreo-speaking people organized social space. Ayoreo people used the word *uniri* to refer to the areas belonging to a certain band or people, and it was often translated by them as *territorio*. When I asked Totobiegosode men to draw maps of the *uniri* their people ancestrally controlled and that they themselves occupied before contact, they invariably depicted their land as comprised of distinct and separated sequences of circles in a wider empty space. The circles represented large named and fixed regions called *pamite dateode*. People said these regions—which they drew like disconnected beads—were the salient features of their *uniri*. They usually traced one meandering striation of sandy soil, or *ugaraijnai*. The borders between these places—*paminone erueode*—were recognizable by the location of particular named watering holes, springs or areas where something had happened and that "everyone knew was part of that *pamite datei*."

The *pamite dateode* regions, in turn, were subdivided into smaller named regions, *paminone dei sona*. These were productive spaces, which included defining topographical features; a dizzying array of named, permanent villages, or *idaiode*; temporary dry-season camps, or *deguiode*; and hunting areas. Favorite spots where people routinely camped but didn't build a house were called *yocapaingai meque*, the same term used to describe a place where turtles make a shallow depression to sleep but do not build a nest. The *paminone dei sona* of any *pamite dateode* could also include specific areas set aside for hunting and gathering activities, such as three-day *pimoi* trips, or the longer *pachabie*. These hunting areas rotated throughout the year. The empty spaces on the dust maps were also an integral part of each *uniri*. These were places that were *puyaque* and off-limits but symbolically rich. They included burial places, sites of murder or death, and places where rituals associated with blood, such as *taboidi* or *ditai*, or renewal, such as Asojná, were held. These were not empty spaces but places of ambivalent spiritual power, whose significance exceeded any simple economic or utilitarian function. Furthermore, each *uniri* and the areas between them also contained places of human absence. There were places through which people passed but did not stay for various reasons and the well-defined roads that "break our land," *yocuneone ahugeusori*.

In this system, borders were not fixed lines that divided one area from another. It was impossible for each person to visit all known places within their lifetime, and those who hadn't did not necessarily know which particular place corresponded to which *pamite date*. I was told that this was common when people moved far away from the areas they were accustomed to into zones that they had only heard about from their grandparents, relatives, or others. However, my teachers also strongly emphasized that Ayoreo people were rarely lost in the forest, because they knew where they were at all times. "We knew it because it was our country." How, I asked, could they know where they were if they didn't know which zone they were in? I was told that in such cases, people could recognize the places they encountered on the basis of something we might call intuition. Certain individuals were able to discern their passage into another zone or *uniri*, even if they had never been there before, by "searching their minds," or "their insides."

These *pamite dateode* were not only shared by different bands of the same *uage*, but the margins of large spatial divisions were often shared by different Ayoreo-speaking groups, such as the *garai* prairie on the southern edge of Manenaquide'uniri. Because of historical alliances between the Totobiegosode and Iñojamuigosode, bands from the two groups ranged into each other's *uniri* even though other bands of the Garaigosode allied with the Iñojamuigosode sought to exerminate the Totobiegosode. Such fluid boundaries and agreements created a number of wider joint-use areas around the margins of fiercely defended territorial cores. In times of peace, Ayoreo-speaking groups could send out messengers and request permission to visit or cross their *uniri*. I was told that such *dajetagode* visits from other groups, even from distant *unireone*, were common.

Place names, as well as borders, were also highly fluid. Places were descriptively named after events that happened there. This system of naming meant that a single group of people for each place could have various names, and that the group's knowledge of place names and social events was continually evolving depending on its composition. It also meant that events could potentially delimit a new place. As Yoteuoi told me, "There are very many places in our country. There are so many places that even if we all sat here and helped one another we would never be able to tell you all the names of the places where people used to camp."

A single individual only knew part of these place names, depending on his or her social relations. That person could never know all the names of the places within his or her own *uniri*, nor were the referents of such names entirely fixed. This multiplicity was not seen as contradictory or confusing. Rather, place names were constantly being generated, applied, cited, resuscitated, forgotten, or disputed. Such names were—are still— often invoked to show solidarity or index a shared experience that ties an individual to the group. This meant that the landscape or *uniri* space did not contain named places that imbued it with meaning but, rather, that the practice of citing and naming places patterned time and space in ways that always had the potential to reorganize it.

This fluidity also applied to Ayoreo ways of grouping people, from band to *uage* village confederations. Each *uage* was named after a particular place, such as the People-from-the-Place-of-the-Village, or *Guidaigos-ode*; People-from-the-Place-of-the-New-Day or *Direquedenejnaigosode*; and People-from-the-Place-of-the-Hole-in-the-Ground, or *Jnupedogosode*. Yet neither these names nor group membership were fixed. Genealogies I made in the Totobiegosode villages suggested that such circulation was pervasive. Although Totobiegosode say they have not had sustained peaceful contacts with Guidaigosode since around the 1930s, nearly all living Totobiegosode can count someone who originated in a Guidaigosode subgroup as a direct ancestor within the last three generations.[27] The *uage* group, in other words, was not historically conceived only as a function of its individual members, nor did group identity or territorial range define the limits of the individual or the family. Echoi was perhaps the most elaborate expression of a general socio-spatial system, in which group affiliations were counterbalanced by wider bonds of shared belonging. These bonds were emphasized in face-to-face encounters between groups that were as mutually suspicious as they were deeply related.

Colonial Coordinates

Echoi also played a central role in colonial cosmologies of space and value. My explorations in the archive revealed that the association of Ayoreo-speaking people and the area of Echoi organized a variety of colonial projects. The success of early missionary efforts to collect Ayoreo souls depended on the historical link between Ayoreo and Echoi. Like the Jesuits and Padre Cardus before them, twentieth-century evangelical missionaries learned of Echoi's significance from Ayoreo slaves. Armed with this strategic information, missionaries soon began to concentrate their "first contact" efforts on the vicinity of the salt lakes.

These contacts were the final stage in a much longer process of dispossessing Ayoreo-speaking people of land and life. In the 1870s, the Paraguayan government sold off large portions of the territories of southern and eastern Ayoreo-speaking groups on international markets to pay for the disastrous War of the Triple Alliance, which killed nearly 75 percent of Paraguay's total population.[28] Much of this land was bought by tannin companies seeking the red axe-breaking wood of *quebracho colorado* trees used to tan the hides of feral Argentine cattle for later shipment to Europe. For instance, land occupied by Totobiegosode and Garaigosode was included in the holdings of the quebracho baron Carlos Casado, who at one point owned more than eleven million acres of land through which his own private railroad extended 145 kilometers inland from the Rio Paraguay. To replace the extreme war casualties, Paraguay also encouraged European immigration, particularly from Germany.[29]

Tensions between Bolivia and Paraguay over control of the northern Chaco were already simmering by 1900. As part of Paraguay's claim to the area, the government brokered a deal with Casado in which part of his Chaco holdings were granted to a group of Canadian Mennonite colonists who had emigrated from Russia in 1874 but, whose pacifistic stance had created a hostile environment for these German speakers in post–World War I Manitoba and Saskatchewan. In 1921, the government began negotiating with Mennonite delegations, and Mennonite colonies were established in the central Chaco in 1927. In exchange for "populating, cultivating, and civilizing" the so-called Green Hell, Mennonites were promised special rights (including political autonomy, exemption from taxes and military service, and permission to establish their own German-language schools and ban alcohol in their communities). These positions were formalized in 1921 through Law 514. By 1928, eighteen Mennonite villages and 255 farms had been founded on sixty-five hectares of land in the central Chaco. Two other waves of Mennonite immigration in 1930 and 1947 added additional colonies, each comprised of numerous villages and corresponding farms laid out in the *Strassendorfer* administrative style imported from sixteenth-century Prussia, with its spatial emphasis on community order and church supervision.[30]

Salt is also a key indicator of the presence of hydrocarbons. Oil was discovered in the Andean foothills in 1919, a crucial factor in the subsequent Chaco War (1932–35). Standard Oil, which already had producing wells near Villa Montes, Bolivia, began competing for leases in the northern Chaco with Shell Oil, which was affiliated with the Paraguayan government. Both governments took a sudden interest in this area, which included a vast amount of ancestral Ayoreo territory, and both sides began making preparations for war. These preparations included building roads and forts. In 1913, the Paraguayan army surveyed a road straight through the center of the lands occupied by Ayoreo-speaking groups, north from Fernheim to Madrejon, some fifteen kilometers east of Cerro León. In the late 1920s, this road was built and forts erected along its length, almost always near the few permanent water sources in the region.[31]

Echoi again figured prominently in these geopolitical skirmishes, precisely because of its earlier importance for the Jesuits. In the buildup to the Chaco War, history was instrumental. At the center of this was a debate about the precise location of the "lost" mission of San Ignacio de Zamucos. Government officials from Bolivia and Paraguay eagerly sought historical precedents for their claims to the northern Chaco. Thus, Paraguayan scholars placed it north of Echoi, while Bolivian sources from the 1930s regularly placed San Ignacio far to the south, near the site of present-day La Gerenza or Fortín Ingavi.

These contested historical geographies became military coordinates. In July 1931, the Bolivian Lieutenant Coronel Angel Ayoroa set out south from Roboré to the Salinas, with the objective of "reincorporating the region of the Zamucos" into Chiquitanía and the Bolivian nation. Ayoroa followed "the old Jesuit road" straight to the base of Cerro San Miguel. He and his men worked for several months opening the road and constructing a number of forts near water sources found along the way, including San Ramon, Ravelo, and Ingavi. Their route passed directly through the center of the lands controlled and favored by Ayoreo-speaking people, pivoting near the base of Gososo or Cerro San Miguel, twenty-five kilometers to the east of Echoi, continuing south past Nakaje, or Palmar de las Islas, and terminating to the west of Cucaani, or Cerro León.[32]

This route became one of the main transport roads for troops and supplies in the Chaco War. Between 1932 and 1935, more than one hundred thousand soldiers, most of them Quechua and Aymara Indians from the Bolivian highlands, moved through these roads in the northern Chaco, along with military transport trucks and airplanes sent by German firms for field testing prior to World War II. More of these soldiers died from thirst and exposure than combat. Warplanes regularly flew over the villages of Ayoreo-speaking peoples.

The invasion of Ayoreoland only increased when the war straggled to a close in 1935. The peace treaty that ended the Chaco War included a provision that enabled Bolivia to access the sea via the Rio Paraguay, essential for shipping hydrocarbons. Because of that agreement, construction on the Santa Cruz–Corumba railroad began in 1945 and was completed ten years later. It was laid west to east across the paths to Echoi, effectively sealing the Direquednejnaigosode and many other groups from their southern territory, as well as facilitating the efforts of those eager to track down and enslave Ayoreo-speaking people. Furthermore, the highway and the railroad opened the area for intensive cultivation and permanent settlement. In accordance with the 1943 Bohan Plan, the railroad was seen to be an ideal route to promote the colonization of lowland Bolivia by highlanders, and colonies of domestic and foreign immigrants were quickly established along the tracks. The railroad also played a crucial role in efforts by missionaries to follow and contact northern Ayoreo-speaking groups.[33]

In the southern Ayoreo territories, the road from the Mennonite colonies north to Madrejon separated the regions known in Ayoreo as Ñacore'Abode from those of Tamocode, Amotocodie, and Chungupere'nate. The Totobiegosode called this road Dajei'date or the "mother road." They say this road "cut our land into parts, *yocuneone ahuguto.*" In Bolivia, the road south from Roboré established permanent *Cojñone* settlements near Echoi, Nakaje, and Cucaani. This road skirted

Echoi and later became the principal route for clandestine trade in the Chaco, beginning with smugglers moving Bolivian rubber to be sold to Nazi Germany during WWII and, later, stolen automobiles and cocaine paste. Mennonite colonists and immigrant ranchers also began pushing farther north along the Dajei'date in the 1950s. From these routes, farmers, ranchers, soldiers, and hide hunters moved into areas that had long been extremely difficult to reach.

At the same time, petroleum exploration in the Chaco began in earnest after the Chaco War, and oil companies expanded military roads. The first well in the Paraguayan Chaco was drilled in 1944 by the US-based Union Oil Company, under a decree by the dictator Morinigo. Throughout the 1940s and 1950s, international companies such as Texaco, Standard, Pennzoil, Placid Oil, and Victory Oil drilled forty-seven additional wells in the Paraguayan departments of Boquerón and Alto Paraguay. The Pure Oil Company of Chicago, then one of the largest companies in the United States, obtained leases in the middle of the home territories of the Ducodegosode and Ijnapuigosode subgroups, to the north and east of Cerro León. The company established its base of operations at Madrejón in the late 1940s. By 1948, Pure Oil was drilling north of Fort Pitiantuta and had brought in heavy machinery to build water tanks and roads. The line of wells began in Teniente Martinez and continued to Madrejón, Cerro León, and near the Bolivian border at Mendoza, along Dajei'date and straight through Ayoreoland. From their base, Pure Oil Company workers had repeated sightings of Guidaigosode groups, with whom they established a series of contacts in 1958 at a drilling rig abutting Cerro León.[34]

Violence as Antihistory

Ayoreo groups could not defend their ancestral territories in the face of such a massive invasion. Unknown diseases came from the metal and clothing of the *Cojñone*. As early as the late 1920s, measles, influenza, and smallpox devastated Ayoreo camps. Ayoreo *Daijnane* shamans could not cure their people and were blamed for the epidemics. In response, northern Ayoreo groups decided to massacre all shamans around this time:

When they killed the shamans I still lived with the Direquednejnaigosode. I was about 14 or 15 years old. I think that in each group and in the old camps there were leaders of the shamans. Because there were shamans in my village I will say some of their names: Ujnohihai, Jochabiadacode. When the people killed all the shamans all the people from all the groups gathered together. That was because there was no longer anybody to make another person sick. Everyone was calm. Some said that the shamans were the ones who cured people who were sick, and others said, "I am against the shamans because they were the ones who were killing all the Ayoreo people."[35]

Simijáné told me about this time in the following story:

The people said they were going to find all the shamans and kill them. Ajnisidai was frightened because he knew that he was a shaman. The son of Ajnisidai told his father that other Ayoreo bands had already killed all of their shamans. The Jnupedogosode and Uechaemitogosode had also. All of the Direquednejnaigosode Ayoreo were going to kill all of the shamans. The people were searching for Ajnisidai to kill him. He knew that there were warriors coming to kill him. So his family dug a grave for him to enter and die in. He told a woman named Diguere to dig him a grave. But before he entered the grave he had something to tell. He said, "I'll get into the grave. But I'm going to put on my *ayoi* [jaguar headdress], and my *cobia* [vulture feather collar] will go with me to the grave." He said, "I'm going to finish telling what I have to say and then I'll go to the grave. I'm going to tell about the time I shot a white man with an arrow. But I'm going to enter the grave." He hadn't tied his hair up when he was telling the story. . . . But he tied it up [as if for war] when he was ready for the hole. He was angry, already thinking about the earth that is black as if it were burnt. He was angry with the overturned soil of the grave. The sickness of the black earth had grasped him, it had grasped his eyes. Then, he began to paint himself. He painted his entire body, his shoulders, his stomach.

His children told him, "Papa, the grave is ready." When he had entered the grave, his son said to him, "Father, don't hurt those of us who are still living after you go in." When they had finished putting the soil over him, the earth trembled. And trembled. "Father, don't hurt us. You are the one who asked to be buried!" After he had died, the men who wanted to kill him arrived. They said, "There is a grave here! Someone has recently died." And again the earth began to tremble. An *uitaque* seer among them told his companions, "Let's go right now because his *pujopie* still exists and can hurt us." Suddenly, a strong wind arose. It made a noise like *uuuuwaah* when it blew. The warriors were frightened and they went away. The son of Ajnisidai said, "I want to be like my father." He knew that his father had a heavy soul, he had power, and that is why his son wanted to be like him.

I discovered that—alongside memories of peace and unity—Echoi also articulated this darker story of violence for many of my elder Ayoreo teachers. Death, in the form of murder and massacre, converged around Echoi in the first half of the twentieth century. Precise testimonies of these events do not exist. Witnesses often could not put such violence into the confines of narrative or words. For many Ayoreo-speaking people, the violence of the past was tangible. It did not take the form of firsthand reports but rather secondhand stories, rumors, and bodily dispositions. The residues of violence eluded categorization. They refused to become history.

Most of the violence against Ayoreo-speaking people, and that which occurred at Echoi in particular, was impossible to piece together. It was reported, for instance, that Bolivian soldiers at Fortin Esteros in the late 1920s regularly hunted Ayoreo-speaking women near Echoi, who were taken back to the fort and sexually abused until they died. We do not know the names or numbers of these victims.[36] In 2002, Ayoreo elders pointed out to me more than a dozen middle-aged people who "once were Ayoreo." These people had been captured by slave raiders as children, sold and raised as domestic and sexual servants. They often did not remember or acknowledge that they had been Ayoreo. We do not know the scale of this slavery, but it was pervasive and minimally involved hundreds of Ayoreo captives.

The few archival residues of violence against Ayoreo were incidental, footnotes to other projects. Some, like the account of an Ayoreo man murdered in Roboré Bolivia, were simply filed away and forgotten. Others registered as traces of effective knowledge. Missionaries often noted a local "hunger" for Ayoreo slaves in the towns of eastern Bolivia. Missionary Joe Moreno recorded one of two firsthand descriptions of an Ayoreo captive in 1944 ("They couldn't keep clothes on this woman, she would tear them off. . . . When I saw her four months later she had gotten over her leg wound but took malaria and died two days later"). But even this record of a woman captured on a path to the salt lakes was no more than a casual footnote.

The fragments that remain bespeak a saturating violence whose degree has already escaped documentation. Killing an Ayoreo-speaking person in the 1940s and 1950s wrote missionary Jean Dye Johnson, was considered less morally compromising than killing a dog. At least two soldiers in the Chaco War recalled killing entire groups of Ayoreo at the Salt Lakes with machine guns and light artillery. The official policy of the Paraguayan government well into the 1960s was a "state of war" against Ayoreo-speaking people. Paraguayan military conscripts at some Chaco forts could reportedly gain exemption from service if they presented the severed head of an Ayoreo person to their commanders during this time. After two cowboys roped a twelve-year-old Ayoreo boy in 1956, he was put in a cage and exhibited in downtown Asunción as an "authentic savage Moro Indian" before being sent to the zoo. How did the upper-class Asunceños, who each paid a dollar to see him, perceive this slight figure huddled behind bars?[37]

Perhaps the most vivid records of all were the tall tales and rumors still told about Ayoreo-speaking people in small towns along the Chaco frontier. "They used to take children and eat them," an old man in northern Paraguay once told me, "I can show you the place." Some remembered them as incestuous cannibals. During my fieldwork, they were sometimes still called Moors (an odious name imported from the Iberian peninsula) or simply barbarians. I was warned about traveling to their villages. About being too near. "You must take a pistol." One rancher told me not to go out at night in the Chaco, that savage Ayoreo were still lurking nearby.

The violence against Ayoreo by outsiders was amplified in warfare between various Ayoreo subgroups. Ayoreo elders agreed that sometime in the 1930s, the peace established at Echoi was broken. While there was no consensus on which group started the war, all concurred that a man from one group speared another in the chest at Echoi in a dispute over a woman. Ayoreo groups soon turned against one another. The main aggressors were the newly allied Guidaigosode and Garaigosode, who launched devastating raids against all other groups. Demoralized and defeated, Direquednejnaigosode and Jnupedogosode bands initiated contact with evangelical missionaries in 1947. From the missionaries, they obtained shotguns and machetes and launched their own revenge attacks against their enemies. The result was that by 1977, nearly all Ayoreo-speaking bands had made contact with missionaries from one of the six Christian denominations that were pursuing them.[38] The boundaries of territories and *uage* membership hardened. The southernmost village confederacy, called the Totobiegosode refused to make any contact at all. And, their former Guidaigosode and Garaigosode enemies continued to hunt them down, armed with shotguns and machetes obtained on the missions.

The decades of these conflicts are now called simply *pocaningai*, or war. This warfare by all accounts was intimate and increasingly brutal. By the 1940s, it had escalated to an unprecedented scale. The aim was not to demonstrate bravery and prowess but to exterminate entire groups of the enemy. No quarter was asked nor taken. Men, women, and children were equal targets. Ambush, mutilation, and gang rape were common practices. Battles were usually highly unequal, and suicidal defense actions the norm. On several occasions, massacres occurred at Echoi.

People usually spoke of such things in fragmented images when they spoke of them at all:

"We found them at daybreak. They screamed when they saw us. We ran after them. The guns were loud. They fell, they fell. I shot a woman who was running away. The bullet hit her by the kidney. She fell."

"Under a bush. The warriors came. They came. They killed my mother, my father, my brother. They cut their bodies. Two days later I climbed a tree. I wished I was a bird."

"The roots cooking sound like guns. The bullets bit my mother's body. She dropped me. I passed out. Here I am."

"The earth turned black there. Blood. It soaked into the earth. Do not go back."

Upoide's Song

Elders say that during this time of war and dispossession, Echoi changed forever. Salt gathering trips began to take on the form of a raiding party. "When people gathered salt after the war began, they would go at night and put it in their bags. They were fast. They didn't stay there long." It was during that time, they say, that Echoi was lost to the Ayoreo. "We couldn't go back. We began to think that Echoi was going to vanish. . . . It will no longer be like it once was, when Amomegosode put their camp so close to Echoi."

The residues of violence can unmake places in ways distinct from the conscious redefinition of the past demanded by conversion or anthropology or activism. A history of violence is no history at all, and it defied a single interpretation for us all. One of the few explicit explanations I heard of how Echoi was lost came in the form of a song, sung to me by Yoteuoi in Arocojnadi when I asked a question about that place:

It is the fault of the shaman Upoide that the *Cojñone* arrived to those important places like Echoi. Satan told the shaman that he should make a road to cause problems for his people. That shaman was angry at Uejnai, who was more powerful than he was, and whom he could not defeat. He knew Uejnai was too strong. So he went to Heaven to bring back something cold to hit Uejnai with, something cold to calm him. But he didn't kill him. So his spirit (*ujopie*) went running. That shaman couldn't do anything, so he had to go far away to make problems for his people. He sent his spirit (*ujopie*) to the edge of the earth (*Jnupe'ureo*) to see, and from there he called to the ants that were working to make roads. That shaman did that to make a road and bring the *Cojñone* into his territory and against his people. All of this has already happened and there is no way to get that land back because the *Cojñone* made a big road there. That shaman was a Guidaigosi. That shaman had a song. His spirit went to the ants. When he speaks of the ants, he is referring to the *Cojñone*. The spirit of the shaman makes the *Cojñone* and the ants the same. He told the *Cojñone* to make the road to that good place. They first passed a place called Piogoto. Before, those places were for all of us. This shaman had a song:

I am going far, to the land of Earth's End
I spoke with the warriors and I took out their righteous anger to fight
I went and brought back cold water lilies
I used them to hit my enemies
But I could not overcome their will
I went and I spoke with the ants under the Earth
And I sent them here
I went and we ate up the forest
I drive them to all corners of the world
I drive them to all corners of the world

It was this shaman's fault that we had so many problems before, that the people split up. This is the trunk of all histories (*cuchade erode udi*). It tells about *yocuneone togode*, our dead lands. Do you understand?

I said I did but now I am not so sure. I cannot understand why Ayoreo-speaking people never returned to Echoi, why it was abandoned so abruptly, so completely, why this ambivalence they held so lightly still seemed paralyzing. The memories of its significance were not told to outside researchers but neither were they passed on to younger generations of Ayoreo, even within the same family. I struggled to reconcile this apparently self-conscious collective decision to let Echoi vanish with the nostalgia that tinged elders' stories about their experiences in this place.[39]

As I continued my pursuit of Echoi, this mystery deepened. It seemed somehow related to the common sense that Echoi was not only impossible to reach in the present, but that any attempt to do so was full of real dangers. Although many elders shared their memories of Echoi, others refused to speak about it. None dared to visit, and many whispered about the vengeful spirits that inhabited it. I was told several times that Echoi is now *Dajegeo*, a place to be avoided.

Dajegeo is an old phrase, used to describe places where one could not stay. The word is applied to sites of murder, shame, or burials. The past lingers in such places. There, the past is contagious, capable of rising up within and overwhelming bodies and minds. I was told by the Totobiegosode elders in Arocojnadi that *Dajegeo* is "a place where humans hunted other humans," and if one were to visit, "there is a danger that someone else will be killed." If someone goes back to a *Dajegeo* place, "they will die quickly, very quickly. Breaking a bone, maybe."

More than anything, a *Dajegeo* place is a site contaminated by spilled human blood. Ayoreo-speaking people reportedly never considered blood to be a biological substance that simply disintegrated or rotted away. Rather, it was capable of moving through time to replicate or copy the events of its production. That is, blood was the supreme substance of sympathetic magic for Ayoreo-speaking people.

Dajegeo or Blood as Witness

The power of blood figured prominently in Ayoreo life. In the past, all Ayoreo men who were not proven warriors or *dacasute* were known as *ayore poitade*, or worthless ones. To become a *dacasute* warrior—at least within living memory—it was necessary to kill an enemy and survive exposure to its ritually polluting blood, *iyojna*. The blood of *Cojñone* was considered the most potent and thus the highest prestige was associated with its shedding, whether the enemy was a man, woman or child. Second in potency was the blood of an Ayoreo-speaking person, followed by the blood of a jaguar. Much less powerful but still significant was the blood of cattle and horses.

When a man vanquished an enemy, he was expected to touch the corpse and the blood that welled from the wounds he had caused. Upon doing so, he and his blood-spattered weapon entered into another state of being known as *ditai*. A *ditai* man was *puyacho*, or taboo. He could not travel with anyone who was not a *dacasute*, and he could not enter the camp upon the party's return. The blood of the enemy that covered him was contagious and this blood transformed the killer's metaphysical being. The *ditai* state lasted until he was cleansed in an all-night ceremony.

Fellow *dacasutedie* made a circle around the man who was *ditai*, and a *daijnai* shaman traced a rectangular shape in the dust with single lines representing arms and legs and dots for eyes. The warriors did *chugu'iji* there and sang songs of rage and death. The killer jumped over the figure and the shaman wiped it out. At that precise spot a hole was dug and the death weapons broken and then buried. The earth that comprised the figure was turned over and used to cover the broken weapons, and thus the soul of the dominated enemy vanished into the underworld of death, *Jnaropie*. The men then shared food that only they could safely eat, as the food itself represented the flesh and blood of the victims and would kill any noninitiated person who dared to consume it.

For the *ditai* man, blood was more than the second-order residue of a primary violent event. Rather, spilling and consuming human blood was transformative because blood was considered to be an agentive sign of life itself. A *dacasute* proved his capacity to dominate existence and to realize his full human potential through surviving, vanquishing, and incorporating the very life substance of his enemy. If this was his first kill, the *dacasute* was given a special warrior name known only to his peers, usually the name of a carnivorous animal or carrion-eating bird, part of a complex vocabulary exclusively used during journeys of war. He was then eligible to don elaborate feather *potaye* and *cobia* and the *ayoi*, the jaguar-skin headdress that symbolized the spiritually potent blood of the jaguar and marked his status as a man of consequence and potency.

Yet this state of being a *dacasute* warrior itself was fraught, the power of blood a dialectic. "It is as if a *dacasute* always has a sickness," Simijáné said, "because he has spilled blood. The blood of enemies is in all of his things, they are filled with blood and no one else can touch them or they will get sick and die." The dangerous power of blood was in its mimetic excess, its capacity to act against the very life force it signified.

Disiejoi, a Direquednejnaigose elder bent with arthritis and the leading pastor of the village of Zapocó, recalled visiting Echoi as a child and then killing his Guidaigosode enemies with a shotgun there later. He responded to my questions about Echoi in the following way. "It is true what you say. Echoi used to be a good place for all Ayoreo. It was our trunk. Our grandfathers and their grandfathers always went there. They were happy there." He paused, and looked away down the road where we were sitting after a hot afternoon's work in his garden.

"But in my opinion we cannot go back there. Many people killed each other in that place later. There was a lot of blood. The soil there, in that place, it swallowed up the blood. It swallowed up all of our blood. That blood is still there, in the earth. We cannot go back. We have lost that place. It is like Cain and Abel. When one brother killed another, he could never go back. The blood is still there."

"Do you understand?" he asked.

Journey to Echoi

After months in the field, I began to suspect that I had missed my elder teachers' fundamental point all along. Echoi was a constellation of binaries that refused coherence. It ultimately defied all of my attempts to give it a stable, singular form. As soon as it was evoked in one register, it slipped into its opposite. This was the source of its power, its ghostly presence, and even, perhaps, its real political potential. It was uncanny because it mirrored precisely the fragmentation of a meaningful Ayoreo past and turned contradictory desires to find a redemptive history against one another.[40]

Even so, I still yearned for some kind of closure. I thought that visiting the former site of Echoi might offer a sense of order or magic that eluded me in the disrupted rhythms of everyday life. A friend and I decided to make the long trip, and we invited several people from Arocojnadi to come along.

Traveling north from the Mennonite colonies in central Paraguay, the road was worn by cattle trailers and oil trucks into deep ruts and *arenales*. It passed a hundred miles of land stripped bare, pastures for estancias called Betty's Ranch or Light to the Moros. It went by several Ayoreo and Manjuy Indian villages, the mud huts and people waiting in shade, and then Last Drink, a clapboard shanty with a hand-painted sign that sold cold Brazilian beer.

The road forked at Madrejón. Sixty years ago it was headquarters for the Pure Oil Company. In 2007, it was a wooden store with canned peaches and gingham tablecloths and a compound of brick buildings where tourists to the Defensores del Chaco National Park could sleep on dusty foam if they ever arrived. We followed the narrow double-track to the west and north where a string of contacts were made in the 1950s and where Guidaigosode Ayoreo speared the Mennonite missionary Isaak Cornelius. Like the concealed Ayoreo bands, the low forest along the road was a palpable presence with its smell of sun-baked spice and hidden sounds.

The northern Chaco was a place of extremes where the margins between reality and fantasy were stretched thin. Official maps of the area showed settlements and roads where none exist. At a military outpost, a handful of adolescent soldiers harassed travelers and told stories of savages. White-lipped peccaries rooted in the road's puddles and its bends glinted with shell casings from archaic machine guns. More than seventy years since the end of the Chaco War, the entire northern Chaco remained a militarized zone requiring special permissions and inspections to traverse. Despite the bombast and a giant concrete cavalryman outside Fort La Gerenza, the frontier itself was a vague no man's land where national borders were only formally fixed in 2009. It was a place where local strongmen were the rulers and the whims of their hired hands became the rules.

We followed the road through the twenty-five-thousand-acre ranch of a retired army colonel, who met us in his yard in a threadbare undershirt. He was clipping his nails and he did not look up. "I know where the savage Ayoreo are, I know where they live and I know many things that you would like to hear," he said unprompted. "But I cannot tell you anything." As he walked back inside, we continued down tunnels in parched brush.

After hundreds of monochrome miles, the road curled around a bend of white sand, and I knew we had arrived somewhere that should be named, if it is possible to fix names on a place and a time at all. An immense blue lake lay in front of us, encircled by green grass and swaying palms, dotted with flocks of white and gray water birds. Behind it a flat-topped mountain loomed blue in the distance. The road ended at its base in Fortin Ravelo, where two dozen teenage Bolivian soldiers ringed us with machine guns, then invited us to join a soccer game played behind a single mud wall with flag-mounted battlements breached by a hole they made to facilitate the retrieval of stray balls.

Near the mountain the forest changed. The breeze was cool. Hardwood trees towered to twenty or thirty meters, twice the size of any trees to the south. Many of their trunks were marked with ancient rectangles: the precise cuts of Ayoreo honey hunters and their two-handled tension axes. The ground beneath them was fragrant with fallen leaves. Bees and tapir and jaguar drank from springs that trickled down gashes in sheer basalt.

"I am sorry," the lieutenant said. "But no one can go near the salt lakes. They are inside Kaa Iya National Park. It is prohibited to go there. Tell us why you really want to go to that place in the middle of the *monte*. You are looking for the savage Ayoreo, no? I recommend that you go back to Paraguay before there is a problem here."

But what, I asked him, about my Ayoreo companions? Surely they could go alone, another day. It had been their land forever.

"Impossible," he said. "They have no permits. Would you like some Kool-Aid?"

The boys had stopped playing soccer. We were assigned an escort with a pistol in his belt. The salt lakes laid twenty-five kilometers west beyond a dry riverbed, but they may as well have been pictures from a satellite or a book. From the top of the mountain they looked like a half-remembered fever dream, impossibly white among the heavy mottled green.

THREE

Hunting Indians

"Well, we have worked with a lot of Indians before but this is about the Indian-est yet." Later he said to us, "And this is about the Indianest I want to get, too!" NORMAN KEEFE

The story of the Areguede'urasade is inseparable from stories about earlier Totobiegosode contacts. These, in turn, are invariably about missionaries. The most widely told story about Totobiegosode contact begins with some version of the following events.

In late December 1986, the New Tribes Mission pilot Dean Lattin spotted a column of smoke. On December 26, he made another flight to the area and located a camp of forest Totobiegosode with a communal house and eleven fresh garden patches. On December 27, missionaries took a group of thirty-four Christian Guidaigosode Ayoreo eighty-six kilometers to the northwest of the mission Campo Loro, to the nearest spot reachable by truck. On December 30, the mission group neared the camp of the forest Totobiegosode, who had seen the airplane and prepared defenses: a thick thorn barrier, dug up ground.

When the mission group approached the next morning at dawn, the leaders of the Totobiegosode, Aasi and Ducubaide, sent the eight women and the children to hide near the gardens with one man, Yoteuoi, to protect them. The other nine men lined up, painted for war. The mission group yelled that they came in peace and proceeded to walk into the village. They were led by Jochade, a Totobiegose who had been captured in 1979 along with the rest of Pejeide's band.

Upon entering the camp, one of the mission men grabbed a young Dejai and claimed him as his captive. The others began to do the same thing. But when the man let go of him to pick up his cassette recorder, Dejai grabbed his spear and began to fight. Others soon joined him in attacking their captors. In the ensuing battle, five men of the mission group were killed, including another Totobiegosode captured in 1979, and another four were badly wounded, including Jochade.

The mission group retreated and continued to shout to the others that they came in peace and wanted to take them to the *Cojñone*. That night, an uneasy truce had been brokered and the Guidaigosode raiders slept in the camp. By the following afternoon, most of the Totobiegosode agreed to go back with them to the mission. Some of the captors walked to a nearby ranch and radioed the mission. The mission plane dropped medical supplies and airlifted one of the wounded men out. The contacted group of eighteen captives arrived at Campo Loro on the back of a tractor-trailer. It was the evening of January 4, 1987.

What few could agree on—in 2007 or in 1987—was the ultimate significance of these events.

Once news of the contact became public, a small group of Paraguayan *indigenistas* mounted a vigorous public critique that resonated across the social divides of the military dictatorship and internationally. This critique built on several prior denunciations: international accusations of state genocide against Aché groups, Luke Holland's exposé of the New Tribes 1979 pursuit of Pejeide's band, and the fallout from human rights scholar Richard Arens's 1977 visit to the New Tribes Mission (which he likened to Nazi concentration camps, controlled by death-dealing missionaries and focused on routine patrols to hunt down the forest bands).[1] The 1987 critique, which called the work of the mission "offensive to our nation," was widely circulated in books, articles, and newspaper stories. It became an organizing point for the nascent Indigenous rights movement in Paraguay.[2]

The *indigenista* critics denounced the central institution of the New Tribes Mission: actively pursuing and capturing so-called uncontacted bands of Ayoreo. They argued that such Totobiegosode contact expeditions were manhunts. Moreover, according to these critics, the Totobiegosode "human hunts" were best understood as a form of "ethnocide," or the active extermination of culture and a collective psyche:

This process is simply and straightforwardly called ethnocide; institutionalized ethnocide, systematic ethnocide. More than the declared redemptive objectives, the

consequences of the dismantling of cultures is the resigned submission of the indigenous person and his integration, always degraded, to the Western civilizational model.[3]

According to this narrative, evangelical ethnocide against Totobiegosode destroyed them as surely as their physical extermination:

The most tragic effect of missionary ethnocide is that it breaks the spine of a people,
it converts their members into caricatures of Westerners and later, as it does not have
any reasonable project to offer them, it pushes them irresponsibly to sell hides, to hand
over their forests and symbols, and to direct exploitation or it sends them to a marginal
underworld of begging, prostitution, alcoholism and petty delinquency where they
end as beings that have no place in their culture nor the culture of others. . . . Once
the community is dissolved, the pact that united it with nature according its common
designs is also dissolved; soon, the Indian finds himself in opposition to a universe with
which he once identified himself and he sees his relations with his ecological environment adulterated; at the end, he himself is turned into a predatory destroyer. Collective
tradition is also dissolved: the anathema launched against his ancient beliefs creates
an artificial wall of forgetting that interferes with the transmitting metabolism of his
culture: a phobic negation of the past and a schizophrenic fracture of time invents a
pure present without ghosts or memories.[4]

In this evocative critique, Ayoreo humanity was irrevocably lost
through contact and a pathological internalization of oppression. The
Totobiegosode became global icons of such ethnocidal violence. This
narrative of the human hunt was repeated so often—"The Facts: Hunters,
Hunted," "Chronicles of Human Hunts," "The Hunted Ayoreo"—that it
has become inseparable from their political visibility.[5] For many outsiders, it is the foundation myth of Totobiegosode humanity.

This narrative is also an origin myth for the political anthropology of
indigeneity. "Genocide assassinates people in their bodies," wrote Pierre
Clastres in his famous 1974 essay on the topic, "ethnocide kills them in
their minds." For Clastres, ethnocide was at once "the dissolution of the
multiple into the One" and the "normal mode of existence of the state."
The reduction of difference, he argued, was the shared aim of capitalist
productivity and the modern state. The task of the political anthropologist
was thus clear: to prove the historical contingency of this ethnocidal state
form, by showing that primitive society offered a radical alternative.[6] This
possibility, or so the story goes, was destroyed at the moment of contact.

Predictably, the New Tribes missionaries offered a starkly different interpretation of what was at stake in these events. In early 1987, they

wrote and circulated a short public response, titled "Fear and Hope in an
Ayoreo Contact!" In this remarkable document, the missionaries claimed
their actions were not genocide but a kind of ethical humanism.

The document stated that forced contact was, in fact, helping the forest people: "In a world of mistrust and confusion along with civilization
moving into your area . . . what hope of the future does one have?" At
the same time, it argued that the lack of contact was a violation of Ayoreo
self-determination. "Why should educated man be allowed to force the
Indian people into something they don't want?"

The document gave six examples in support of the mission's argument, including the existence of shared kinship ties between Jochade
and the forest people, the fact that the Ayoreo mission did not shoot
any of the Totobiegosode, that the Christian Ayoreo planned the entire
operation, and that the forest band "went willingly" to the mission. "Is
it not reasonable for them to want to contact their family and relatives
and to offer them the opportunity of a life without fear and a better life
for the future?"

To further complicate the debate, Christian Guidaigosode at Campo
Loro also stridently rejected the human hunt narrative. They joined the
missionaries in articulating contact as humane. Tensions ran so high
that when the most prominent critics sent a delegation, the aptly named
"Commission of Indigenist Entities," to interview the recently captured

Totobiegosode, they were denied entry to Campo Loro by a crowd of hostile Guidaigosode. Some say the critics were physically attacked. Instead, they sent a supposedly secret cassette with questions for the Totobiegosode, who then sent tape-recorded responses back. One of the Guidaigosode leaders summed up the situation to a journalist at the time:

These people only make problems for us. They want to film us and compare our way of life with that of our ancestors. When we say something, they twist the meaning of our words. We believe that they are exploiting us. They themselves should go to the forest and live there, like our ancestors lived. They should take off their shoes and their clothes and walk around naked, just like the Ayoreo did before. We want to be integrated into society and live this new life that we have now, and we do not want anybody to act against our purpose. We are tired of being the object of discussion for those scientists.[7]

In the aftermath of the contact, there was no formal resolution of such tensions. The government dictatorship took no serious action against the New Tribes Mission, which it had long supported. The Totobiegosode remained subordinated in Campo Loro, until Jochade and Yoteuoi initiated their land claim in 1993. The *indigenistas* all became involved in various NGOs. Mennonite efforts at mediation failed, although Wilmar Stahl and his brother Uwe did produce a decidedly surrealist video, complete with staged reenactments of contact by the actual participants several months after the traumatic event. Its unintentionally ironic title: "Reconciliation."

My fieldwork began after many believed this story of the Totobiegosode had ended. But it was soon impossible to ignore how the events of 2004 depended on and demanded the reinterpretation of earlier contacts. Making sense of the Areguede'urasade and their subordination meant reopening old wounds and resignifying the past for everyone involved. And it also brought us all face to face with the images of Ayoreo humanity historically cocreated on evangelical missions. Because it was obvious that conflicts about contact were always debates about the meaning and value of Ayoreo life, this interpretive labor was distinctly political. There was no middle ground. Witnessing meant taking a side.

When I arrived, I thought nothing could be clearer than which side I was on. It was perhaps fitting that Totobiegosode voices themselves unsettled any easy narrative of victimization, ethnocide, or redemption. If I was perplexed by the stark disconnect between an imagined process of culture death and actual Totobiegosode lives, I was even less prepared for how contact survivors interpreted the "human hunts" some twenty years later.

What was I to make of a quiet consensus that the outcome of being captured was deeply ambiguous? To be sure, most people refused to discuss their contact experiences in any depth with me. They were deeply distressing memories, often reduced to an oblique image or partial allegory or quickly covered up. I heard enough fragments to know that the transition was acutely painful. I was told that a shockingly large number of Totobiegosode had died of "sadness" afterward, by refusing to eat or drink.

Edó was at the center of the most complete story I ever heard about these events. She was a *daijné* shaman married to Yoteuoi and was Dasua's mother. Yoteuoi told me she had a vision a month before the contact and knew that enemies were coming. She took various precautions and she stressed to all that a black *Kiyakiyai* bird would appear in the middle of the village before the attack and it would lead them to safety. No one should touch it or scare it away. If the people did as she said they would survive. The bird came and landed where a warrior was building the defenses. Without thinking, he scared it away. The bird took flight, and there Yoteuoi's story ended. Edó, pregnant at the time, starved herself and died on February 17, less than two months after her capture.

Because of such arresting images and the drama they implied, I initially presumed we all agreed that the conditions of the present were much worse than those of the past. Like many outsiders, I was constantly searching for a Totobiegosode critique of the *immorality* of what had been done to them. But this was not forthcoming from contact survivors. In fact, many people subtly but firmly rejected my framing. It was hard for me to accept that we did not agree even on this. I struggled to understand how Totobiegosode survivors made sense of being hunted down and captured, years after the events.

Aasi, the leader of the 1986 group, was Jochade's brother and Dasua's husband. He took the unprecedented step of renouncing his status as a *dacasute* after being captured and was widely known as an *ayaajingaque*, or peacemaker. He was a slight man who rarely spoke but who radiated a calm strength. It was hard earned. He had killed enemies and jaguars alone with a spear. He often told the story of the time he fell from the top of a tall honey tree onto his face, breaking several ribs, bleeding internally, and losing consciousness. Eventually he crawled back to the camp, only to discover that everyone had decided to move that day. He did not wish them to wait on his behalf and so he said nothing about his injuries. He told them to go ahead and without a single word of complaint he picked up his heavy bag and followed as best as he could, stopping only to pass out from time to time before continuing his journey. That was the kind of man Aasi was.

He and I spent much time together in the twilight of his life. I like to think we were close. In our conversations, he always framed contact in a series of implicit and enigmatic contrasts. "We were happy when they captured us, because we no longer were happy in the forest, there were not many of us left." Or, "Life in the forest was very difficult. It was hard to hunt in the rain. We suffered from hunger and thirst." Or, "Before we were always scared." Or, "Now, I am happy, I live in peace." Or, "I am not a child who fears death." I could not share the logic but I became convinced such statements were sincere. However, what I could never understand was the unsettling fact that Totobiegosode seemed willing or even eager to participate in the pursuit of forest bands so soon after their own traumatic captures.

Jochade had helped to hunt down his brother, Aasi, and the others. And Aasi stated on the cassette returned to the *indigenista* commission in 1987 that "if they organize a trip to search for the other Totobiegosode, I am going to participate as well because I have relatives in the forest: a brother, a daughter and others."[8] During my fieldwork, this was a typical attitude rather than an exceptional one. More than one of the former Areguede'urasade said they too were in favor of tracking down and capturing the band of their relatives who remained in the forest.

Were they lying to appease the missionaries? Was life in the forest actually so hard? Was this ethnocide speaking? None of these explanations seemed entirely plausible. Yet I could not come up with anything better. The only way I could even begin to imagine how this desire to track down and capture relatives springs so quickly from starvation and sickness and ridicule and death was if the pursuit of the forest bands was never a hunt for the human at all but a reaction to the breakdown of knowledge and being through colonial violence, a doomed effort to transgress savagery by creating it and conquering it anew, to terrify terror into going away, to domesticate the wildness of death. Yet this formulation raised more questions than it answered. How did the murderous hunt for the uncontacted appear as the profane dissolution of human society and the script for its salvation at the same time? If human hunts created ethnocide and a fetishized image of it, where did the ongoing pursuit of the wild man lead us all?

The images of Ayoreo alterity cocreated on evangelical missions were the logical starting point for backtracking this peculiar and nonsensical drive to hunt forest Indians and by their violent subjection to care for and save them. And there was only one place this reverse pursuit could begin.

Bobby

On a white-hot day in late 2007, I made my way to the New Tribes Mission of Campo Loro, hoping to interview the missionary named Bobby. At the time it was the largest Ayoreo settlement on either side of the Bolivia–Paraguay border. With nearly a thousand inhabitants, this square mile scar of grit and dust contained approximately one-sixth of all Ayoreo-speaking people.

Experiencing the sheer scale of misery in Campo Loro was like a physical assault until you began to get used to it and I was never sure which was worse. It was a searing scene. No trees broke the sun's relentless heat on patchwork roofs of tin and tarp. Clutches of anonymous figures and skeletal dogs sat in the scant shade of mud huts and plastic-walled shacks. An incongruous brick church loomed in the center.

It was an unruly place where hunger lurked. Where the hollow trunk of the death tree yawned. Where crops did not grow well in the hot sandy soil. Where there was no game left in the shred of forest nearby, few jobs in the mission brick factory and fewer still in the Mennonite colonies forty miles to the south. In the Totobiegosode villages far to the north, we heard of elders starving to death on the mission. A cruel irony had led these people, descendants of the great Guidaigosode confederation, from their ancestral territories south to this place in 1979—nearer, they believed, to jobs, prosperity, and inclusion into moral modernity. And it was to this place that the Totobiegosode bands had been brought in 1979 and 1986, clutching bows, covering eyes.

In the early 2000s, an enterprising young teacher converted his ramshackle hut into a combination bar, pool hall, and movie theater at night. Crowds of teenagers paid the entrance fee to drink cheap soda or beer and watch looping videos. Among the most popular were martial arts films by Jean-Claude Van Damme. Campo Loro was divided into two unequal parts. The Indian section of the mission was bordered on the west by a neat airstrip, the small missionary neighborhood a safe quarter mile beyond. It was a short distance from the devastation of the Ayoreo encampment, but it seemed like a different world.

There a pale teenage girl with braces and American clothes walked alone down the planed gravel, playing with her fat black bottle calf. I passed a handful of shady recessed houses, garages, and a generator station, before stopping in front of a large brick house with a picket fence and flowers shaded by palm trees. A Great Dane greeted me at the gate, and a half-grown pet anteater raised its sticklike nose to examine me.

Soon, a smiling woman with pale skin and auburn hair invited me to sit on a padded wicker chair on the cool porch. She spoke English, her words clipped and polite. She served me a glass of cold lemonade. I held each sip in my mouth before swallowing. It was cold and sweet. It just couldn't be, but so it was.

I had arrived hoping to interview her husband, Bobby. We chatted while we waited. She was born in Thailand, the child of New Tribes missionaries. She asked for news of the US, which she imagined as a place of gang violence and childhood obesity and video games. She told me they kept newspaper clippings of all the accusations of genocide that anthropologists had leveled against her husband's family and that she was ambivalent about the recent news that her nephew was taking anthropology classes in college.

Like his wife, Bobby was also a second-generation New Tribes missionary. His parents were famous in the Chaco. In 1966, they had established the first permanent mission with the Guidaigosode at a remote camp near Cerro León. His father had been involved in several contacts with forest Totobiegosode. Bobby and his brother were raised among Ayoreo people. It was Bobby who had the closest friends among them and it was Bobby that Ayoreo-speaking people respected, over and above all other missionaries. He spoke impeccable Ayoreo, bow-hunted peccaries in the forest, and gave oranges to the children when he came to preach to the New People in Chaidi. They said it was Bobby who brought messages from the Son of God.

One of these messages, which began shortly after I arrived, was that I was opposed to God, that I was a helper of Satan, that I would lead the Totobiegosode to *Ngahu Pioi*, the "lake of fire." For more than a year prior to my arrival at his house, Bobby and I had carried on a debate by proxy. I put out word that I didn't hate God. It was disconcerting to find myself imitating preachers I had known as a child in western Kansas, where my grandfather often took me to services at the evangelical Church of Christ he so loved. To argue with Totobiegosode that they were not intrinsically sinful, I ended up quoting scripture and discussing Bible parables at length.

I tried time and again to interview Bobby, but he always refused. My meeting with him that day was more the result of Ayoreo lobbying than anything else. Many people said they liked us both and that we should meet to reconcile our differences. I hoped he would give me a clue about missionary perceptions of Ayoreo humanity.

He arrived in a faded baseball cap, a lean sunburnt man in his early forties who looked like my own midwestern kin. After such buildup, the conversation was anticlimactic. He preferred to talk in Ayoreo with English sprinkled in and didn't have much to say. He compared Campo Loro Ayoreo to the recently contacted Totobiegosode at Chaidi. Those at Campo Loro, he said, were really "straddling the line" with their faith, while those at Chaidi were more focused on Jesus, because they had less distraction from "outside influences." But he was tense and abruptly ended the interview with an observation. "Outsiders show up and think they understand the Ayoré," he said, looking away. "But they get it all wrong. There is no community here. Each person is out only for themselves."

I left disappointed but not sure why. Had I expected some easy way to reconcile devastation and death with the faith of this guy who wasn't as different from me as I wanted to imagine? A man who could have been me and was my opposing double, instead. I wondered if Bobby might be frustrated too. Struggling to distinguish the doomed and the redeemed in a shady house over cold lemonade, did he also yearn to fight a clear battle against a Devil incarnate who appeared one afternoon on a borrowed motorcycle with tape recorder in hand?

As I motored slowly back through the desolation of the Ayoreo camp in the evening shadows, I saw clusters of faceless figures silhouetted against cookfires. It was like riding through a battlefield in which incantation was the weapon and the terrain to be contested was life itself and we had all already lost something whether here at its end or in fealty to our creed or before the journey began.

The noninterview with Bobby meant I had little choice but to backtrack the allure of the Indian hunt further, to the missionary imaginary of Ayoreo humanity stored in the archives. Perhaps in their writings I could find some preliminary clues about the images these evangelicals constructed of themselves, their labor, and their ideal object: the Ayoreo soul.

The Bride of Christ

The story the New Tribes Mission tells about itself goes something like this. Three middle-class men founded the New Tribes Mission in August 1942. Their leader was a young charismatic man named Paul Fleming.

His peers described him as filled with "consuming passion," to see the last "unevangelized tribes," reached for Christ "in his generation." "[Fleming] did not play around," as one of his contemporaries put it. "He was obsessed with this one great command of our Lord, to preach the gospel to every creature."[9]

That Ayoreo-speaking groups were the focal point around which the New Tribes Mission became institutionalized in 1943 was not due to any particular interest in them. Missionaries, instead, "eagerly anticipated reaching" Ayoreo groups because they were a "savage tribe that had never heard the Gospel."[10] The point was not their humanity but their savagery.

They lived deep in the "Green Hell" of the Chaco wilderness and were a "savage, nomadic tribe who had never had a chance to hear of the love of Christ." From its inception, the New Tribes Mission specifically focused on "the hardest tribes," those considered to be the most difficult to contact. "Our first missionaries made the Ayore tribe their immediate goal because they felt that other missionaries would never get to them. The Ayores were infamous for their hostility."[11]

Fleming and other New Tribes missionaries believed that, like all "unreached people," Ayoreo-speaking people had to learn their critically important role in God's universal plan. Inspired by Revelations 5:9 and 7:9, Mark 16:15, and Matthew 24:14—which agree that "the end shall come" after the gospel has been preached "unto all nations"—the New Tribes Mission philosophy is to hasten the apocalyptic return of the Messiah by bringing all of the world's human groups together "under the sound of the Gospel." The salvation of Ayoreo was merely incidental to this larger project, "the completion of the body of the Bride of Christ."

The body of the Bride of Christ is composed of the souls of all believers. But it will only be formed when every "kindred and tongue and people and nation" on the Earth has heard the Word of God. "We believe that the purpose of the Lord is that from every tribe and language must come those who are to complete the Bride of Christ."[12]

According to the foundational covenant signed on August 1, 1942, on the kitchen table in Paul Fleming's Chicago home, the foremost task of the New Tribes Mission was to "work toward the completion of the Bride of Christ until death, and to measure all our efforts in the light of this task." The rapture of the faithful—including missionaries themselves—could only occur when Christ "comes for his Bride" and consummates his union with the church.[13]

From the beginning, the New Tribes Mission imagined a world in which taming savage Indians was essential for transforming the power of their difference into the salvation of all Christendom. Yet dark forces opposed them. They imagined themselves in "spiritual warfare" against a diabolic savagery that was "trapping" and "holding prisoner" the "world's untouched fields" of Indian souls.[14] It was no coincidence, then, that the Bolivian director of colonization Viador Moreno Peña himself, granted the New Tribes Mission a state mandate to "civilize the Barbarian Indians" of the Chaco in 1946.[15]

Brown Gold

The first aspiring New Tribes missionaries stepped off the plane in Roboré, Bolivia, in 1942. Or so the story goes. They were told by one army official that Roboré was "the most strategic spot in all of South America to evangelize savages."[16] A small town of whitewashed adobe houses, wooden vigas, streets of red earth—a place cooled by scented breezes from the taller forests of the gentler northern edges of the Chaco. "Women with long black single-braided hair smiled their *buenos días* as we passed. Barefoot children stared."[17] The striking impression made by these gringos, with their brass instruments, penchant for public prayer, and ardent faith, lingered in local memories for years.

Many of the male townspeople of Roboré and neighboring towns were veteran Indian hunters. The Indian hunt was a venerable backwoods institution. These hunts were based on a simple formula: track down a band of forest *Indios bárbaros* with dogs, attack at dawn by rushing *en masse* into the middle of the camp, shoot any adult men, and capture as many women and children as possible. The captives were roped together and taken back to the towns, where they were sold, divided among townspeople as gifts or sent on to relatives in Santa Cruz. These Indian slaves were called *criados*. They worked as domestic servants, sexual servants, clerks, shopkeepers, ranch hands.[18] Having one was a mark of status and prestige.

Ayoreo warriors raided back, killing a cow or a lone peasant from time to time. Military commanders advocated exterminating all of the *Bárbaros* with regular bombing raids. Not surprisingly, some residents of Roboré doubted that these tall, well-fed gringos were willing to risk their lives to befriend and convert the *Bárbaros*. But everyone knew that Jesuits and Incas—and who knew who else—had left behind hidden hoards of gold, concealed in the damp tangled forest. Jean Dye Johnson was one of the early missionaries. In her remarkable memoir *God Planted Five Seeds*, she writes that locals at first thought the missionaries were gold seekers.[19]

Gold. What else could these men be seeking in the Green Hell so far from the temperate soil of Indiana, or Michigan or California or wherever it was from whence they came in the DC-3 plane with a stenciled face of a "tribesman" on the tail—a plane only recently returned from the killing fields of Europe? That alluring substance was the only thing that could compel these men to behavior like aerial pursuit of the barbarians. In eastern Bolivia, as in many other places, gold was considered dangerous, capable of creating desire so excessive that men would risk their lives to acquire it. In such cases, it was considered to be a substance not of God but of the Devil.

The missionaries laughed at such attitudes. Yes, they answered. "We are after gold—but brown gold."

For so did these missionaries refer to Indian souls, "worth far more to God than any nugget on earth." This image became a serious mantra for missionary work among Ayoreo-speaking groups. One Ayoreo soul, they wrote, was "more precious even than gold that perisheth," and "one brown tribal person is worth all the monetary wealth of the world."[20] Indian souls were the only coin by which eternal salvation could be purchased. As such, they required the intertwined labors of transgression and extraction. Fleming named the official New Tribes newsletter *Brown Gold* because he was convinced that Indian souls were as valuable as spiritual gold to their finders seeking to achieve heavenly paradise[21]: "If there is any such thing as eternal gold, these brown men and women are it!"[22]

What is the value of a soul? Seldom do we as missionaries ourselves sense the great value, but Jesus Christ paid a great price to win these men! Truly these men were "brown gold" that would not fade or canker, and there were eternal values for the winning of such men for the glory of God . . . ["Brown gold"] seemed more and more to be the most adequate way to describe these men—like a priceless treasure that had not been found.[23]

The Indian-soul-as-gold metaphor is found over and over in missionary writings from the 1940s and 1950s and then cited until the late 1990s. Cecil Dye, a former Michigan businessman, even wrote a song about brown gold. He and his brother Bob sang the song during prayer meetings in Roboré. They were singing it when they walked into the jungle in late 1943 to contact the barbarians, and one wonders if it was on their lips when a Jnupedogosode band speared them and their three companions to death several weeks later:

Gold! I know where's hidden treasure
Souls who've never heard of Him
I will leave earth's shallow pleasure
And search for gold in paths of sin
Gold from every tribe and tongue we'll gather
From fields that are white we'll glean rich treasure
By His Spirit we will win souls from doubt and fear of sin
Each effort's worth it all a thousand measure
(Songs of Challenge, no. 38)[24]

But why gold? Regardless of missionary intent, the metaphor of gold was uncannily apt. Gold is deeply woven into Christian symbolism as a potent and sacred substance. Yet gold is also widely recognized as a dangerous material capable of driving rational men mad. It is the material form of a polluting temptation. As Taussig describes it, along with the stuff of sacrifice or offering, gold is "the epitome of evil, a veritable code word for all we want but in our innermost hearts know we must not have."[25]

For Marx, gold by its nature was money, and vice versa.[26] In his analysis, gold was a universal measure of value for several reasons: it was scarce, it never corroded, it could be infinitely divided and reassembled, and it was the condensation of labor.[27] The magic of gold lay in its imaginary or ideal nature, which in turn atomized human sociality and imbued things with ghostly spirit.[28] The universal value attributed to gold, in turn, made it intimately linked to transgressive desires.

Marx most cuttingly described this excessive power of gold through the figure of those who hoard it.[29] Because gold promised to contain everything within it and, thus, erase all forms of relational distinction, the hoarder's limitless desires were unleashed. Marx quotes Columbus himself on this point, "Gold is a wonderful thing! Its owner is master of all he desires. Gold can even enable souls to enter Paradise."[30]

The pursuit of brown gold presupposed the alienation of Ayoreo-speaking people not from their labor power but from their vitalism and the nuggets of soul content it presumably covered. Such missionary attitudes were strikingly similar to those of tradition-seeking anthropologists. Both presumed a disembodied substance of universal value resided inside degraded Native bodies, which a properly industrious labor could extract and collect.

Yet weren't missionaries also dazzled by brown gold? Indeed, the entire mission order was explicitly predicated on a fundamental disorder: the irrational and excessive desire to hoard Indian souls. Missionaries wrote that the primary criterion for selecting new recruits was that they be "seized" by a "consuming" and "consistent passion for souls."[31] This desire ideally "overwhelmed" the self; it was "something that should monopolize us completely."[32] That is, the ideal missionary was not a rational collector but someone who had surrendered his intellect and reason in the headlong pursuit of souls.

But this was fraught ground. It meant coming face-to-face with the Devil, the original owner of yellow and brown gold alike, who threatened to beguile the faithful with lightning-quick shifts between the apparently human, the divinely valuable, and the eternally doomed.[33]

Slavery

Dealing with the Devil for Indian souls meant abandoning any clear distinction between dark and light. Missionary work depended on acquiring reliable interpreters. And there were plenty to be had for the right price. Seventy years after Padre Cardus was loaned an Ayoreo slave to guide him to Echoi, American missionaries discovered that captive *Bárbaros* abounded in towns like San José, San Juan, Santo Corazón, and Roboré. The missionary passion for Indian souls was mirrored in the thriving market for Indian slaves among the townspeople.

Jean Dye Johnson, widowed when Ayoreo warriors killed her husband and his four companions in 1943, wrote of the inhabitants of Roboré that they "were shameless in their desire to get their hands on some Ayoreo who would become a laborer without pay." She described how ranchers familiar with Ayoreo slavery eyed the recently contacted Ayoreo bands and "looked them over calculatingly, picking out likely prospects." This hunger for captive Ayoreo-speaking people was so pronounced that missionaries feared that truckloads of armed townspeople would raid the fledgling missions to slaughter or enslave Ayoreo.[34] And after the initial contacts with Jnupedogosode and Direquednejnaigosode bands in the late 1940s, missionaries went to elaborate lengths to hide *Bárbaro* groups and transport them to the missions before word got out: "Joe is working desperately to contact these savages before they show up again because many of the people now want to capture some of them. Others hope to shoot at them."[35]

Is it surprising, then, that local residents initially mistook New Tribes missionaries for slave traders and contact for enslavement? Jean Johnson complained that the same people who supported missionary efforts could not seem to distinguish between these two activities. "People," she noted in the October 1948 edition of *Brown Gold*, "again and again criticize us for not handing these people out among the townspeople for them to raise as slaves!" When in town, missionaries commonly filled the roles reserved for slave owners. Missionary Joe Moreno reported in the August 1946 edition of *Brown Gold* that he was "developing the contact with some captured jungle people in San Juan," and missionaries had identified "a captive *Bárbaro* woman in San Jose," who believed that a particular band who appeared on the railroad tracks "is the one from which she came. She is of the opinion that if she could talk with her father, who is the chief of her tribe, we could make friends with them."

This Ayoreo-speaking woman was called Inez in Spanish and Aroide or Guto'date in Ayoreo. A rancher named Ignacio Paz captured her in 1929 near San Jose with a group of other women, including her two teenage daughters. Paz gave Guto'date and her lone surviving daughter to his son, Ubil, shortly thereafter. When Jean Johnson met her in 1945, this "Ayore woman servant whose master would permit me to take time with her," lived with her daughter, who by then had been impregnated by her captor, in a "rude shack of poles, covered with woven palm leaves." She was a "short, chunky, dark-haired woman nearing forty years. She seemed awed and puzzled" at the missionary's interest but agreed to teach Jean her language and eventually came to live in her house.[36]

From Guto'date, missionaries learned that the *Bárbaros* called themselves "Ayore" or "Ayoreode." Guto'date recounted "customs, living patterns, and the likes and dislikes of [her] fellow tribesmen." From her, Jean learned how to lure Indians by gifts of steel-leaf spring, instead of mirrors, and that the best twisting footpaths for hunting Indians were those that led to the salt pans.[37] Despite the love that Jean expressed for Inez in her book, she soon returned to the United States. Guto'date married a man from one of the recently contacted bands and reportedly died of tuberculosis at the Catholic mission of Santa Teresita sometime after 1956.[38]

There are so many of these contradictions—flashes of light within darkness that created darkness within light—that it is difficult to imagine them as evidence of a missionary failure or hypocrisy. Rather, it is more likely that such ambiguity was precisely the aim, technique, and motor of the missionary project.

The Dangers of Proximity

According to missionary writings, one of the first tasks of the missionary in the field was to create a sense of internal sin and the idea of Hell: "It has been so hard to get words to give them the gospel. Little by little the idea of hell seems to be getting across to them."[39] It meant communicating a moral order by parsing humanity into binaries of opposed elements. Yet this order was constantly unraveling, and the elements never stayed in their proper place:

Jean has never been able to get hold of a word for sin. The word she thought she had, turned out to be fish, for the slave girl did not know the difference between the two Spanish words. (Fish, *pescado*, sin, *pecado*.) The captive believes all Jean Dye has told her of the Bible, etc., but so far she seems to have no sense of her own sin. To her the bad people are those who kill people.[40]

What is most striking about early missionary self-reporting is the degree to which Ayoreo-speaking people were seen as already dying or as the walking dead. They were "those whose manner of living is more like animals than men."[41] Their bodies, language, and practices were a veneer of apparent health concealing a fallen state and a profound spiritual degeneracy. "Always we are amazed how normal human beings could sink so low as to be so much like the forest from which they came."[42] "In the natural, it would be hard to love them, but when we realize that they are wandering around in darkness and dying without Christ, his love fills our heart and goes out to them."[43] Indeed, the physical life of the Indian was irrelevant.

The terrible irony, of course, is that missionaries had to first intensify or create the savage realities they aimed to alleviate and transcend. They thanked God when he "opened doors" through devastating epidemics of smallpox, measles, and influenza that swept missions in the 1950s and 1960s. "Sickness entered as before and even caused the death of some who had so confidently carved out the little poles. It has given us the opportunity to witness to him because of their act, and we have likened their poles to the graven images that God spoke against in Exodus."[44] Ayoreo people began to take Western medicine when they saw "that those who died were those who wouldn't take it."[45] Missionaries noted the "miraculous" effects of antibiotics and even aspirin on the *Bárbaros*, as well as the extremely rapid deaths, often in less than twenty-four hours, from diseases such as the common cold or measles. When medicine was scarce, they reportedly gave first attention to Christian converts.

Missionaries used these epidemics to stage demonstrations of the power of God's Grace over satanic witchcraft. Yet stopping Satan at times required mimicking him. Bill Pencille led the Ayoreo work of the South American Indian Mission in the 1950s and 1960s. His outsized role in these early contacts was mythologized in a 1967 book called *The Defeat*

of the Bird God: The Story of Missionary Bill Pencille Who Risked His Life to Reach the Ayorés of Bolivia. In it, Pencille boasted of relying on the very same practices he considered satanic in order to convince Ayoreo of the power of Jesus:

They considered it all witchcraft. I don't purposefully deceive them, but neither can I explain to a savage how an antibiotic works. So I don't try to explain. I let them draw their own conclusions. And the conclusion is that just as their witch doctors have power over the spirits, so this great spirit, *Dupade*, who sent me, has bestowed his power on me.

I picked up a little bottle, opened it, and took out a tiny black pill. I dropped the pill into the can of water. The water turned blood red. Potassium permanganate was all it was—a disinfectant. But of course, it was magic. . . . They oh'd and ah'd at that. . . . I took out a couple of white lumps, ground them to powder and sprinkled this powder carefully on the wound. . . . Then I opened a bottle, and they all spit on the ground. Alcohol was what it was in the bottle, but they didn't like the odor and spat to show their disgust. As I washed her arm with this, they had quite a gagging time of it. Then with all the magic of sterilizing a syringe and boiling it and putting it together and washing the top of a bottle and sticking a needle in and filling the syringe with penicillin—you can imagine how that would impress a raw savage! And then as they gasped, I plunged this thing into her arm![46]

101

Direquednejnaigosode elders often told of how this former Minnesota farmboy began to strip bare and don the jaguar-skin *ayoi* headdress reserved for *dacasute* warriors during his contact work.

Slaves and converts, missionaries and slave owners, shamans and farmers, Christians and barbarians, sin and fish, illness and piety, hunting and saving: every image created and threatened to blur into its own opposite. The paradox is that the power of missionaries was not diminished by these contradictions. Rather, it seemed to grow from this tension of creation and transgression and collapse of the limits between savage and civilized, sacred and profane, Jesus and Devil, White and Indian, human and nonhuman that were both immutable and instrumentally unclear for everyone involved.

One gets the sense when reading missionary writings that this dialectical game of terror and salvation, however, almost immediately spiraled out of control. Missionaries arrived in the field expecting a certain kind of confrontation with Satan. But what if conversion, and hence the entire economy of salvation, required becoming complicit in a mortal sin?

We had a real shock this morning. A woman killed her little girl, probably about five years old, just because she didn't want her anymore. When we found out about it Evelyn and some of the others dug her up, but she was dead. We sent the woman out of the camp, a hard thing to do, and told the others in no uncertain terms that we were going to have nothing like that here. We know that unless we stand out firmly against their evil practices that they will not see the sin in them, though it would be easier to do otherwise.[47]

The horror occasioned by the common Ayoreo practice of burying unwanted infants threatened to overwhelm missionaries with rage, impotence, and shame.[48] Missionaries of both genders who confronted such scenes reported turning white, vomiting, or fainting. They later wrote of suffering nervous spells, insomnia, and long sessions of uncontrollable "bawling."[49] Such experiences could be so upsetting as to plunge missionaries into confusion about their capacity to control and guide the endeavor they had worked so hard to create:

We're here too late for many of these and only those living here and seeing these people day after day and these things happening could quite understand the distress in this heart and life for more of Christ. I don't know how I could ever take seeing some of these things we have had to with these folks. . . . I know Satan is dragging as many as he can into Christless hell before they can hear. . . . As the Devil stands and sneers, we as sons of God are failing so pitifully in declaring and living Spirit-filled lives of power.[50]

Why was infanticide so upsetting if everyone was convinced the Ayoreo were truly spiritual degenerates? Such imagery confirmed Ayoreo savagery and called for the intensification of missionary efforts. Yet it legitimated further contact work only if the mission itself had nothing to do with creating a situation that was so extreme that mothers felt compelled to bury their own infant children alive, which by all accounts occurred on a dramatic scale. The figure of savagery in need of contact was thus amplified not because the missions functioned smoothly but precisely because they did not. This leads to the uncomfortable conclusion that the ever-increasing drive to hunt Indians was at least partly a response to the overwhelming contradictions of running missions that resembled death camps.

Likewise, the pervasive missionary horror and fear of Ayoreo was articulated not primarily in terms of their own constant fears of being murdered by the Indians as they slept but as the fear they attributed to the uncontacted *Indios bárbaros*.[51] This notion of Indian fear was repeated over and over again in missionary writings until it became the guiding chant of the Indian hunt. "They lived in constant fear of each other and of their terrible Bird God. . . . Was there no escape from Asojna and her anger? Must it always be this way? Would she always reign supreme in the country of the Ayores?"[52]

The point, of course, is that this imagery described not the savages but the missionaries who had to terrorize the terrifying barbarians in order to tame terror itself. Ayoreo had to first become the missionary image of the walking dead to become savable life. Only then could Indian souls be reaped and redeemed.

Indian Hunt as Colonial Becoming

Does this all mean that the pursuit and capture and enslavement and conversion of forest Indians can be considered the crucial ritual by which the missionary project was sustained? New Tribes missionaries may have claimed that they led regular wild man hunts because they wanted "to give the Pig People the chance to hear the Bible, cause if they don't they'll all go to hell and suffer eternal damnation," but the stakes were never limited to Ayoreo well-being.[53]

It was only by the constant creation and extermination of savages through the creation and alleviation of terror that the staged scene of death could figure as an ordered collection of Indian souls and conjure lifesaving Grace. This nonlinear system worked not through its predictability but its instability. Wild man hunts promised to restore a functional and redemptive order even as they invariably amplified a dysfunctional and profane disorder, left as they were in the feeble hands of mortal men. The appeal of the Indian hunt derived from contradictory meanings, half-truths, interrupted images, the constant play of opposites from which new images of difference rushed out in all directions.

What are we to make, then, of the fact that Indian hunts were also a central ritual of the civilizing project more generally? By the time that Aasi and the others were captured, Indian hunting was a long-standing tradition. Paraguay only outlawed "hunting and selling Indian children" in 1957 as a response to the whole-scale traffic in Indians, especially of the small pale-skinned Aché, commonly known as Guayaki. German anthropologist Mark Munzel lived at the "National Guayaki Colony" in 1971–72. He described the official overseeing the colony, Manuel De Jesus Pereira, as a former Indian hunter who ruled the national reserve like a slave colony, selling Indians or giving them away, raping the girls and young women at whim or offering them to visitors, beating Indians to death, shooting them for sport, punishing them by placing them in crude stocks or the *tronco*, starving them to death.[54]

There were five girls between the ages of six and twelve in the group. De Jesus Pereira kept four of them in his house at night. He continued to hunt down, capture, and sell groups of forest Aché well into the 1960s. These hunts usually involved fatalities, the female children were sold or given away, and the captured women were customarily taken from their husbands and turned over as rewards to the small group of Aché "hunters" who had carried out the slave raids with Jesus Pereira. "The Aches are being convinced that it is a shame to be an Ache," wrote Munzel in 1973. "The only way to escape from this shame is to become a hunter of Indians like Jesus Pereira."[55] In 1972, "Papa Pereira" was arrested due to international pressure. His replacement? New Tribes missionaries.

Slave trading and soul collecting were fused through the institution of the Indian hunt. Moreover, at times the Indian hunt also blurred into the ethnographic search for blood and tradition. Jehan Albert Vellard, a prominent French ethnologist and founder of the French Institute for Andean Studies, wanted to collect physical samples of supposedly archaic Indians like the Aché and Ayoreo. To collect such samples, however, he had to literally collect Indians. "True manhunting," he wrote, "was the only way of meeting them." In the name of science, he sponsored a 1933 Indian hunt. Dr. Vellard and his group of local Indian hunters tracked down a band of forest Aché and charged into their camp. In the chaos, his men shot one Indian and captured a child, later taken by his men.

The Indian who was shot "groaned on the ground and did not survive more than a few minutes. . . . As we could not take the corpse with us, I contented myself with measuring it. The little Aché boy looked at us with his astonished eyes, without a shout, without a tear, without a gesture of emotion in front of the dead; he obeyed our signs without saying a word."[56]

All this can be reduced to a single question: what does it mean if being civilized and being Christian is the same as hunting Indians and being an Indian the same as being hunted?

Sequelae

Missionary work progressed more slowly among Ayoreo-speaking groups on the Paraguayan side of the border, not least because the Mennonites largely halted their attempts to contact the so-called Moros after the spearing death of Kornelius Isaac in 1959. Bobby's father was one of the first New Tribes missionaries to establish a settlement in the mid-1960s near Cerro León with various bands of Guidaigosode. He promptly put them to work building roads and trapping big cats for their hides. From the beginning, the New Tribes missions of the northern Chaco served as bases for armed expeditions to hunt down and exterminate or capture the Totobiegosode bands remaining in the forest, or "the Pig People" as they were referred to by the missionaries.[57] These raids were commonplace.

Bobby's father and the other missionaries involved in the 1987 contact were unrepentant about their actions in reports sent to fellow believers. The frustrating outcome of the contact with Aasi and the others was quickly subsumed into the call for more Indian hunts. Missionaries Robert and Cheryl Ketcham described the contact to their prayer partners in a letter titled "God Plants Five More Seeds." The physical deaths of Indians were seeds planted for the spiritual salvation of everyone else. The letter concluded with the following call:

There is yet another group of Totobiegosode in the bush; they are a break off of the group just contacted with 7 men plus women and children in that group. This last group was last seen by the group just contacted about 3 years ago. They are still friends with this group and we hope that some day soon this last group of Totobiegosode people will be found and brought out of the bush so they can hear the Good News also! It may take more seeds, but we are reminded that some plant or are planted, others water, but God gives the increase!!!

Missionary prayers were answered when the Areguede'urasade appeared, on their own, in 2004. Bobby had been waiting a long time for this, his own opportunity to capture Indian souls and make New People. By October of that year, he was making regular trips to Chaidi. His visits were described to New Tribes believers and prayer partners through posts on the New Tribes Mission website like that entitled, "Pig People Clan Hearing God's Word."

As [Bobby] taught about the Ten Commandments he held up a mirror, showing the Ayores how he could look into it and see himself. Then he took mud and spread it all over his face. The people thought it was hilarious, but Bobby brought out the seriousness of the lesson. He told them how, in the mirror, he could see the dirt all over his face and that God's Law was like a mirror. It showed people how they are dirty (sinful) before God.[58]

We're suddenly back to savagery and gold and an anxious drive for more Ayoreo souls to collect and consume and discard, a race for the last Indian hunts to once again recreate and unsettle the boundaries between Indians, missionaries, and anthropologists. It remains to be seen how Totobiegosode people are dealing with the legacy of being hunted, the topic of the following chapters.

For now, we can conclude that the mission was animated by two contradictory energies. One was the drive to create a neat economy of salvation, to reproduce "the romance, the ecstasy, the catharses of the fantasy of order by which the conquest of the New World has been so constantly rendered" and effected.[59] Like the tradition-seeking *Abujá*, the soul-collecting missionary rendered Ayoreo life intelligible only to the degree that it was reduced to a single, valuable, and interchangeable image with magical powers and to a preordained function in an already-established whole. The moral humanity of missionaries and *Abujádie* was the function of their capacities to control the terms of categorization and objectification and, thus, the reproduction of the cosmic order by which their own self-transformation was ultimately possible to realize.

Yet the drive to impose this hermeneutic order on the violent chaos elicited by their encounters was a second-order effect of the more potent breakdown of images and constant unraveling of sense. Indeed, the mission system was animated primarily by its systematic dysfunction and epistemic failure on all sides. As Lévi-Strauss reminded us, the foundational myths of savages and missionaries and contact and Indians and ethnocide exist to mediate the binary contradictions they reify. Yet each attempt just as surely engenders its opposite, an inverted excess that defies and demands further images to represent it and create another kind of excess that demands further images of difference.

These tensions found their clearest articulation in the drive to hunt Indians. Repeated time and again, this operation aimed to track down and capture and enslave and convert small groups of forest Totobiegosode in order to save and care for them by touching and then eradicating their difference, yet in doing so the Indian hunt created and amplified the savage figures of forest Indians to be hunted anew. Orgiastic representation flashed into another Indian hunt, which flashed back into another limit of humanity to be transcended, which flashed back into another Indian hunt, and so on. Yet neither wild man hunts nor my own reverse pursuit of them offered any positive synthesis for anyone involved. Instead, this lack of synthesis is the point.

In the midst of this turbulence, the figure of ethnocide reappears not as an effective critique but as a crucial pole for sustaining this nonsystem. It is a figure created by the instrumental dysfunction of colonial violence and inseparable from the allure of hunting Indians. It does not question Indian alterity but reifies it anew. That is, the notion of ethnocide is a key colonial metanarrative that inverts and naturalizes the operations of colonial subjugation. Through it, Indian difference is imagined as a preexisting multiplicity forcibly reduced to the singular, and this singularity is imagined to make Indians identical to the model we impose.

Yet in practice Ayoreo missions instantiated the very opposite dynamic. Indian hunts did not eradicate Indian difference but tamed it and through taming it they unwittingly reproduced the difference in need of being hunted. Only by having a barbarian to hunt and tame and displace back into the forest could one truly become Christian, civilized, and saved. The drive to domesticate multiplicity, in other words, did not imply its rational reduction to a singularity but rather a much more chaotic tension between the amplification and homogenization of multiplicity itself. In this case, then, the narrative of ethnocide substitutes cause for effect and gives new momentum to the apparatus of eradication, salvation, and alterity that was indistinguishable from colonial violence. It emerges from the same nonsensical logic as the Indian hunt.

Taking this dynamic seriously obviously charters a different kind of political anthropology than did the ethnocide envisioned by Clastres. Such a project can only start by recognizing that the notion of primitive society as "the conceptual embodiment of the thesis that another world is possible" is fused to the appeal of the Indian hunt. It risks diverting our attention from the more pressing question of how such images work through and against the substances of life to create vertical hierarchies of human kinds. In doing so, they are masked and given new power to loop back into the very ontological senses of being in the world that they purport to describe.

Such incomplete images exerted their scorching force on retired missionaries too, who, after years of hunting Indian souls in the Chaco, went back to the middle-class suburbs and megachurches of South Carolina, Florida, California, Minnesota, and Ontario. I heard that one missionary named Asi'guede was so powerful he could make machines move with his spirit. And I heard he wept when he last returned to Paraguay as an old man and saw the squalor and disease of the Ayoreo camp on the outskirts of Filadelifa.

Surely he was no less haunted than Bill Pencille. I found him four years before his death in an assisted living home beside a rushing highway in Rochester, Minnesota. The three-story building was made up in faux colonial style, complete with pillars and synthetic white siding stretching out into two impersonal wings like a budget commuter hotel.

He answered the door like he had been waiting, a tiny nonagenarian in worn slippers and a shirt stained with green Jell-O. The room smelled like chemical air fresheners and urine and the plastic of its generic furnishings. There was no trace of his past, nothing to indicate the outsized role this shrunken man had played in setting the course of Ayoreo history. I tried to interview him, but he tired easily and his memories had faded.

He initially mistook me for a young missionary, asking for advice about how to work with the Ayoreo.

"Oh man," he replied. "I haven't thought about that for a long, long time."

He said he had dragged the Ayoreo from the Stone Age to the modern world. His most vivid memories were of his fear at being speared, his affection for the slave boy that he had acquired, and his meeting with the great Guidaigosode chief Uejnai.

"I'm not a tall man but I was just the right man. I was God's chosen man to do that."

I couldn't quit staring at the interchangeable and sterile surroundings. Struck by the frailty of human designs, the list of probing questions died on my lips. Without thinking, I impulsively offered the old missionary my woven *guipe* bag. At this, his face changed. He lit up, slung the bag around his chest, and then, to my astonishment, recited John 3:16 in perfect rapid-fire Ayoreo: "For God so loved the world that he gave his only Son, that whoever believes in him shall not perish but have eternal life."

Just as suddenly, he again fell silent, as if emptied. We sat for a long time in quiet, the air-conditioning whooshing in the background. When I got up to leave, he stirred and imparted one last nugget of wisdom.

"Just trust God and love them," he said. "Be a man of faith. It is the only thing that lasts."

FOUR

Mediating the New Human

No dream experience, no ancient religion ever separated spirit from flesh more effectively than the electronic media. **EDMUND CARPENTER**

The sound of a two-way, shortwave radio transceiver is unlike any other. It is equal parts static and voice, recognizable utterances bracketed by blips of pure noise. When used Ayoreo style, it resembled a stretched echo chamber wherein dozens of distorted voices moved and resounded all at once in the sparsely consonanted tones of the Ayoreo language, the words overlapping and interrupted and recombined in a schizoid montage. Meaning was anything but stable. The background was as richly layered as this mobile foreground, galaxies of buzzes and crackles and moans and shrieks, the electronic signature of aging solar panels and dried up batteries and atmospheric conditions, of sunbursts and storm clouds and seasons and winds. When you pressed the button to add your voice to the cacophony, all sound ceased as if severed. The handset clicked into a dead weight, the silence flat and empty and solitary. As soon as the transmitter button was released, the noise burst forth again and you were never sure exactly what your voice had conveyed. It was a sonic surreality.

This sound defined Ayoreoland during the early 2000s. Before 2010, I never spent a day in any village on either side of the border without hearing it, flowing from this person's mud house or that wooden shack or brick school. I learned to tell time by it and by the cross-border, all-community radio sessions held without fail at 8:00 a.m., noon, and 5:00 p.m. every day. Sometimes it was the first sound I heard when waking and one of the last before I slept. Plans to establish new settlements had been abandoned due to the lack of a radio set, or more precisely, the existence of a village was explicitly predicated on this sound. In my memory, it has become more than a constant backdrop muting the deeper essences of everyday life. I think of it now as the typical sound of spirit and meaning and becoming and the search for knowledge during days gone by in the Place-Where-the-Black-Caiman-Walks.

It was no secret, no great discovery that this ambiguous and ephemeral sound was the substance of a collective Ayoreo project that hovered like a dust cloud or a mirage over the Chaco—one that emanated from the cheap speakers of radios scattered across thirty-eight settlements and an international border and 180,000 square miles, a haze that evaporated back into the ether just as quickly as it poured out. What was unexpected was the nature of this aural project and that it took me so long to realize that I could not grasp its force through rational means and discursive categories alone. I had no choice but to follow the sound, even though this was not apparent at the beginning.

Jochade, the leader of Arocojnadi, was just as interested in radio as I was. He agreed to teach me to listen to this sound and how to properly use the radio. We sat around the flat Yaesu set wrapped carefully in a stained cloth on top of a rude table in the school building. Cables ran from a small solar panel on the roof to an old truck battery and from the battery to the radio. Six of the *Disiode* children peered back at us, arms and heads pushed through gaping holes in the windowscreen. He said the rules were strict but simple.

To talk, everyone met first on the main channel, 6819.2 MHz. Less public conversations were then moved to one of the other twenty-four channel presets shared by all Ayoreo. If even more privacy was desired, Totobiegosode had created three "hidden" presets or frequencies that were only shared by Chaidi and Arocojnadi. No children or adolescents were supposed to use the radio, and it wasn't seemly for a male leader to talk too much on it. Jochade knew the usual radio operators in the other villages, most of them senior high-ranking women. The primary rule of radio was that conflictive topics were off-limits. He asked if I understood and then he said:

"The radio is only for two things: the Word of God and for finding out if anyone is sick."

With this single phrase, Jochade upended my expectations. Like a good *Abujá*, I had hoped to discover a hidden substrate of continuity and counter-hegemonic resistance within the unregulated domain of Ayoreo radio.

I glanced at him sidelong in disbelief. He was looking down politely and clearly not joking and he was not a man who brooked repetition or incredulity lightly. I tried to condense my confusion into a question.

"Do you talk about how the *Cojñone* are stealing your lands?"

"The *Cojñone* are stealing our lands. It is true. But we do not talk about that on the radio. We want to know who is sick. We want to share the Word of God."

Over the next several months, I struggled to apprehend the full significance of Jochade's lesson. Of course, I knew that Ayoreo radio had long been associated with Christianity. Bill Pencille made Direquednejn-aigosode porters, including a young Jnupi, carry shortwave radio sets on their backs through the tangled brush to Echoi, where he broadcast live updates during his 1960 expedition to contact the band of Uejnai, the great Guidaigose leader. At least since then, missionaries allowed their favorites to communicate via the radio with relatives on allied Protestant missions and even encouraged some of those contacts as part of their evangelizing project.

Yet surely this history was only part of what Jochade intended to convey. For instance, I knew that NGOs were also invested in Ayoreo radio but for the opposite reasons: its potential to preserve tradition, revitalize culture, and create political solidarity. The organization APCOB gave Bolivian Ayoreo radios in the 1970s, and the radio we sat near in Arocojnadi was provided as part of a Norwegian-funded project in the late 1990s. Did Jochade mean that these projects had simply failed, or did he mean that Totobiegosode saw health and Christianity as the primary sources of political agency? Was he speaking about limits or priorities or something else entirely?

At first, I mistakenly thought Jochade was referring to the discursive contents of radio. That is, I was thinking only about words. Accordingly, I set out to chart and transcribe the actual texts exchanged on the radio. The process took months and meant long hours listening to on-air conversations and observing Dasua, Arocojnadi's usual operator. I quickly learned that Jochade had not exaggerated. Nearly every linguistic radio exchange focused on one of these two topics. Rarely was anything else discussed. A common way for Dasua to conclude a radio conversation was "I have nothing more to tell you because we are all healthy here."

Descriptions of a healthy, strong, and resistant body did not usually elicit too much conversation. By contrast, discussing an ill, vulnerable, or threatened body invariably created an intense burst of concern. The recitation of symptoms and stories about infections could last for hours and involve people from dozens of far-flung communities, who expressed their grief or sorrow for the ill person at great and formulaic length.

Likewise, the Christian God *Dupade* was invoked in the vast majority of radio conversations. Common radio expressions included, "Don't worry about us. God helps us here," "We know that God is the one who gives us everything," "I am happy because God gives us strength and health," "God is the one who knows everything," and "God will help us all."

Many people, including Totobiegosode, commonly used the radio to ask others elsewhere to pray for them. Such requests were directed at individuals who were believed to be morally pure and therefore have relatively more powerful prayers. These were usually a handful of recognized authorities on the Word of God living at the New Tribes Mission of

Campo Loro and Puesto Paz. Such spiritual authorities often "gave" Bible verses to one another or discussed parables on the radio, as in the following advice given by a preacher at Campo Loro to a man in Bolivia:

I am happy that you are listening to my words, but you must not forget the teachings I am giving you. The Word of God is the only thing that can save us. The Word of God is true. Before, when our grandfathers lived in the forest, they looked for a spirit that could save us but they never found one. They believed that the *sarode* could save them. They only believed in the *sarode* because they did not know about any other spirit. But today, we have the teachings of the Word of God, and today everything is very good. It is very good because we now have only one God. We now know that it is only God that can save people.

If these conversations were only taken as texts, as I initially approached them, their significance seemed self-evident: Ayoreo radio users were creating a kind of collective belonging predicated on expressions of Christian faith. This kind of belonging did not imply continuity so much as rupture. It was predicated on explicitly distinguishing the moral present of *Cojñone-Gari* from the sinful past of *Erami*. Radio conversations evoked a world through binary oppositions: past, Indian, traditional, satanic, and forest were set against present, civilized, modern, Christian, and moral.

This impression seemed to be further strengthened when I learned that other Ayoreo routinely singled out Totobiegosode as a homogenous group of New People particularly in need of Christian advice. In the following conversation between Jochade and a prominent preacher from Puesto Paz, Bolivia, the pastor suggested the Totobiegosode were morally deficient and should capture the forest bands, while Jochade invoked Christian imagery to bring the conversation back to an acceptable middle ground. Having failed, he ended the conversation:

Ajnocai: What about those who are in the forest? Cambio.

Jochade: We do not know anything about them. We have not found anything yet. Cambio.

Ajnocai: That's how it is. I heard in CANOB that you all there found something from them. Cambio.

Jochade: That is how it is, Ajnocai. We did not find anything, cambio.

Ajnocai: We have not forgotten about you when we pray to God. We never forget you all and we never forget to pray that God arrives to them too. We are always praying so that they arrive together with you all and that they come to know God, cambio.

Jochade: That is how it is. It is good what you say and that you are praying there for them. No one knows the place where they may be walking. Only God knows where they are. He has more power than us and he knows everything. Cambio.

Ajnocai: That is how it is, that is how it is. I say to you that we are all healthy here and that we never met you there. I hope that one day I will have the chance to arrive in your village. Cambio.

Jochade: That is how it is, Ajnocai. We are all healthy here, my brother and I and all of us. Ajnocai, I say to you that I never forget you all there, either, and my sister who is your wife. That is why I will buy something for you all too. I will buy something next Tuesday and send it to you because I am thinking about you all a lot.

Ajnocai: That is how it is. I am always thinking about you, and if a trip comes we hope to arrive there too. There are many words now because we have learned new things about the Word of God, and I will now tell you all there too. I will tell you because we want this news to arrive to all of the Ayoreo people that are in Paraguay. Cambio.

Jochade: I understand, I understand what you are saying and it is good. You should pray to God that some of the forest will be left over, even if it is just a little bit for our relatives to live in. You should pray to God always for our health as well. You should pray that our relatives in the forest can be in a place that is calm, even if it is very small [*ore uja ca yatique*]. Cambio.

Ajnocai: That is how it is. It would be very good in my opinion if they come to live with the *Cojñone* too. I say this because we lived in the forest before and many bad things happened to us there. We always had to run away from the people who were killing us. We were very bad before. We were full of sin and we did not know anything. But now the *Cojñone* taught us the Word of God and we have learned many things. Now our *ayipie* is new. It is the *ayipie* of God. That is how it is, that is how it is. I say this to you now because they are surely close to where you are. Cambio *eee*.

Jochade: That is how it is, Ajnocai. This how it is. You should ask the believers of God there to pray for us because we never know when we may have problems here. I am happy to talk to you and here my words end.

After four months, I had gotten no further and my efforts to puzzle through Jochade's statement ground to a frustrating halt. It was obvious that through radio exchanges, Ayoreo were evoking a collective sense of belonging. Moreover, this sense of belonging was based on a consensus that contact meant a rupture with the past and the wholescale transformation of humanity. But such basic observations raised more questions than they answered. Why were people talking so much on the radio if the contents were so restricted, formulaic, and predictable? Surely, the production of a sentimental community through recognizable scripts was a central part of radio's appeal. But was that all Jochade had been trying to tell me? Why only two topics in particular? How were bodily states and the moral transformation of humanity related and what did radio have to do with it?

By then, I knew just how rude asking direct questions could seem, and I knew that Jochade was particularly touchy about protocol. When I approached him again, I only had one or two shots at getting somewhere with the topic whose grip on my imagination had already begun to slip as I became further submerged in the everyday immediacies of survival and suffering.

I thanked him for his instruction and said I now understood what they talked about but that I still did not understand why people talked on the radio at all. I said I did not understand its power. At that, he smiled slightly and spoke gently to me, as if to a small and stubborn child. "You have to listen to what they say. It is not about their words. It is never about their words, Lucas. It is about their *ayipie*. They are giving their *ayipie* to each other. It moves quickly, here and there, there and here."

I learned that *ayipie* referred to one of the three kinds of soul matter believed to animate human life. The other two—*ore'gate* and *ujopie*—were previously associated with different dimensions of immortal and shamanic power and were no longer discussed except in reference to the Holy Spirit. Not so with *ayipie*. It was associated with the corporeal seat of memory (located in the head), emotion, rationality, and willpower (located in various abdominal organs). That is, it encompassed precisely all those elements believed to constitute the moral human. Moreover, I learned that it was this kind of soul matter that Ayoreo believers imagined to have been transformed by conversion to Christianity. It was a barometer of moral standing and physical health and personal agency. Accordingly, the quality and force of individual *ayipie* was a matter of constant concern: was it coming or going, near or far, expanding or diminishing, strong or weak?

The relationship between radio and *ayipie* was an ambivalent one. Certainly, *ayipie* figured prominently as a subject of radio conversations. The newly transformed *ayipie* was an ideal type that was routinely invoked. Like Jochade, Ayoreo people often explicitly used the figure of this reconstituted soul matter to articulate the moral imperative of Christian conversion:

Our sins were a big problem for us before. That was our entire life before in the forest. No one was careful for their own life. We always became angry at other people. We always fought with each other. And God doesn't like that. It is better if our *ayipie* in this world is the same *ayipie* as God. The Bible says that our faith is our *ayipie*. It is like our soul. Our *ayipie* is the same as the *ayipie* of Jesus Christ. We therefore cannot forget about him even for one day. God is more powerful than anything else. We cannot forget about him or search for the things of this world. That is all we did before when we did not know that Jesus Christ existed. Before we didn't know this. So we always searched for other wives. We lied. We killed each other. All of the things that Jesus is against. But now we are together under the light. We are in the hands of God and we must do what he is in favor of. Only God has strength. Only God can erase bad *ayipie*. If an Ayoreo makes his *ayipie* the same as God's *ayipie*, it will be saved and live forever.

On the radio and in conversation, Totobiegosode described the transformation of humanity in *Cojñone-Gari* in terms of acquiring *payipie ichadie*, or new *ayipie*. This was the core of what it meant to become New People capable of surviving in a New World where the Christian God set the terms of life and agency. The former slave boy who had been Bill Pencille's key guide throughout the 1950s and 1960s framed this in blunt terms in a message he sent to all Totobiegosode over the radio.

Our *ayipie* were very bad [*yocayipie poitagipise*] when we lived in the forest before. That is why we had to know Jesus. And now we know Him. Before, we didn't know who God was. God knows how to change our *ayipiedie*. He wanted us to give Him thanks. Also, He wants us to thank people like me and to thank Him for "touching" [capturing and converting] so many Ayoreo before. To this day, we are together, all of us, living in *Cojñone-Gari* as God's people [*Dupade-urasade*]. And today, we are looking for the Ayoreo still in the forest so they will join us as God's people. We have left behind the war with other Ayoreo that we always had before. I am always thinking that this life is fair because the work that we did saved all of our souls. We are together now with the Word of God and with our faith. All the Ayoreo and the new people and the *Cojñone* too.

Yet *ayipie* was more than simply a discourse about radio's power: it was also the metaphysical and nondiscursive content of radio sound. This soul matter was a substance that adhered to radio sound and gave voice its force and texture. I learned that radio sound expanded the ways in which the moving breath of the speaker and the vibrations of the vocal chords conjured and channeled a spirit power that could be fused with words and make them capable of moving through time and space to cause the effects they ostensibly represented. This relationship between sound, body, and spirit was crucial to the operations of sympathetic magic and was the source of shamanic power to heal illness in *Erami*. Radio technology involved a double transference of audible sound and disembodied spirit. And it activated twinned circuits of transduction: one which moved from voice to wires and electromagnetic waves and back to mediatized sound, and another from mediatized sound to *ayipie* spirit and back to psychic and bodily interiors.

"When I hear someone's voice on the radio, my *ayipie* goes near them," Dasua told me. This was not merely an expression of empathy. Rather, it also referred to the kind of mechanics through which radio sound became a technique for practical assistance. This pragmatic force was usually rooted in vocalizations of prayer, a practice that was associated with the same haptic mobilization of *ayipie* previously attributed to shamanic healing chants such as the *ujñarone*. Like sympathetic magic of all kinds, the power of prayer depended on being able to effectively channel and direct the movement of one's *ayipie*. Radio sound promised to aid in the expansion and redirection of a Christianized *ayipie*:

> The radio helps us find out about other people like those living in Puesto Paz. We heard recently that someone had bothered one of them and they asked us to pray for them. We can pray for them. We prayed a lot and the situation in Puesto Paz improved. God helped them through the radio. Without the radio, we wouldn't have known, and the strength of our prayers, our *ayipiedie* would not have reached them.

Jochade, Dasua, and the others in Arocojnadi always attributed radio's power to enhance the circulation of *ayipie* to its status as a defining technology of *Cojñone-Gari*. It was suitable for transferring not only new Ayoreo *ayipie* but the *ayipie* of *Cojñone* as well. Through radio sound, Jochade said, some people were able to transform spirit. He implied that radio sound was a medium for the mimetic transference and transformation of *ayipie* itself.

> The radio is useful for someone who speaks Spanish because if a *cojnoi* speaks, the Ayoreo can listen and acquire the same *ayipie* as the *Cojñone*. If Mateo listens to a *cojnoi* on the radio, he might say to himself, "My *ayipie* was not well developed before, but now I will make it grow. I will make it like the *ayipie* of the *Cojñone*."

Jochade's tutelage continued as the days shortened and the rains gradually stopped. By the early dry season of 2007, I was convinced that this mediatization of *ayipie* explained the seemingly mundane content of Ayoreo electronic media practices. The words, as he had said months before, were less significant than this metaphysics of media. Words were often secondary vehicles for the sentiment, breath, and sound aimed at accelerating the beneficial circulation of *ayipie*. When properly deployed, *ayipiedie* could intervene in the Christian metaphysical ecology that now determined physical health, moral well-being, social status, and political agency.

At the same time, Jochade and the others also stressed that these concrete effects were only possible when the *ayipie* that was sent resembled the receiving *ayipie*. *Ayipiedie* were said to be supportive of one another only when they were similar, and this power was cumulative. Thus, groups of people who possessed "one *ayipie*" were considered to be stronger than individuals and more likely to achieve their goal. The process of creating a single *ayipie*, *yipejo yocayipie*, was fundamental to any group action. Weakness, paralysis, and vulnerability were caused by not being able to locate or homogenize *ayipie*. In other words, the power of radio sound also depended on and allowed a collective *ayipie* soul matter to be standardized. It was premised on making explicit a shared consensus about the nature of the new moral human. Indeed, this was the kind of commons evoked and created by the metaphysics of radio sound.

Totobiegosode media practitioners did not locate the power of radio technology in a potential for discursive resistance or the continuity of form but in the potential to objectify and domesticate rupture itself through the particular spiritual harmonics of electronic sound. This process also reasserted the Ayoreo capacity to objectify the terms of their self-transformation and figured the two-way radio as the crucial medium by which a Christian Ayoreo mainstream predicated on the moral value of rupture could be constituted. This metaphysics made radio sound a crucial medium through which the wider Ayoreo project of self-transformation could be evoked and objectified. In many ways, radio sound created this project instead of vice versa. The image of the new moral human and the circuits of aural media seemed inextricably entwined.

Remaking a world and transforming humanity through sound, however, was a deeply fraught process, one subject to its own inversions in The-Place-Where-the-Black-Caiman-Walks. Radio sound was wild and ungovernable and ephemeral. It always exceeded the boundaries of any ideal type or standardized project, no matter how closely guarded the script. The schizoid qualities that imbued it with spiritual force also threatened to evaporate at the very moment their power was materialized.

Jochade was never entirely comfortable with the unruly excess of radio's sound. Instead, he told me he preferred cassette tapes. I learned that a vibrant network of cassette exchanges had predated and prefigured the use of twoway radio. Jochade had hundreds of these cassettes in his house. The cassettes consisted of short greetings from many different people, as well as songs, ordered by rank and prestige of the speakers. He told me these closely resembled the forms of face-to-face greetings considered appropriate in the past, particularly during the great pan-Ayoreo gatherings at Echoi. Jochade listened to these greetings over and over. One of his favorites was a 1989 message from a distant relative of his that he had not seen since leaving the forest and whom he had long presumed was dead:

Jochade, this message is for you. I am one who lived with you all before when you were in the forest. I don't know if you remember what I look like, *ñequenochade*. I am Igue, who lived with you before in the forest, together with my father. My father lived a long time with your people, because my father was one of your people before [contact]. It was long ago. I am meeting you now with my words. . . . I do not know if you remember that when we were children during the war, our parents took us to live with the Garaigosode. We do not know if you remember us. Here my words end. Send me a cassette and some songs, if this cassette of Chogueside arrives to you. . . . We do not know where you all are. We knew before but we do not know now.

According to Jochade, the most important difference between cassettes and radio was how their effects could be contained. "You cannot erase the words of radio like you could erase a cassette. They speak there and it arrives to you. Everyone can hear it. Radio can make problems worse. With cassettes it was better because we could erase any bad words so no one could hear them." I learned that Jochade himself was a notorious tape editor. He would collect and privately listen to any cassette made by the people in Arocojnadi. If he objected to any message, he would secretly edit it out or record something over it. No such control was possible, of course, with the radio voices resounding across Ayoreoland all hours of the day and night.

If the ambiguous power of radio sound derived from its multivocal ephemerality, cassette sound was ambiguous precisely because of its more durable materiality. Indeed, a history of recorded sound was no less fraught than a written one. Profound tensions between remembering and forgetting marked Totobiegosode relationships with aural media, particularly those cassettes deteriorating in the harsh climate. Listening to the sounds of the past and the voices of the dead could be deeply ambivalent and upsetting. It could cause one's *ayipie* to "go back to the past" or become embittered and thus vulnerable to infection and death.

Jochade possessed live sound recordings of the first contacts with various Totobiegosode groups, including his own band in 1979 and the disastrous contact he guided in 1986. These tapes were made by Christian

Guidaigosode from Campo Loro who entered the Totobiegosode villages with portable cassette players in hand.

One morning, Jochade's son-in-law played the tape of the 1979 capture of the Totobiegosode band led by Jochade's brother Pejeide. He made no fuss about it and didn't tell me what it was. But at the first tones of muffled and distorted sound, the handful of younger men sitting nearby knew. One laughed hard, one walked away, two listened quietly. In the foreground were staccato shouts between men, moving in and out of range, punctuated by short screams, sounds of the tape recorder being dropped and picked up and dropped again. After two or three minutes I realized there was also a steady background sound, but one I could not identify, a continuous blanket of distant moaning, something almost human but not quite. Jochade's son-in-law said it was the sound of the women and children who were too paralyzed to run away during the attack. This was the noise they made when they believed they would soon be massacred. Jochade and the others could identify most of the individual voices on these tapes. I asked the young man to turn it off, and we sat as wind rushed through leaves and dragged up dust.

Such recorded sounds, of course, did not catalyze a new memory project for Totobiegosode. Rather, their durability stood in stark tension with the projects of self-transformation effected through the immateriality of radio sound. One night around the fire, Jochade looked up and said that he wanted to use cassette technology to record old-style songs like the one Yoteuoi had just finished singing. In the next breath, however, both men said such efforts were doomed to fail. They concluded that *Cojñone* like me would end up with the only lasting records of their voices, stories, and songs.

"If we don't have a way to record these songs, they will be forgotten," Jochade began. "I don't want this, so I have a tape recorder. If the children don't want to learn these songs, they will be lost. If the *Cojñone* record our stories and songs, they will always have them."

"That is right," Yoteuoi agreed. "We can tape-record our songs, but it will be pointless. We elders have already forgotten some things, the names of some things, because we didn't have tape recorders before when there were still people alive who knew those things very well."

"The *Cojñone* never forget anything," Jochade concluded. "Like *Dupadeuruode*, the Word of God. It is very old but it comes from the *Cojñone* because they recorded it and then they wrote it down."

For Totobiegosode, recording and preserving was never so simple. Precisely at the moment of materialization, the ideal moral self—like the sound and spirit through which it emerged—dissolved once more.

It is no surprise that ethnographers have persistently misinterpreted the ontologically generative qualities of such negative tensions. Most have failed to register Ayoreo media practices at all. The few exceptions—much like my ill-fated initial attempts—have labored to flatten the rich sonic incoherence into the easy textual logics of a traditional culture from which the contents of actual Ayoreo media practices are already excluded.

Fischermann, for example, contends that "Ayoreo have completely subordinated technology like radio to the values of traditional cosmology" and the mythic ordering of the world. Borrowing from Fischermann, Miguel Bartolomé likewise concludes that Ayoreo sonic media "have been incorporated as the material patrimony of tradition."[1] He describes how the use of radios and cassettes has "refunctionalized" the typical communication strategies of hunter/gatherers. "In such ways," he argued, "radio/recorders have been transformed into the transmitters and reproducers of culture, used in accordance with the particular communicative logics of indigenous society."[2] José Zanardini likewise points to cassettes as the reproducers of tradition: "This [circulation of cassettes] proves that the world of the Ayoreo is still united inside, that divisions come from outside."[3] The noted Chaco ethnomusicologist Jean Pierre Estival is the only scholar to develop a deeper analysis of Ayoreo cassette and radio use, but he too describes it as a "medium of cultural resistance" that "reproduces past values and symbols" in the present.[4] For each of these authors, the significance of Ayoreo radio and cassette media lies in continuity with ontological alterity and counter-hegemonic resistance.

The problem, it seems, is that such descriptions reduce the significance of Ayoreo aural media to an impoverished notion of essence or being without taking into account the relationships between sound, spirit, and body that animate these media exchanges. Moreover, they ignore how all textual representations of radio sound are poor imitations. Such representations flatten the remarkable ways that Ayoreo media practitioners deployed the nonlinear qualities of haptic meaning making to agitate, extend and attempt to channel the ontological murkiness of the colonial encounter. This reflexive capacity of radio sound—how it allows for reflection on meaning itself and thus demands a reobjectification of the incoherent objecifications on which colonial subjection depends—is the truly resistant ontological labor that resounds through the fecund distortions and surreal montages of radio, a potential that uneasily reverberated through the simultaneous playback of colonial divides and differentiations. Radio sound voiced precisely the contradictions through which Ayoreo projects of self-transformation could coalesce and collapse.

Is it any surprise, then, that the new moral human evoked through sound during those bygone days was ultimately illusory? That it was an ideal impossible to realize, one constantly interrupted by blunt-force trauma and a more insidious hermeneutic violence that measured such creative and disordered Ayoreo projects of becoming against a colonial fantasy of order to which they were already illegitimate, spurious, diminished? Like radio sound itself, the new moral self was fractured and ambiguous. As quick as it could be conjured, it dissipated back into the ether, just out of reach of us all.

On my last visit to Chaidi and Arocojnadi in 2013, it was clear that radio technology was on the way out. Cheap cell phones had replaced it like radio had replaced cassettes. Everyone had one. They said the phones were better because they were beautiful and more private and you could call anyone you pleased and no one else could listen. They said the two-way radio was only for old people.

Arocojnadi was one of the few villages without cell signal. For the time being, the young people climbed a tall quebracho tree in the evenings to catch a single bar or walked a couple of miles to a nearby ranch house equipped with a microwave relay. Even the New People had phones, with a constantly shifting set of numbers and very few incoming calls.

Jochade, however, couldn't seem to break his daily habit of checking on the radio. The last time we turned it on to the main Ayoreo channel, we sat side by side in the evening dusk listening as the radio buzzed and crackled with the sound of empty static.

FIVE

Apocalypse and the Limits of Transformation

The only philosophy which can be responsibly practiced in the face of despair is the attempt to contemplate all things as they would present themselves from the standpoint of redemption. THEODOR ADORNO

This does not imply, however, that for the Jews the future turned into homogeneous, empty time. For every second of time was the straight gate through which the Messiah might enter. WALTER BENJAMIN

During the dry season in northern Paraguay, the bulldozers never stopped. Manned by rotating crews, the massive caged machines with stabilizer bars and halogen lamps rode over and crushed the low forest, too dense to undercut, the metal worn smooth by the plants. Working-class Paraguayans said that driving one makes you *embrutecido*, brutish and mean, that it damages your kidneys, spine, brain. As the days lengthened and the heat of the sun intensified, ranchers burned what the bulldozers pushed into mile-long rows. Smoke covered thousands of square miles for weeks, enough to make the streetlights in the Mennonite colonies far to the south turn on at midday. The sky was twilight gray; I woke with the taste of ashes and a thin film of white on my tongue.

The harsh drone of the bulldozers was audible all day and all night. In 2006, they were working south of Chaidi, carving a grid that pushed to the very limits of the Totobiegosode land claim. The Areguede'urasade had occupied this forest until the day of contact. The land being bulldozed

was intimately known. It was dotted with old camps and houses and gardens and trails and carefully stewarded reserves of honey and game. The bulldozers uncovered a traditional Ayoreo *iguijnai* house in the path of the first planned road. The Nivaclé Indian drivers walked off the job when they were ordered to bulldoze the dome of mud and limbs. It stood defiantly for months in the center of the planed earth until it too was finally crushed.

Such incongruity, at times, could seem typical. If I always felt welcomed at Arocojnadi, this wasn't the case at Chaidi. Although I divided my time equally between the two Totobiegosode settlements, my relationships in Chaidi were uneasy from the beginning. Most of the people had only recently arrived from Campo Loro and had not met me during earlier visits. Few of them were willing to be interviewed. The evangelical preacher and his family, newly installed by Bobby as his proxies in the village, were openly distrustful of my motives. The New People, too, were highly surveilled and controlled. Although I did not know it at the time, they had been warned against talking to me.

Yet the temptation of my truck—the Giant Armadillo—proved hard for anyone to resist. The people of Chaidi kept me in constant motion hauling firewood or sick patients or *dajudie* plants or wage laborers or supplies from the new gas station the Mennonites put up some twenty miles away. Eager to escape from the monotony and heat and relentless brinks-manship, I was usually happy to oblige. I especially looked forward to *baaque*, one- or two-day hunting trips. They meant strenuous exercise and the possibility of meat or honey to give away. Moreover, hunting meant a chance to talk privately with Siquei, Emi, Asôre, and my hunting partner, Cutai.

Invariably, these hunting trips were aimed at the same lands the Mennonite bulldozers targeted. Often the sounds of growling chains and exploding trunks grew louder as we neared and parked. At such times, my grim-faced companions spread out silently in pairs and disappeared into the soon to be flattened brush. It was hunting without restraint, an orgy of killing and extraction before the great machines arrived to crush the ancient tortoises in their sandy dens, to frighten away the peccaries, to pulverize the secreted honeycombs, to spill the rainwater trapped in the hollows of the *najuane* leaves. Once we ran alongside the fearsome machine itself, close enough to feel its heat and almost close enough to touch, four abreast completely concealed not fifteen meters from where it was ripping the forest a new edge like invisible spirits of destruction stalking and grabbing and racing to dig up turtles and cut out hives before the bulldozer could grind out another inexorable turn. We dripped sweat and blood from a thousand scratches and our ears rang from the noise and afterward there was no sense of pride and nothing to say at all.

During this time a religious fervor gripped Chaidi that was extreme even by Ayoreo standards. Performances of Christian faith soon dominated public sociality in the village. The change was spurred in part by Bobby, who had waited years for the next first contact. In the beginning he tried to keep his visits to Chaidi a secret and he wouldn't come if I was there. But by February 2007, he was visiting openly, every two or three days, his SUV loaded with oranges for the children, each visit followed by a short peccary hunt. Backed by Bobby's influence, the aspiring preacher Achinguirai organized a faction of people to oppose the authority of Dejai, the more secular leader of the village. Achinguirai encouraged daily discussions of evangelical doctrine aimed mostly at the New People, and began to flaunt his power. Bobby overlooked such earthly power plays and even paid for the construction of Chaidi's first church building. It took the ambiguous form of an open-sided tin awning because, as he explained, Ayoreo Christianity could only succeed if it were hidden from the NGO and visiting anthropologists.

Under this awning or at Achinguirai's hearth, the growing Christian faction held religious services twice a day. The New People, in particular, were required to attend. They were publicly interrogated about Bible verses and scripture and most of them seemed to take it very seriously. Siquei was discredited as a leader. Bobby took two of the men on preaching trips to Campo Loro and even to the New Tribes Mission of Puesto Paz in Bolivia. There were rumors he was trying to convince Jotai and the others to hunt down and contact the last Totobiegosode band remaining in the forest, led by Jotai's father, Jotaine. Twice a week, the New Tribes missionaries hosted a program in which Ayoreo preachers from Campo Loro broadcast fifteen-minute Ayoreo-language sermons on the Mennonite radio station. When they aired, the village hushed. Adults in every household huddled near their battery-powered radios. I learned that these sermons were strikingly creative interpretations of Christian scripture, such as the following by Refresco'daye:

We know that the stars, the moon, the sun, the night and the day are very different. We know that up to this day. We know that because we have these stories in our hands [in the Bible]. It will be the same for the Ayoreo. Those that die will come back to life again, even though they were dead. When they live again, they will have another body. Their old bodies will stay underneath the earth. . . . They will have a New Body when it happens. . . . Our old bodies are very ugly and dirty. . . . Our new body will never die. It will never die because we believe in God. The dead ones will come back to life again. They will have a new body. They will change to another body. Their new body will be very beautiful. It will be clean and white.

Such themes, I discovered, were not the exception but the rule. Indeed, the performances of Christian faith so common in Chaidi invariably invoked the end and rebirth of social time. Many Totobiegosode were convinced that believers would soon be transformed into prosperous, white-skinned, celestial beings. Redemptive transformation was an appealing vision in this land of bulldozers and blood. As Siquei put it, "We have to leave behind all those bad things from before. Only then will Jesus take us to His village when the world ends. They say our *ayipie* souls have to become new."

Anthropologists of Chaco peoples have long reacted to such statements with skepticism or despair. Widespread Ayoreo professions of faith are either lamented as evidence of evangelical ethnocide or dismissed as a misleading appearance of change that conceals an unchanging core of continuous and pure cultural difference. That is, it is often presumed that Ayoreo traditional culture and evangelical Christianity are fundamentally opposed and mutually exclusive. In such models, "indigenous

Christianity" can only be intelligible in relation to continuity or rupture with the past and it is imagined to be the result of collisions between two bounded and incompatible "regimes of value."[1] Like anthropologists working elsewhere, I gradually came to realize that this model did not adequately describe local faith practices. Moreover, as Tom Abercrombie painstakingly shows, the common perception that indigenous Christianity represented only a "thin veneer" of change concealing a deeper essence of alterity is itself a colonial logic long deployed to morally justify the violent dispossession of heterodox Indians.[2]

Yet it was precisely through such creative heterodoxy that Ayoreo believers also attempted to render their new subject position as "indigenous peoples" inhabitable. That apocalyptic futurism was the primary drive and reflection of this creative agency added a further complication.[3] Ayoreo apocalypticism created a limit situation in which continuity implied rupture and vice versa.[4] It was never a "fully articulated horizon" that "freezes our view of the reality that immediately confronts us" or a simple evacuation of the near future by the far.[5] The opposite appeared to be the case: the capacity to alter the experience of time and to reclaim the possibility for self-transformation is what made apocalypticism an appealing and commonsense way for Ayoreo people to understand the end of past worlds. Perhaps more disturbing, the widespread adoption of apocalyptic futurism as a source of optimism showed how the fusion of Western eschatological models and global political economies created a world where hope for marginalized peoples was restricted to the same terms as their imminent destruction by forces they did not control.

Space of Terror

For Totobiegosode, bulldozers became both vehicle and sign for the end of time. They called them *eapajocacade,* a word that likens them to "attackers of the world." Siquei told me that bulldozers were his people's greatest fear. The sound of one—impossible to pinpoint in the complex acoustics of the forest—caused Totobiegosode to run far and fast.

The machines of industrial agriculture haunted the forest of the concealed Totobiegosode throughout the 1980s and 1990s. The forest bands were constantly harassed by their noisy incursions. On several occasions in 1994, 1998, and 2001, Totobiegosode warriors attacked bulldozers with spears and arrows. They remembered these acts with pride. Usually they fled:

Before, we walked from here to there, around the land. We were very afraid of the bull-dozers, and that is why we walked to every place. We walked at night, from one place to another. . . . We did not know where to go that would be safe from the bulldozers.

These people concluded that the bulldozers—like dogs, chickens, and trucks—were monstrous beings sent by *Cojñone* to follow their scent and consume their land. As Siquei put it,

We did not know before that people were inside. We thought that machines moved by themselves, but it turns out that there is space inside where people sit. We did not know what was inside of them. We thought that they had eyes and could see where to go. We thought that they obeyed the words of the white men [*Cojñone*]. That is what we thought before because we did not know. We thought machines were like dogs. . . . We thought the bulldozers were looking for us because they knew how to find the places that had always been ours. We thought they were following us. Every time we made a village a bulldozer would come.

They imagined that these noisy beings communicated with the *Co-jñone* in a foreign language and told them where to find the forest people.

"We kept running and we did not know where to go. The *Cojñone* and their bulldozers had destroyed the world. We sat in the darkness."

The former Areguede'urasade each pointed to this pervasive terror of bulldozers. They depicted their recent life in the forest as a series of encounters with bulldozers, each one more demoralizing than the last. By all accounts, these were catastrophic events. In their terror and headlong flight, people left behind their homes, possessions, food reserves, gardens, and reliable sources of water. Often, people were separated from one another and sometimes left behind, putting the survival of the entire band at risk.

One particularly traumatic encounter occurred sometime in the mid-1990s near Arocojnadi. At the time, the forest Totobiegosode were living together in a large encampment. They knew a bulldozer was working nearby but were caught off guard when the terrible machine suddenly appeared in the middle of their village. Siquei described this event in the following way:

I was working to prepare my garden when the bulldozer arrived in the village. I began to run back toward the village. But I forgot my *cobia* [feather collar for war] near my garden. I went back and put on my *cobia* of stork feathers and went to fight the bulldozer. It was very loud, and it came closer and closer. It seemed like it was angry with us, that it wanted us to go to another place so it could have our beautiful gardens. We loved that place very much. We all ran away, the women first. Then the men organized to return [and fight]. . . . But at that time [the elder] Areguede said, "It is acceptable if the bulldozer pushes us out of here." So it was finished. We [did not fight and] returned to join the women. They had already run far away and we followed them north. We met them and we kept running. We traveled by night. It was very dark but we left that place. Our grandfather Ugaguede followed behind but he walked slowly because he was old. He lost our trail in the night and his daughter had to go find him the next day. After that we suffered thirst. We almost died when we went to find water. Everyone in the camp was very thirsty. No one had strength to move. I cut some cactus trees and we squeezed the juice from the pieces. That was in the place called Bajoite-ajuqueode. We were very thirsty and we almost died. We decided to go back to our old camps. We though the *Cojñone* and the Ayoreo were still at war. We didn't know that they had stopped looking for us to kill us. We stayed there in the place called Cuguedodie. But the bulldozers came again and again we went back to the north.

They linked a solar eclipse in 1994 with their attack on a bulldozer and as a sign of their impending demise:

We came back to the east and we shot at the bulldozer that was bothering us. Then we ran out of fear of the *Cojñone*. We were those who wanted to kill the *Cojñone* but we ran. Jutaine and I walked behind the others. I looked at the sky and Sun had died. Because we shot the bulldozer. The world went dark. The women began to cry. We did not know where to run. The women cried. We men became enraged. Jutaine is a powerful shaman. He said, "That truck said to me that it would kill the Sun." It would empty the Sun and make it red." A shaman cannot speak directly about his vision. I also saw that those *Cojñone* had been in our other camp and learned a lot about us from looking at our things and then they shot their guns in all directions. My eyes were full of spots. We thought we would die but we blew away the badness.

The forest bands said they were also afraid of the pure white Zebu and Brahman cattle that soon appeared to populate the vast clearings left in the bulldozers' wake:

We were afraid of the noise of the trucks and the bulldozers and we ran. We were afraid because the *Cojñone* always killed our people before. We went to a road to look for the things of the *Cojñone*. We saw some Cattle, *Aquiejnanie*. We thought that Cattle knew how to speak. We began to speak to the Cattle. We spoke to them but they did not speak back. They did not understand. We thought that those Cattle would speak with us. We spoke to them but they did not answer. We said to the Cattle, "Cattle, we are your Fathers too!" We thought that the *Cojñone* were like Fathers to the Cattle and that is why they obeyed them. We wanted to tell the Cattle to leave us alone. But we left the Cattle behind there because they did not speak to us. They did not understand our language.

The bulldozers made roads that cut up Totobiegosode lands. These roads were particularly difficult to cross without being seen or leaving tracks. To cross, the entire band would hide near the road and wait until no trucks were visible. Siquei followed behind to wipe out their tracks. Cutai described one crossing.

We arrived before dawn. We did not cross because we were afraid. We wanted to cross but we did not know where to cross because there was much movement on the road. I waited because I also wanted to see a truck. The trucks are very fast; they do not walk slowly. We waited there for the trucks. I wanted to see one. We lay there under the brush at the side of the road. They said that one was coming. We prepared ourselves. But it passed so fast that we could not see it. We had waited for no reason because we did not see it. So we waited for another to pass and caught a glimpse. . . . At dusk the women said we are going to cross the road. We were afraid because there were Cattle behind us and in front of us the road. The Cattle made sounds like *ooo ooo ooo*. The women stayed behind with one man to guard them. The men said, "Let's cross now because the moon has come out." We crossed the road. Siquei erased our tracks and we walked all night to the east.

The forest bands also spent long hours observing the *Cojñone*. They entered unoccupied ranch houses in search of clothing and metal:

We hid for a long time behind a cukoi tree, watching. I said to Jutaine, there is no one there, it is an empty house. There were only sheep there. We went into the houses. I picked up something there. I picked up some things but it was very dark inside. I touched something there and it felt like the intestines of an anteater. Chicoi picked up another. It was big but there was nothing inside. Around the house we found glass bottles. Very big . . . one of the sheep came into the house. I did not want to kill it because I wanted to kill a *Cojñoi*. I said to Chicoi, "Let's go, there is nothing here." I shot a sheep with an arrow and it cried a lot. Jutaine shot another. We climbed on top of some vehicles. Chicoi found black clothes in one and gave it to us. We took the thing we know is a bucket and some wire. We tore some of the clothing. We found a tarp and we took it. We called it Parojnai. It was very dark but we walked back to our camp. When we arrived to the women we told them of our trip and divided everything up.

Such stories mock the conceits of the "uncontacted" and "isolation." Belying the stereotypical primitive aped in earlier descriptions, the perceptions of the forest bands were not determined by a cyclical time of myth but by their relationship with global political economies.

Siquei told me his people began to imagine that new kinds of evil spirits were persecuting them. Among the most fearsome were those he called the *Cojñoque chaguide*, or ravenous Strangers. These giant blue-eyed beings consumed nature and society. They were said to be responsible for the neatly cut stumps of *quebracho colorado* trees often encountered in the forest and they had an insatiable hunger for Totobiegosode flesh.[6]

Often, the Areguede'urasade were forced to camp in the ten- or twenty-meter-wide margins of brush and shade left around vast empty pastures. They told me of their sadness when they returned to find favored places flattened, burned, bared to the sun. Such places, they said, were dead. More than anything else, they remembered running from the people and machines they thought were trying to kill them. As Siquei put it,

We were afraid of the Guidaigosode [an enemy Ayoreo group], we were afraid of the *Cojñone*. They had killed many of our people before. We were afraid that they would do the same thing again. We were afraid of the noise of the vehicles, of the bulldozers that moved us out of our land. . . . We didn't know if the war with the Guidaigosode was over. We didn't know that any Ayoreo were left; we didn't know if the *Cojñone* had killed them all.

Some people became so terrified that the slightest noise could provoke paralysis. Stories are laughingly told of instances in which the sound of someone walking nearby caused all of the women in the group to enter a kind of hysteria, crying and running, leaving behind all of their possessions, stripping off their skirts and running nude to go faster, only to later discover that the noise had been made by one of their own hunters returning to camp at the expected hour.

The fear of bulldozers recalled past fears of warriors from enemy Ayoreo groups. The Totobiegosode, in particular, had been the frequent targets of raids aimed at their extermination. Genealogies I compiled suggest that more than 80 percent of Totobiegosode were killed by enemy Ayoreo groups between 1940 and 1979. All Totobiegosode adults over forty are survivors of more than one attack by raiding parties from the missions, armed with shotguns and machetes. Most of the men bear traces of battle on their bodies. A fleshy knob conceals a bullet lodged against the skull, crooked spines and bent limbs, indecipherable scars and puckered craters speak of spears, machetes, bullets, clubs. Cutai's mother, Ajidababia, told of one such attack in the 1960s:

We died there. I was a girl, around nine or ten years old. The enemy arrived. My mother called me to her and carried me over her shoulder. I do not know why. She carried me and ran. But she tired because I was heavy. When she stopped to let me down they killed her. I never ran but that time I ran. I was afraid of the guns that wanted to sting me and take the blood from my body. Something hit me and I fell. They cut me there with a machete. I ran but fell again. I got up then passed out again. I got up and heard someone running and it was Pejei'daquide. I arrived at a clearing and saw them and they saw me. I was thirsty but could not look for water. It seemed like I would die that afternoon. We went back to the place they killed us. No one was there. We arrived at our camp and I drank water. I put much dirt into my wound. The dirt stopped the bleeding. I drank water and began to feel better.

One man named Poaji was overcome by bouts of terror after he was the lone survivor of such an attack. He hid under a bush and watched silently as his mother, father, siblings, and other relatives were killed and dismembered. After months alone, he found a friendly group. Decades later, the slightest sound would cause him to vanish again for months or years at a time. Siquei and his group found him living alone in a hollow tree, and he stayed with them for two or three months in 2000 or 2001. They remembered him as one moment taking delight in the children and the next trembling with fear. When they heard a chainsaw one morning at daybreak, Poaji disappeared. As far as anyone knows, he is still hiding alone in the forest.

The terror was so pervasive that the *uitaque* seers began to have visions of group death, a prophecy with historical precedents among Ayoreo-speaking people. Ethnographers who visited the Guidaigosode and Garaigosode-Ayoreo groups shortly after first contact in the 1960s noted "a collective delirium of the end of a world by a group massacre"[7] and "a great fear of the civilized . . . they expected at any moment to be attacked by some fantastic machine."[8] Such fears only intensified among those Totobiegosode groups who remained in the forest, as the death of their world was evoked in images of endless night without fires, children who could not speak, a warrior's club that could not be lifted, and crushed and dead lands.

"There were shamans who knew our land would be destroyed," I was told by Jochade, a leader of the Totobiegosode band captured in 1979. "One shaman had a dream. He only saw darkness in the forest, because there was not a single fire. He knew that the land would disappear. It is like today, the lands of Uejnai and Manenaquide, they are gone, finished." His brother Aasi added, "The spirit *Pujopie* spoke to him and told him to wake up and sing his vision. Eotedaquide woke up and told his wife Gapuome'dacode what he had seen when he was sleeping. He said, "Do not believe that we are still Humans. The day will come when we will all disappear and the world will disappear too." That is what Eotedaquide said to his wife long ago, that we will all die." Jochade agreed. "Many places have died. When there are no Ayoreo living there, the land dies. The *Daijnai* said one day we will all disappear. The world will be dark without any children to light a fire."

Totobiegosode living in concealment in the rapidly dwindling forest of northern Paraguay concluded that there was little hope for survival. "Before we said to one another, it appears that we will all be killed," my adopted father Yoteuoi recounted. "We thought that other Ayoreo or the *Cojñone* would kill us all. That is what we thought would happen." Genocidal violence and ecological devastation reverberated in bodies and psyches. These dynamics inverted the existential conditions of precontact concealment and transformed the forest into a space of imminent death.

The Failure of Mimetic Magic

In the face of such violence and fear, established theories of causality
and agency began to fail. Totobiegosode groups roaming the dwindling
Chaco forests in the 1980s and 1990s turned to long-standing theories of
sympathetic magic and the mimetic faculty to seek alternatives to group
death. But the object of mimetic magic shifted from spiritually animated
natural elements to the superior power of the strangers.

After the former shaman Guedeuejnai had a vision in the early 1980s,
he put his entire group to work digging a shallow ditch about a mile in
length along one side of a familiar path. If anyone stepped into the ditch,
he said, he or she would die. He told his people that if they dug the ditch,
then they would never die, that death would stay in the ditch. "We dug
that ditch," I was told twenty years later. "But he was a liar." Others attempted to reproduce the specific material forms they believed gave the
Cojñone spiritual power. Yoteuoi tried to make metal. He put the men of
the group to work digging a deep pit, in which they mixed soil of five different colors and water. "It did not work," he told me. "But almost."

Former members of the Aregued'urasade also recounted how they tried
to access the power of the *Cojñone* through mimicry. They broke into isolated ranch houses and took clothing and items of metal. They also spent
hours hidden under thick brush, observing the *Cojñone* inhabitants. They
hypothesized that these outsiders must have access to an unseen magical substance, which was explained to me as taking the form of a tubular
object about twelve inches in length that generated light. This sticklike
thing, Totobiegosode reasoned, was impregnated with a form of *pujopie,*
or supernatural power, distinct from the one possessed by their own shamans. It provided the foreigners with sustenance, for they never seemed
to eat, and gave their machines the force to move with great speed. "We
thought they could rub it on their bodies," I was told. "And they would
fly around the world." Totobiegosode entered several houses looking for
this source of superhuman strength. When they did not find it, Jutaine
attempted to fabricate such a magical stick out of hardwood and incanta-
tions. "He rubbed it on us, but it did nothing."

These stories are told with a gentle bitterness, when they are told
at all. Mimetic magic ultimately failed to summon a source of power
great enough to provide an alternative to group death. The former
Areguede'urasade interpreted this failure of transformation as indicating
a pervasive weakness relative to *Cojñone*, a loss of control over the means

to summon moral and physical power from the set of forces that hovered near Human Beings. This implied that the boundaries of humanity were becoming increasingly porous, leaving people at risk of affliction, a sense that only intensified in the aftermath of contact:

We suffered a lot in the forest and that is why we decided to live with the whites. We thought that the food of the *Cojñone* was given for free. But now we know that it costs a lot. We were hungry before. But it turns out that they sell their food. Everything is very expensive. Even beans are expensive. We thought that things were free before. We thought that the *Cojñone* always gave away everything. We thought they would bring us melons, beans. We thought that life among the *Cojñone* was very easy. We thought it was easy to learn their language. We didn't know about money.

Within four months of contact, the men of the New People were working as wage laborers alongside bulldozers, fencing off pastures for four dollars a day. With their meager earnings, they bought noodles, cookies, and Coca-Cola. On such a diet, their bodies were often weak and sick. Ayoreo-speaking people do not usually distinguish between physical health, moral well-being, and social agency. Daily experiences of disease and social marginality, then, are seen to indicate a moral weakness intrinsic to Ayoreo and neighboring Indigenous peoples. Many people claim that they cannot remember much of their former lives.

In the "experience of world-annihilating violence," Veena Das writes in the context of postpartition India, the "grammar of the ordinary" fails, and the criteria relating causality, justification, and action may be abruptly ended or rendered opaque.[9] For contemporary Totobiegosode-Ayoreo, the aphasia of suffering and terror has become a routine part of everyday life. Yet this silence was distinct from the cases described by Das and other theorists of trauma, in which violence caused the failure of language and the social.[10] Among Totobiegosode, apocalypticism reclaimed the capacity for human transformation within such extreme experiences. Colonial violence generated the social conditions under which apocalypticism became appealing and intelligible, even as apocalyptic sensibilities offered a way to reclaim life from death or violent events from the domain of the unspeakable. Totobiegosode believers thus solidified the sense that the present occured within a radically different moral ecology than the past did, one that was structured by the future return of Jesus and the bodily transformation of true believers.

Touching and Time

We are very afraid of the Word of God, because the Bible says that he will come again in the near future, and it is a sure thing that he will come. That is why we read the Bible often. It says that he will come again, and we are very afraid. It says that he will come again and he will kill all of us who don't believe in him. He will take only those who are very faithful with him to his village, like Norman and those that believe in him a lot, like Carodi and Cadui. That is what will happen when God comes to punish the Ayoreo.

Ayoreo-speaking people converted en masse to Christianity in 1975, and outside observers have noted their rapid adoption of evangelical Christianity with surprise. "What is suggested by the data is that the Ayoreo conceive of their past life in the woods as extremely hard and very contrary to the will of God (Our ways in the woods were extremely hard. We suffered much. God hated our ways)," wrote a Mennonite committee charged with evaluating the New Tribes Mission at El Faro Moro in 1977. "They have now, as a group, set their sights on moving toward the civilized way."[11] David Maybury-Lewis and James Howe also reported an "extremely rare" 100 percent conversion rate among Guidaigosode-Ayoreo at El Faro Moro in 1978, and even Salesian missionaries were astonished that, by the mid-1970s, "the majority of Ayoreo are fervent Christians."[12]

This impression of fervor was cemented by the particular form of Ayoreo Christianity, which emphasized public professions of faith even while imagining true faith as a difficult-to-realize ideal. Common expressions such as "God hated us before" or "we were worthless and ignorant" emphasized that contact created a New World in which Jesus was the ultimate arbiter of power and knowledge. This New World, called *Cojñone-Gari*, was associated with a distinct metaphysical ecology—a distinct relation between cause and effect—than that which is attributed to the past, precontact forest world, *Erami*. In this shared recognition, Christianity became synonymous with both modernity and an Ayoreo morality. Believers developed a unique version of Christian faith that both cited and exceeded missionary visions of Ayoreo humanity.

During my fieldwork, the conversion of the former Areguede'urasade was carried out by their Totobiegosode relatives, provoked in part by Bobby's visits every three days. Witnessing the violent conversion of the new group, based largely on ridicule and domination, never failed to bother me. Siquei, the former leader, was especially targeted. I could not understand why the New People did not object to it or simply leave the village to avoid it. Part of my confusion, it seems, was that I perceived conversion as a cognitive process aimed at dominating reason. This was not necessarily the primary plane of belief and contact among Totobiegosode.

Many Ayoreo-speaking people said that touching, -*isa*, is a particularly significant and violent act. Rather than only a tactile exchange, it also involved gaining possession of another being's willpower. When hunting, the person who touched an animal first rather than the person who delivered the fatal blow was the one who "killed" it. According to the terms of traditional warfare, a warrior could take a member of the enemy group captive by touching that person. The touched ones, *isagode*, were incorporated into the family structure of their touchers, *isasorone*, as servants, subordinates, or slaves.[13] Many *isagode* intentionally starved themselves to death, but I am not aware of any cases in which an individual, once touched, attempted to escape.

The first Direquednejnaigosode-Ayoreo groups contacted in Bolivia in the late 1940s interpreted the relationships established in contact with missionaries as a form of touching, in which contacted groups became the *isagode* and the missionaries unwittingly became the *isasorone*. This idea was extended by converted Guidaigosode-Ayoreo (directed by missionaries who, by that time, had learned the potency of the concept) to apply to the Totobiegosode groups they captured in 1979 and 1986 and brought back to the mission of Campo Loro, where the families of the newly contacted were divided up among the touchers. In 2004, the Totobiegosode who were touched in the human hunt of 1986 added another layer to this term by applying it to close relatives. Thus, the Areguede'urasade became the *isagode* possessions of their own relatives.

Touching became a key idiom by which Ayoreo-speaking people made conversion to Christianity intelligible. During my fieldwork, many Ayoreo people used the same term, *isagode*, to refer to individuals hired as wage laborers. Each mode of touching—being captured by an enemy, contacted by missionaries, hired by a boss, and converted to Christianity (touched by God)—caused a process of fundamental change, of *chinoningase*, the internal transformation implied by becoming subordinate to another willful being. In the case of conversion, I was told, the phrase meant that "God grabs everything that is inside and it then belongs to Him." The concept can refer to the complete destruction or alteration of an object (i.e., *yinoningase yasore*, "I turned my spear into a tent pole"), the effects of a terminal illness (i.e., *uajedie chinoningase yu*, "my cough has destroyed me"), or the process of adoption into a clan (*Pukoi'date chinoningase yu enga chiquenoi uyu*, "Pukoi'date transformed me into her clan and now I am Chiquenoi"). God was also said to *checae* the convert, a term that likened the transference of God's willpower into the body of the new believer to pouring a liquid from one container into another.

When God *chinoningase*, or converted, someone, he was thought to create a new kind of person fit for life in *Cojñone-Gari*. Conversion was associated with a radical transformation in human substance. Ayoreo-speaking people said that faith in God *chieta bacajeode*, literally fills up your insides, will, thoughts. This process of filling up meant that God erased the convert's *ayipie*, the soul matter that encompassed the corporeal seat of memory (located in the head), emotion, and willpower (located in various abdominal organs). Through *chinoningase* and accepting the Word of God, the insides and *ayipie* of the convert were reconstituted anew along the sentimental and temporal axes that were pleasurable to *Jesui*. It was this erasure, *-iro*, of the convert's memory and its replacement by God's spirit that supposedly restrained what was possible to express about life before conversion. This was why, Totobiegosode said, so many of them could not remember too much of what happened before contact. "Your *ayipie* must be erased before you can be saved by God."

However harsh this process of becoming New People may have been, it was considered essential for human survival in *Cojñone-Gari*. This position presumed that the contemporary terms of life itself were so radically altered that the only hope for survival resided in the death of older human forms and the internal transformation of bodies (in terms resonant with the ancient transformations of mythic beings). Yet Ayoreo believers did not simply "internalize" a colonial image of themselves as "quasi-men ignorant of the truly true."[14] Rather, they imagined their present humanity to be a form of immanence that was a precondition for their transformation into fully human beings of another kind. Such Ayoreo beliefs, then, were a testament to the generative force of violence. They exceeded both missionary frameworks (with their particular notions of Native souls and bodily interiority) and creatively reorganized Ayoreo ontologies, catalyzed by a potent touch.

Segregated Memory

You all can use the spears that God gives us in the Bible to defend against Satan if he arrives to your village. We are in a war again, but not against other Ayoreo. Now we are fighting against Satan. Satan wants to attack those of us who believe in Jesus. Don't forget that we are bothering Satan, but remember that God is stronger than him. The person who remembers this will be taken with God when he returns to end this world.

Apocalyptic futurism evacuated the time and terms of the past. Totobiegosode regularly used the word *nanique* (lit. "a long ago time"), to refer to all events that occurred precontact. This was the case in 2004, when their relatives taught the former Areguede'urasade to use the word *nanique* to refer to events that had happened only six months earlier, when the group was still in the forest. The relatives used words for the near past, such as *dirica* (lit. "yesterday") or *irica* ("a while ago") to refer to events that happened up to twenty years prior, after they had been contacted. The same applied to *casicaite* and *casodica*, words that meant "something dead or absent a long time," which were used to describe precontact social forms, even if contact was recent. This shift also corresponded to the flattening out of precontact time by a general discontinuance of those words—such as *naninguejna*—capable of marking stages of the past.

Precisely such a rupture was also hypothesized in the idealized Indigenous subject position: Whereas NGO officials and anthropologists often desired to preserve or resuscitate the homogenous time of tradition, many missionaries strove to expunge it from the present. Yet Ayoreo-speaking people marked this disjuncture in distinctive ways. One way they did so was by applying words that once referred to ruptures between the time of myth and the time of sociality to the ruptures of the pre- and post-contact periods. For instance, the term *Jnanibajade* (Original Men) could refer to mythical beings; Old Testament biblical characters such as Noah, Jonas, or Adam; and Ayoreo-speaking people who refused direct contact. *Taningane*, a word literally meaning "the beginnings," could refer to mythical time, biblical time, and precontact life. The overall effect was to emphasize that human–nonhuman divides applied as much to processes of contact as they did to the original differentiation of nature and culture.

The religious conversion of human substance realigned the boundaries between the sacred and the profane, the moral and the immoral, or the healthy and the sick along the temporal divide of contact. Such concepts and the reworking of memory they implied were publicly performed in village *culto*, or church services, in which a central component was renarrating past events as evidence of God's interventions for or against the narrator's actions. A man told about a rock that mysteriously fell but narrowly missed crushing his head as a warning from Jesus to repent. Another talked of recovering from an intense fever as evidence of God's love for him or of an illness that compelled him to accept Jesus as divinely inspired. Someone might have talked of reaching shoulder-deep into a turtle den and encountering a rattlesnake that moved along his arm but did not strike or of finding a job when his family was on the brink of starvation. These miracle narratives reconstituted and mystified the links between cause and effect in the time of memory. Aasi was fond of telling the following story:

Now I am going to tell of the time I went to bring firewood, and a rattlesnake almost bit me. I did not see it until I was very close. I saw something, and it was the rattlesnake. You know that when the rattlesnake is ready to bite, it coils its body. I could not move my foot; it seemed very heavy. I wanted to jump but I could not. The snake was ready, and I could not understand why it didn't bite me. God protected me so that the snake did not bite me. Instead, it moved away. There are many things that have happened to me like this, in which God protected me. Because of the protection of God, I was not bitten. I thanked God because the snake did not bite me. Because of God, I am still alive today.

The distinctions between past and future human kinds were also spatially inscribed. Totobiegosode, for example, at times associated *nanique* and its animating forces with certain places in the wilderness. This use was consistent with a general schema in which Ayoreo words for place and direction usually referred to experienced time as well. Thus, *jogadi* could refer to a specific location as well as a specific time; the future was referred to with words that also meant before, in front of, above, and on the other side; and the past was expressed with words that meant behind, after, and what has been passed. Both the future and the past could be near, far, a medium "distance," between, or otherwise located spatially vis-à-vis other moments. Up and down, however, referred to parallel universes that represented distinct modes of nonlinear social time. This spatialization of social time may at times have threatened the strict segregation of memory that Ayoreo said was a necessary condition for apocalyptic redemption.

The Christian God, *Dupade*, was said to absolutely control the terms of human life in *Cojñone-Gari*.[15] But the extent of his power was less clear for Ayoreo-speaking people who encountered the material remnants of the past or found themselves back in the places they used to inhabit. Most of these zones and the occasional potsherd found while hunting were treated with no particular reverence. This was not the case, however, when we happened on one of their more permanent village sites, *ore idai*, with its dome-shaped communal house, *ore iguijnai*. Because such houses were often abruptly abandoned and are very durable, they could seem jarringly contemporary. Firewood remained neatly stacked and ready for use, reeds for arrow shafts were tucked into the roof, tins and blankets and bones were strewn about, favored paths were still clearly visible. The sight could trigger deep emotions. In one such place that has since been bulldozed, Siquei and I walked through overgrown gardens. "This was my father's garden. We were happy here."

The Rotation of Apocalyptic Horizons

Everyone alive now will surely see the end of this world. This will happen when this generation is still alive, it won't happen to another generation. It will happen in this generation. No one will be able to distinguish Heaven and Earth when God comes, only the Word of God. The things that we see here, like food and motorcycles and jobs and money, seem to be important things. But they will all disappear when God comes. We will be in the village of God that will never die. No one knows what time he will arrive, this afternoon, tomorrow, or the next day. At midnight or noon or six in the morning. I don't think it will be earlier than six. . . . We must be prepared because he can arrive any day.

Impending bodily transformation and the destruction of the world were not much of a logical stretch for people who believed they had already experienced such events. Many Ayoreo said they were waiting for Jesus. As in Kenelm Burridge's description of collective myth dreams among cargo believers in New Guinea, Ayoreo believers animated this "structure of waiting" with half-articulated expectations, conflicts, and rumors.[16] Stories of speaking dogs, satanic animals, and cannibalistic white men were common. Yet these visions and hopes were also frequently frustrated by the precarious conditions of post-contact life.

Once every three or four months during my fieldwork, a rumor swept through the communities that a foreign, white-skinned man with mysterious powers had arrived in the Mennonite colonies. Usually, the man had departed, walking away down the road with a band of Indian disciples, just before the Ayoreo source of the rumor found out about him. Poverty-stricken, monolingual Ayoreo families saved money for months to travel to distant churches in search of miracles.

These miracle searches usually bore no fruit, as in the case of my acquaintance Juan, a Ñamocodegose man, who took his wife to find a place in Asunción where they had heard a man could expel Satan and make people vomit out their diseases with his touch. The sickness, Juan told me, would come out in a hard, multicolored ball that the man could destroy. "He talks to God, saying, 'Come here God, cure this person.' And then the person is cured. Surely there are miracles happening there."

I found Juan and his wife in Asunción camped in an empty lot behind a Catholic church. Juan produced an indecipherable map, consisting of a single line and two words drawn onto a tattered napkin, and asked me to guide him to the miracle place. A week later, the couple returned to the Chaco.

Like miracles, clear causes and clean ends were elusive for Ayoreo people, many of whom were starving or chronically ill. In *Cojñone-Gari*, Ayoreo were urged by the terms of lumpen capitalist exchange to embody scenarios freighted with senses of an ending. They were often compelled to be coparticipants in the destruction of the natural environment that was previously the source of the hidden sacred.

The old people would run up to a recently arrived visitor in the New Tribes Mission village of Campo Loro, ten or twenty pushing to the front of a larger crowd, all soft eyes and ravaged gums and wild hair, pulling up tattered rags and pressing the visitor's hands to their ribs so that he or she would know that they were not lying when they said they were starving and asked for food. When the price of Paraguayan beef and uncleared land surged in the late 1990s, when the bulldozers worked twenty-four hours a day, someone decided to donate food to "the poor starving Indians," subsidized in part by humanitarian aid from a European government. Once a week for several years, a truck carrying unusable bovine entrails from the Mennonite slaughterhouse would drive through the settlement, make a slow U-turn, and dump the wet offal on the dusty ground. The news was shouted among the households and the race was on. Old ladies and children trampled and pushed down other old ladies and children, fighting over the small shreds of foul viscera, covered in bloody mud.

These efforts were replaced by a scheme to have Ayoreo make charcoal, also sponsored in part by international development aid. This scheme required the inhabitants of most Ayoreo communities—people who once believed that every plant, insect, and animal in the universe had been a member of their tribe—to clear-cut the trees and woody vegetation remaining on the small plots of land they still control and burn them in underground pit ovens. They were paid approximately six cents per kilogram of charcoal produced. The grain and texture of the wood was visible in the fossilized black lumps, which were sold to Mennonite middlemen and eventually fueled backyard barbecues in Germany. One of the highest-selling brands of this charcoal featured a half-naked cartoon Indian as its logo. The workers emerged like half-remembered dreams, pallid skin visible in sweat lines through black soot. The fragile alkaline earth was left bared to the sun.

Scenarios that may provoke frustration, confusion, and despair (not to mention sympathy, nostalgia, or shame) made a sort of hopeful sense in apocalypticism's inverted logics. The more intense the suffering, the nearer the new beginning.

And yet the conditions of post-contact life also prevented the closure of such horizons and threatened the integrity of the new kinds of humans Ayoreo believers ideally imagined themselves to be. There was a pervasive sense that *ayipie* were particularly vulnerable and at risk in *Cojñone-Gari*. During my fieldwork, it was a matter of daily concern whether someone's *ayipie* had been taken out, left behind or recovered, moved by God or Satan, diminished or grown, gone in this or that direction. Someone's *ayipie* could be dominated by another (*yui*), leaving its original owner senseless or dumb. Even worse, one's *ayipie* could die (*toi*) from fear, weakness, or extreme emotions of sadness and anger, processes that *chejna dayipie*, overwhelmed one's *ayipie*.

In *Cojñone-Gari*, remembering the past was a potentially dangerous activity, capable of causing infection or illness. If one's *ayipie chicaji te*, or goes in an opposite direction from the group, it could return to the *cucha bajade*, or past practices, opposed to modern life (*ayipie echaji cucha bajade; ayipie chajesa daquigade*). This threatened to unravel the seat of moral humanity in the present. That was why many Ayoreo said it was important to keep one's *ayipie* focused on the future, *payipie chicaji piquei*. Thus, the proper moral human no longer had any *ayipie* for the past, *ijnoguipise yocayipie ome cuchabajade iji Cojñone-Gari*. Yet the space of the present was marked by existential contradictions that prevented the full realization of the new moral human.

Some younger Ayoreo people did not even aspire to become true believers. As my young friend and assistant Yakayabi put it, "The people of my generation are between the old things and the Word of God. We don't do what God's Word tells us to do, but we don't know anything about the old things and no one will teach us. We respect what the older people or the missionaries tell us but then we forget it." Rather than threatening apocalyptic horizons, this dangerous lack could be interpreted as a necessary catalyst for individuals to cycle out and back into narratives of impending crisis and loss, of life beyond humanity. And this ability to reborder the human and the inhuman as a viable space of life—however constricted and fraught with despair—was what made apocalyptic futurism an appealing state of being and form of intelligibility for many Ayoreo-speaking people.

Life within the Labor of the Negative

In such ways, apocalypticism illuminated a set of emerging ontological sensibilities that escaped the confines of either tradition or Christianity.

Ayoreo believers and doubters used such imagery to turn the colonial situation inside out. Apocalypticism was appealing precisely because it asserted the capacity of Ayoreo people to control the terms of their self-transformation within the profoundly disturbed conditions of post-contact life. Present contingencies and unspeakable traumas constantly realigned the past and the future and, in doing so, produced an infinite number of causal possibilities and events.[17]

Moreover, apocalyptic sensibilities cited and extended notions of the originary differentiations of human/nonhuman. They interpreted contact and conversion as implying an equally fundamental transformation of human life. This was articulated, both implicitly and explicitly, in the transition from the forest/past of *Erami* to life in the modern/present of *Cojñone-Gari*. In such ways, apocalyptic futurism creatively reconciled long-standing Ayoreo notions about human/nonhuman divides with evangelical eschatology and colonial discourses of subhuman savagery. Like missionaries and local ethnographers, Ayoreo believers imagined contact as a profound rupture between two mutually exclusive forms of human life. Apocalypticism was a form of moral reasoning that mediated this divide and transformed the fundamental contradictions of the colonial situation into the principles by which moral life may be inhabited and reproduced.

Accordingly, missionary attitudes and disciplinary practices alone were not sufficient to explain the adoption of apocalyptic futurism among Ayoreo-speaking peoples. Yet neither could apocalyptic reasoning be explained as the simple continuation of past ontologies or mythic cosmologies into the present. The distinction, rather, rested on the particular meanings and values attributed to contact in its guise as a radical transformation of moral human life. Indeed, contact was envisioned as the latest movement through several contradictory regimes of biolegitimacy in ways that closely mirrored colonial logics while inverting culturalist ones. We may predict that a future transformation to culturalism lies ahead. Many Ayoreo-speaking people concluded that modernity and indigeneity were both regimes of life only inhabitable for a radically transformed human—thus, the common wordplay among Ayoreo people, that "we are no longer Ayoreo (Human Beings), we have become *Ayore-Cojñoque* (a phrase that implies both white Ayoreo and human–nonhuman)."

Through these apocalyptic inversions of the colonial labor of the negative, Ayoreo-speaking people recast such negation of their past humanity as the space of modern life. In doing so, they reclaimed a fundamental capacity for metaobjectification, the agency to speak and act beyond these limits of the human. This implied nothing less than reclaiming the terms of their own subjective immanence. At the same time, they also became negative outsides to the zone of "traditional culture" and, thus, political subjects of a certain kind.

It was a predictable irony that this creative heterodoxy reappeared to culturalist outsiders only as evidence of an essential Ayoreo difference. Many insisted, without any evidence whatsoever, on the existence of clandestine rituals or shamanic practices carried out in secret by Toto-biegosode. (It is likely, of course, that such practices will eventually be created in response.) This created a situation inverse to that documented by Tom Abercrombie in the Andes, in which sixteenth-century Counter-Reformation techniques aimed at extirpating Indigenous heterodoxy produced a bifurcated cosmos comprised of clandestine "Indian" elements and public "Christian" ones.[18] Among contemporary Totobiegosode, it was an earnestly felt evangelical apocalypticism that was produced as the clandestine private in creative tension with public declarations of culturalist agendas and the preservationist rituals of Indigenous rights.

Such was the colonial crucible which contemporary Ayoreo ontologies and moral sensibilities were coproduced. No stability was possible. Rather, they could only be understood in relation to the global political, economic forces destroying the Chaco ecology at breathtaking speed, the violence of internal colonialism and techniques of missionary control, the long-standing moral value given to self-transformation and mimetic magic, and the recent rise of culture as a regime of legitimate Indigenous life. Ayoreo ontologies were always politico-moral ontologies written into the past by the ever-changing concerns of the present.

Rain

The bulldozers must stop when it rains or risk becoming entombed in the mud for weeks. In a village at the edge of the forest, other sounds emerge when the motors are silenced. Brazilian disco music thumps from a speaker held together with fiber twine. Wind rolls like a wave along the treetops, raindrops patter against leaves and tin. A child giggles.

"I am the beautiful buds, returning in the forest," the old man sings. "I am the first fat rain. I am the scent of the flowers carried on the wind."

"I sound like *mei mei mei mei!*
Ti ti ti ti ti.
Joooo! Joooo! Joooo!
Se se se . . ."

SIX

Shame and the Limits
of the Subject

It was that shame . . . that the just man experiences at another man's crime,
at the fact that such a crime should exist, that it should have been introduced
irrevocably into the world of things that exist, and that his will for good should
have proved too weak or null and should not have availed in defense.
PRIMO LEVI

It was not that I was finding febrile coordinates in the world. I existed in triple; I
occupied space. I moved toward the other . . . and the evanescent other, hostile
but not opaque, transparent, not there, disappeared. Nausea.
FRANTZ FANON

When I met Siquei in July 2004, he was a barrel-chested and
supremely self-confident man in his late thirties who had
killed several enemies. He whispered to the trees when he
walked in the forest. He had a quick laugh and a disarming
smile and he always brought back meat and he often stayed
up late at night singing with his rattle, his deep voice rolling
defiantly through the black brush.

In September of that year, Siquei and the Areguede'urasade
decided to settle permanently in Chaidi, over my ineffective
protests. When they completed the move back to Chaidi,
Siquei told a story in the old way, with elaborate panto-
mimes and imitations, of fending off the repeated attacks
of a spirit jaguar. Shortly afterward, Dejai assembled the
Areguede'urasade. "Now that you have settled here permanently, you should know that there is nothing that is taboo among the *Cojñone*. You should not obey the *puyaque*

taboos," he said. "Here, we only listen to the Bible and *Dupade'uruode*, the Word of God."

When I returned a year later, Siquei was thin and sick and had developed a severe stutter. He was among the most marginal people in Chaidi. I could not understand why he did not simply leave. When I tried to ask him one afternoon, alone at his house, he just laughed nervously and looked away. The others said something was wrong with him, that he was ashamed. "Ashamed of what?" I asked. No one answered.

During my fieldwork, I learned that Siquei and the others used shame to explain a variety of reactions that I found difficult to understand, not least because of my own ideas about loss. For instance, the Totobiegosode communities were perched on the southern edge of a territory claimed in one of the most ambitious land claims cases in the Southern Cone. Yet outsiders routinely failed to recognize their Aboriginal title to this territory. The Totobiegosode communities were unique in that they were not founded by missionaries. Yet Totobiegosode self-consciously replicated the forms of the mission settlements where they had been held captive, complete with church, store, and their own set of New People to subject. They sold wild honey to buy white sugar. Men preferred makeshift firearms to bows and spears. Some refused to eat all but the most common traditional foods. All ignored the delicate and detailed *puyaque* taboo restrictions that had formerly regulated life with such remarkable precision. When I asked people to explain each of these behaviors, the answer was the same: they were ashamed.

This answer seemed straightforward, yet I struggled to understand all it conveyed. On one hand, it appeared so predictable as to be vague. Indeed, much anthropological research foretells such a response.[1] Scholars have documented how moral sentiments like shame reflect ethical concerns, manifest culturally specific moral orders, mediate universal tensions, and serve as the crucial affective gradient through which hegemonic and counterhegemonic projects alike are instantiated.[2] On the other hand, this Ayoreo explanation raised more questions than it answered. What did it mean to become ashamed of the very practices that once defined moral humanity? And what did it mean if a group of "ex-primitives" felt ashamed but in ways that were almost precisely the opposite of how we think they should feel? What kind of social orders and what kind of colonizing violence might this dialectical tension between self-objectification and frustrated attribution allow? If ontological sentiments created New Worlds and if shame was the defining ontological sentiment of becoming

New People, what kind of world was the cosmos of shame and what kind of life could inhabit it?

The Ayoreo word for shame was *ajengome*. In the stories I heard about *ajengome*, it was linked to the reddening of certain internal organs and the resulting sensations of heat. It was a bodily state that could carry shades of obedience and piety as well as impropriety and cowardice. It always implied subordination. More than anything else, I was told, *ajengome* was an indication of individual weakness.

Ajengome, they said, arose from attribution by another person. It came from a witnessed transgression. It could be both a verb and an object. You did not have *ajengome* because you steal from another, sleep with someone who is not your spouse, or run from enemies; you have *ajengome* when others find out.[3] *Ajengome* was most often described as a function of failing to act, speak up, or assert one's will, but it could also arise from acquiescing to the negative judgments of others. Like the Ayoreo words *-itodo* ("to be afraid of") and *-angari* ("to listen to"), to have shame was an implicit acknowledgment of one's subordinate position.

I was told that the opposite of *ajengome* was *ajingaque*, or righteous anger. Like *ajengome*, it was associated with a hot bodily interior. This kind of anger was said to overpower the *ayipie*. It was the appropriate way to refuse the attribution of *ajengome*. As Siquei told me, one day when we were out hunting, "If someone is very angry, his gallbladder moves. It sounds like *tucu, tucu, tucu* because of his anger. He trembles, his face changes. He doesn't know anything. He doesn't recognize the world." He paused, looked away. "He is capable of killing anything." *Ajengome* was not felt equally by everyone, and this uneven distribution provided most of its intimate force. It was said that a very strong person never had shame for any reason, including the most flagrant transgression. Those aspiring to high status must contest the label.

To illustrate this point, the adults in Arocojnadi told me a story about a warrior I'll call Asi'de. This man lived so long ago that his grandchildren would now be old people. He was a huge man, a head taller than the next tallest warrior. He was also very strong. No one ever beat him in wrestling contests and he killed many enemies with his bare hands. When he was still a young man he separated from his people and made a new band. He led raids on his enemies that were ever more daring and risky and each time he returned stronger. He began to violate *puyaque* taboo prohibitions, the minor ones at first and then the more powerful ones. If he wanted the wife of any of his followers, he would take her. If someone of his own group offended him, he would kill the person. This reached the point where he killed one of his own wives for not agreeing with him. The other wives were terrified. His people wanted to leave but were too frightened. One night while he was sleeping, they fell on him with dozens of clubs and spears and killed him. This was because Asi'de had no shame.

In this story of a long vanished past itself now considered shameful, shame and anger articulated two opposed trajectories of the ideal moral self. One is that of "respecting," "fearing" (-itodo), or "listening" (-angari) to the moral boundaries formerly established by puyaque prohibitions and now by Dupadeuruode or the Word of God. Those people that respected such boundaries were said to be paaque, both quiet and kind, and caniaque, generous. This was premised on the voluntary subordination of individual will to group norms and an ecology of metaphysical forces that imposes limits on human behavior. (Thus, Ayoreo people commonly described the immorality of themselves and their ancestors prior to contact as the result of being duped or ignorant about the true boundaries of the moral human rather than an intentional violation of them.) The second trajectory of moral agency was the agonistic rejection of those limits through unfettered strength, or etotiguei.[4]

Dominance was considered a virtue necessary for human survival. Although metaphysical forces were more powerful than human will, all was lost without the capacity to resist. The greatest moral failing was to be weak and afraid. Both were synonymous with death. It was no coincidence that the word for "pious respect" was also the same word for "fear," or that "listening" also meant "obeying." This, elders said, also applied to the adode / puyaque / ujñarone complex. Survival in the forest, I was often told, meant not only following taboo prohibitions but also dominating the willpower of other beings—game animals, food plants, enemies, allies. This was marked with the suffix/morpheme -sori. When added to a noun, it meant the object in question was taken as the subordinate possession of a more dominant other. Even mundane phrases retained a martial quality. The words he yuque yu referred to being tired but literally meant "I have become the victim of another." Likewise, the phrase ore surei yibai was how to say that you have lost at a game but it literally meant "I am the one others burned the skin from."

The moral weight of dominance found its clearest expression in the ideal type of the dacasute warrior. Only a proven dacasute could lead others, marry more than one wife, and speak with authority. They were said to fear nothing, and the prototypical dacasute never had shame for any reason. Men who did not achieve this status because they failed to dominate others were known as *ayore poitade*, or worthless men. In other words, elders recounted a world composed of two countervailing forces that were inseparable but often pitted against one another: the moral limits imposed by an ecology of metaphysical forces, and the individual human drive to status, dominance, and strength. It was, of course, nearly certain that this theory of immanence arose as a direct response to colonial violence in the last century or so. Regardless, the world—past and present— appeared before Totobiegosode contact survivors in these terms.

This agonistic sense of being was explicitly focused on the body as well as the soul. Sentiments were embodied moral states that reflected and created an individual's position of strength in relation to this metaphysical ecology. Ayoreo developed an extensive vocabulary to describe the bodily seats of specific emotions. Such feelings were divided into two classes: those associated with the *pajei*, or generic insides, and those associated with the *agute*, or gallbladder. In the first class were positive and generative feelings. These included *pajei sereringane*, "calm insides," or peace and relaxation; *pajei chietaringuei*, "thick insides," or joy and happiness; *pajei dotaningai*, "fevered or heated insides," or anger; *pajei omi*, "beautiful insides," or peaceful and strong. Feelings associated with actions of the gallbladder were often negative and destructive. These included *-ajuque agute*, "to cut the gallbladder," for fear or fright; *agute cho jora*, "the gallbladder resembles its companion," for having a bad feeling of unease or guilt; *cucha agute*, "gallbladder thing," for a bitter feeling; *agute cho tucutuguji jmainie*, "the gallbladder is not firmly planted, wobbly or pliable," for being in the throes of panicked terror; *agute tutaji jmainie*, "the gallbladder is moving away," for being upset or frightened by an event. Someone who was very strong or beautiful may inspire the sensation of *pagute tutubai*, "a painfully full gallbladder," in those who feel inferior. In other words, this was a system in which moral virtues (including domination), metaphysical forces, social sentiments, and bodily capacities were causally intertwined.

Totobiegosode turned to these theories of being to interpret the process of leaving the forest world of *Erami* and becoming New People in *Cojñone-Gari*. The former Areguede'urasade were taught that *Dupade*, the Christian God, and his helper spirits controlled life and agency. "It is from *Dupade* that all strength and health now comes." While the *puyaque* prohibitions established by *Asojná* and other spirits had made past life in *Erami* possible, those very limits were resignified as dangerously immoral and antisocial in *Cojñone-Gari*. Indeed, the moral human of the past could not survive in the moral ecology of the present. *Asojná* and other spirits, as well as the practices associated with them such as *adode* myths and *ujñarone* curing chants, became associated with Satan. "According to *Dupade*," Siquei later told me, "the *puyaque* prohibitions are worthless and bad." The transformation of contact, in other words, realigned the cosmic binaries through which human becoming was charted. In the bifurcated cosmos of the New World everything associated with past morality became resignified as satanic and displaced spatially and temporally into the forest/past. It is no surprise that living in *Cojñone-Gari* also implied an inversion of the moral sentiment of *ajengome*. "Before, we had shame of the things that were *puyaque*," Siquei told me in 2010, reflecting on how he then perceived the world. "Now, we only have shame of *Dupade*." He told me that those who did not have the proper shame relative to *Dupade*, those who did not follow God's rules, were sure to get sick.

Yet the audience for this conversion—and the intentional expansion of becoming it entailed—was not only limited to *Dupade* and his missionary helpers. Indeed, it included a diverse cast of *Cojñone*. Ayoreo quickly learned that there were many types of Strangers, and that they rarely agreed. Surviving in *Cojñone-Gari* also meant being observed by *Cojñone* such as the *Abujádie* believed to oppose Christian morality. It was a world in which self-transformation was routinely interrupted and you never knew quite what to feel about it.

Disgust

A long line of *Cojñone* have worked with Ayoreo people in some capacity, and they routinely arrived unannounced in Totobiegosode communities to check on this project or that initiative in a shiny four-wheel-drive truck, to teach the poor Indians something, to spend part of some huge international development fund on water tanks and cisterns or tin roofs for the houses. Many of these projects targeted the hygiene of Ayoreo people, like the decades-long missionary attempt to separate extended families and construct single-couple homes with indoor bathrooms in Bolivia. (One typical example was the construction of expensive communal laundry stations in several communities in 2005. Based on a "Tuscan model," the idea was that these square concrete block buildings were "culturally appropriate" because they would encourage group activities, and would bring the added benefit of cleaning up the dirty Indians, who presumably could go about in their freshly laundered rags. A year later, they were rarely if ever used, mostly because of conflicts over who could

control access, lack of water in this drought-stricken region, and the failure to properly install piping.)

Maybe it was the sight of so many brown bodies together, or something about the way people would avoid eye contact that spurred them to it, but even the best *Cojñone*, decent people, acted in a way that can be described in the most generous terms as violent. Totobiegosode people were continually being lectured to by this rotating and impermanent cast of characters for being dirty, for not washing their bodies or their hair or their children, for not picking up trash, for living in such close proximity to one another, for not making food properly, for sleeping close together, for how they cared for their children. Not too many were convinced by any one of these outside plans imposed on them, but over time it was glaringly obvious that no matter what the *Cojñone* found the Ayoreo to be insufficient. They were too cultural or too modern, too Christian or too pagan, too primitive or not primitive enough.

One morning, I woke to find Dejai—a proud man who by 2006 had triumphed over his aging rival Jochade to become the principal leader of the Totobiegosode—on his hands and knees in the dust, where he had been working for hours to pick up the tiniest scraps of twigs and plastic.

"The *Cojñone* are coming. They will be against us if they see any trash." And soon three dozen government officials and hangers-on arrived in an air-conditioned caravan of new 4x4s, many of the men dressed in safari suits, one with a pistol tucked behind his belt, swilling cold bottled water.

The Totobiegosode leaders got up and each sang a song to honor the occasion. The officials smiled and promised nothing but their sincere friendship. After an hour or so they were all sweating and ready to leave, but first they wanted to make a stop at a traditional-style Totobiegosode house that was near the road. They asked me to guide them there, and the Totobiegosode leaders agreed. At the sight of the house tucked into the forest, far from the village, the group's anthropologist—who had been notably silent during the encounter with actual Ayoreo—lit up. Panting from the heat and exertion, his canvas vest showing sweat, he proclaimed with a smile, "It's beautiful! Just like a photograph in a book!"

This distance between degraded Ayoreo bodies and any single one of the contradictory images of moral life they were supposed to embody was filled by violence. Many *Cojñone* seemed to desire close encounters withTotobiegosode only if they were sterilized and abstracted from the flapping plastic, the biting insects, the delirious heat, the nonchalant bodily processes. Local experts often complained that working with Ayoreo was impossible because they were so difficult.[5] William Miller has described the "intensely political significance" of Anglophone notions of disgust, which is "an assertion of a claim to superiority that at the same time recognizes the vulnerability of that superiority to the defiling powers of the low."[6] He also makes the remarkable observation that "disgust is organized by the laws of sympathetic magic," insofar as it presumes that similarities in form are similarities in substance and that the socially low are not only polluted but contagious.[7] That is, it arises precisely at those moments in which hierarchies of being are vulnerable to being undone and need to be reasserted. Ayoreo people did not have a word that precisely encompasses disgust. But they did have a particularly developed vocabulary for noting shades of hierarchy, strength, and power. The Ayoreo expression most often translated as "disgust" was *etetigai*. This was a derivative of *etei*, which meant one who is hated, despised, and pitted against in opposition or disapproval.[8] In other words, most Ayoreo interpreted such discriminatory attitudes as a reflection of low social status.

At some places in Bolivia in the early 2000s, Ayoreo was considered to be an insult if used between two white people, as in "Don't be an Ayoreo," or "What an Ayoreo you are!" Denigrating words for animals were used in the same expressions, and calling someone an Ayoreo was meant to designate a person who is uncouth, crude, brutish, stupid, and dirty. I was told that if I were called an Ayoreo, "you should be offended," and "speak up for yourself." In Paraguay, the same went for Moro, Indio, and Ava, a Guarani word meaning "man." There was a popular genre of Paraguayan jokes, told even in the finest and most progressive homes in Asunción, about an Indian figure called El Cacique, or chief. This chief usually goes to the city or gets drunk. The humor arises from a situation in which chief causes harm to himself or lets his wife be raped by a Paraguayan because he misunderstands his social station. Similar distinctions between white and Indian figure prominently in the plots of most "national" literature, including the highly esteemed and often cited satire of Paraguayan culture by Helio Vera, the humor of which comes from the author pointing out similarities between "Paraguayan culture" and the degraded culture of Indians.

Ayoreo-speaking people were also commonly barred from restaurants and excluded from respectable spaces of market exchange and commodity consumption. If such lessons were too subtle, routinized violence clarified the marginal status attributed to Ayoreo people by many *Cojñone*. Despite having a continuous presence in *Cojñone* towns and cities for the last half century, Ayoreo people were routinely expulsed from these spaces by armed soldiers as threats to public hygiene.

During my fieldwork, one rancher took both a Totobiegosode woman I knew and her thirteen-year-old daughter as his concubines. When the husband and father protested, he lost his job and was threatened with death. The woman and her daughter were only released when the rancher's wife returned three months later. (The situation was resolved when the rancher and his wife appeared in Arocojnadi and gave Jochade a goat.) There are numerous cases of *Cojñone* murdering Ayoreo people such as the man decapitated and burned near Isla Alta in Paraguay, or a group of whites who made their hired hands dig graves and then opened fire on them, or the soldier who killed a man and locked his wife and child up in his house. Before he could return to rape her, the woman escaped and walked back to the mission. In one example from the mid-1980s, an Ayoreo man named Jorge Chiqueno was shot by a rancher named Urbano Cuellar in the middle of Roboré, Bolivia. He was shot in the stomach with a small caliber rifle so death was slow. The police report mentions that over the next eighteen hours, including part of a full business day, Jorge crawled 156 feet through the streets of the town asking for help, knocking on the doors of various houses, before he finally bled to death. No witnesses testified against Cuellar.[9]

There were so many stories of such violence that Ayoreo people often imagined that national citizens shared a latent desire to kill them. In such an atmosphere, it was no surprise that any unexplained deaths were believed to be murders by *Cojñone*.[10] During my fieldwork, rumors often swept through the communities about such murders. One involved a bizarre case of a man who died after hunting a giant anteater. There was no evidence of foul play, but it didn't make sense to anyone. Ayoreo people were convinced that the man, named Abujei, had been killed by a white rancher and that the police and the lawyers had conspired to suppress the evidence. Many inconsistencies were mentioned and hotly debated for weeks, but his death remained officially attributed to a giant anteater attack (which would make him the first human casualty of the giant anteater in the history of the Chaco).

Alongside such physical violence, there was also a lucrative traffic in the hair of Ayoreo women. Several competing teams traveled in SUVs to Native villages in the Chaco buying the thick black strands. When women were desperately poor, they sold their hair, and many did it more than once while I was there. They came back quiet and subdued, running their fingers through bobbed ponytails. The *Cojñone* paid by length, usually five or six dollars for a head of hair. The hair was dyed and ended up as beauty enhancing extensions in high-end salons in Argentina, Spain, or the US. Ayoreo people consider hair to be a particularly important aspect of feminine beauty, and its coercive commodification reinforced a general sense of low social status in *Cojñone-Gari*.

The category of *ajengome* was expanded to fit these new conditions of Ayoreo life. One way its expansion did so was through changing notions of the public. In the past, I was told that shame was attributed by an audience comprised of those who witnessed the transgressive event. Elders said this had changed. During my fieldwork, Ayoreo people commonly discussed *ajengome* in relation to *Cojñone* witnesses who were not physically present. I was told that young people didn't learn how to hunt, sing, or play the old games because they were ashamed of acting like an Indian in front of *Cojñone*, even in a remote village where no *Cojñone* could see. As usual, it was Jochade who articulated this in the most pointed manner one night around the fire in Arocojnadi. "The young people today do not *ajengome* God and they do not *ajengome* the old things from before," he said as everyone listened. "But now they *ajengome* the *Cojñone*. They want the life of the *Cojñone*, that's why." For Jochade, the nascent sense of being Indigenous arose from the same factors that made shame an appropriate response to the new publics of *Cojñone-Gari*.

Here, *ajengome* reemerged in another, subtler guise as a crucial energy of the contradictory moral economies of life applied to Indigenous peoples in the Chaco. *Ajengome* was the idiom by which violent marginalization and bodily diminishment became linked to ontological sensibilities, moral status, Christian spirituality, and culturalist disgust. Through it, such oppositional standards were simultaneously reconciled and reproduced; a process that lent renewed urgency to the project of self-transformation.

More precisely, *ajengome* articulated the tensions of subjectification and desubjectification that so profoundly defined the contemporary moral economies of indigeneity and Indigenous ontologies alike. For the New People, humanity was a condition both obvious and unassumable. Becoming an Indian subject always meant becoming insufficient, and for Totobiegosode-Ayoreo in 2004, becoming New meant becoming ashamed. In the classic style of negative dialectics, this oppositional framework created a new kind of disordered subjectivity. It was this tension between subjective immanence and the impossibility of subjectivity that shame articulated. This was the stupendously tricky terrain of which the New World was composed.

Political philosopher Giorgio Agamben makes a similar point in his analysis of shame among the survivors of Auschwitz. For Agamben, the shamed subject "has no other content than its own desubjectification; it becomes witness to its own disorder, its own oblivion as a subject. This double movement, which is both subjectification and desubjectification, is shame."[11] This double movement means that "the human being is the inhuman; the one whose humanity is completely destroyed is the one who is truly human." I suggest that the category of indigeneity reorganizes this tension of subjective becoming and desubjectified life in significant ways. The result may well be the creation of a New World, but it is a world of constant inversions where shame is the primary ontological sentiment and where the survivors of contact and of concentration camps may resemble one another in unsettling ways.

Illness as Ontological Insufficiency

According to Totobiegosode, health was an indication of moral well-being. Becoming ill was never a morally neutral event but a reflection of an individual's capacity for resistance and position relative to metaphysical forces. Health and moral sentiments were causally related. Being afraid, sad, or ashamed was the same as being morally insufficient; each state in turn made one vulnerable to bodily infection and death.

Common impressions of moral insufficiency in *Cojñone-Gari* were enforced by the fact that Ayoreo-speaking people were frequently sick with a range of preventable diseases. On their diet of rice, sugar, and noodles—modeled on that of working-class Paraguayans—teeth fell out by twenty-five and young children had the orange hair of protein deficiency. People died from infections that would be minor elsewhere. Teenagers were diagnosed with rare forms of cancer and most recently, HIV. In a cruel irony, the shift to sedentary villages means Ayoreo are now exposed not only to tuberculosis but also to the spores of lung-damaging fungi that are commonly buried in undisturbed Chaco soils. The ubiquitous golden dust itself had become a potentially mortal contagion.

Totobiegosode told me that illness itself had changed. They said that the diseases afflicting them were different than before, that they were particular to the metaphysics of *Cojñone-Gari*. Similar attitudes have long been noted among Ayoreo-speaking people, and were encouraged by early missionaries. New Tribes missionary Asi'guede, reported in 1970 that some Guidaigosode at El Faro Moro agreed to convert to Christianity only if they would also receive injections of Western medicine. "One day in camp," he wrote, "the chief said to me, 'Tell your wife to give shots to all of our people so that *Asojná* won't be able to touch us.' "[12] Western medicine at that time was a cure for the failure to obey *puyaque* limits.

During my fieldwork, Totobiegosode distinguished between diseases sent by Satan/*Asojná* and those sent by *Dupade/Cojñone*. They called these new illnesses *Cojñone Ujatiode*. The phrase implied that the illness originated from the hostility of *Cojñone* and that they were specifically aimed at Totobiegosode. While elders said they were rarely sick in the forest and that they knew how to cure the diseases of *Erami*, there was a pervasive sense that Totobiegosode were fundamentally alienated from ways to cure these illnesses of *Cojñone-Gari*.

Such diseases were commonly attributed to the failure to be ashamed in front of *Dupade*, the Christian God. "*Dupade*," Siquei told me, "will give a sickness to anyone who does not have shame of him." Yet it was rarely clear what precisely someone did to anger God. Illness narratives emphasized this sense of fear and alienation that characterized Ayoreo relationships with illness and health care practitioners. Aasi was infected by tuberculosis shortly after contact. He died in 2013 of the same chronic lung infection that killed Ujñari, Emi, Codé, and a handful of other Totobiegosode in the previous five years:

The sickness became worse after ten years. It began pushing me to the ground. I thought I would die. There was no truck to take me to the hospital. I told my wife, "I am hot and my eyes are swollen. My insides are hot and my eyes are being pushed out. My breath is very hot." They took me to the Indian hospital. They gave me injections. But they were lying because the sickness returned. They let me go but the sickness got worse. They took me again and it calmed. They tricked me again. They let me go. I went back to my community. The sickness grabbed me again. I went again to the hospital. They gave me injections, and it got better. But it was a lie. My sickness grabbed me again; they took me back to the hospital. My wife was angry with me because she didn't want to endure more days in the hospital. But she said, "It's better we stay more days so you get better." But when we were in the hospital they didn't give us enough food. We had no money and we suffered from hunger, and I said, "Go back. I'll stay here." But my wife was stronger than me, and she said, "We will stay." We were suffering from hunger again. We only ate one piece of bread in the morning and for two weeks we were like that, with nothing but our hunger. Some days we said, "Let's go back." Some days we thought to stay. But still we stayed. I prayed to God because that is my way. I asked God to change the attitude of the *Cojñone*. It was at night, and we were desperate. I thought I would die there. That is the story of my sickness. I did not die because I prayed to God. I asked God to give me strength again. I went to the city with Uaque and he said, "They don't know yet what your sickness is." I said, I am not going to be afraid to die. "I am not a child who cries with his sickness or his fear of death. I am a Christian; it is ugly if I cry. If I cry, it appears as though I am a liar, as if I do not have faith in God."

After the fourth month of fieldwork, I too was frequently sick, fungus in my lungs which had been scarred by the fine dust, and infections in the constant scratches of thorns or insects. One particularly bad infection caused my right leg to swell until it was streaked with black and I couldn't walk. I remember gourd rattles and firelight and a group of women who held me down and mercilessly probed the infected spot. In this memory they began to admonish me for my willfulness. "Asojná did this to you. It is because you went to hunt in the forest like an Ayoreo. But you are *Cojñoi*. The forest is no place for you." Just as quickly, they vanished. Did this mean my illness was caused by my insufficient faith in *Dupade* or by my failure to respect the spirits of the forest or both at the same time? Or was this just another fever dream?

As a *Cojñoi*, it was often taken for granted that I knew how to diagnose and cure the sicknesses of *Cojñone-Gari*. In general, Totobiegosode consumed as much medicine as they could find, dozens of pills they classed by color or shape and shared with each other. A mixture of caffeine and painkillers was the most popular kind. Totobiegosode people called pills *semenie* or *bisidode*, and their names were stories: *yajo pororodie*, "we eat white things," was a common way to refer to taking Western medicine. *Bisite utatai*, black medicine, was for stomachaches. *Seme carate*, or red medicine, was seen as a vitamin. Novalgina painkillers were called by Totobiegosode *Nujnanguto'date potedie*, or "that which Nujnanguto'date always wanted," after a woman who frequently took them. "That which Nujnanguto always wanted" referred to Syndol, a similar painkiller. *Chicori-potedie*, or "that which Chicori always desires," was the name for Calmol, a popular mixture of muscle relaxant and painkillers. Needle medicine was the name for antibiotics. Men medicine, or *seme choquiode*, referred to all kinds of pills that were elongated or oval in shape, while women medicine, or *seme chequedie*, referred to all pills that were round.

Many people, even those who seemed healthy, were taking immense quantities of pills. One of my Totobiegosode friends took more than thirty different pills every day, ten of which were prescribed by a doctor. His wife began to get visibly nervous when the supply was low. Several individuals—all contact survivors—were known pill collectors. One woman in her early fifties constantly complained of weakness. She would moan and cry if she was refused medicine, no matter its shape, color, or effect. She was said to hoard the pills and bury them in a jar in her house. Other people refused to take medicine of certain colors and shapes, no matter their illness.

One day out hunting, Cutai asked me if it was true that *Cojñone* had medicines that could make you happy, that could cure one of sadness. I stumbled through a description of mood-altering drugs and antidepressants. I tried to explain in my insufficient Ayoreo that there was a kind of medicine that changed how one feels but that it was only for those who had severe problems. My failed explanation only seemed to confirm what he had heard. He sucked in his breath in amazement. Then he asked me to bring him some of this medicine the next time I came. That way, he said, he would no longer be ashamed in front of the *Cojñone*.

Becoming New, Becoming Ashamed

The 2004 contact with the Areguede'urasade unsettled and tested the tenuous Totobiegosode theories of life in *Cojñone-Gari*. The startling appearance of these people promised an epistemic rupture where everything was up for grabs. Were Totobiegosode contact survivors going to use this as an opportunity to redefine their own histories and the moral value of their past lives or not? Capturing the Areguede'urasade was a unique opportunity for Totobiegosode to reflect on their collective life project. The life project they articulated, however, was based on repeating the same subordination they had suffered twenty years before. By October of 2004, the former members of the Areguede'urasade group were divided up among their *isasorone* captors.

Ritual objects were defiled and broken to show that they had no power. The New People had to provide game to the others in the village who were no longer skilled hunters. Most of their possessions were taken from them and sold by Dejai to a variety of *Cojñone* eager to possess the material artifacts of isolation. The settlement of Chaidi, established at the very edge of the forest, was set up to resemble the forms of a mission, complete with a small store run by Dejai, a school, and Bobby's open-walled church. The first Spanish words the former Areguede'urasade learned were *monte*, for forest, and *sucio*, for dirty.

By 2005, Areguede had died and the group's matriarch starved herself to death. Siquei was morose and withdrawn. He spent most of his time alone in his house, carving bowls of *palo santo* wood to sell to middlemen for the handicraft market. Dejai sold all of his former possession and kept the money from every third bowl he made. The rhythmic pounding of his tools continued far into the night. It was hard for me not to hear rage and loss in the steady procession of staccato knocks.

When I think of that long ago time I spent with the Areguede'urasade in 2004, I remember hunger, a desperate frustration, and Ebedai'date's voice. I can still hear the high whine she made without stopping for three hours straight the second time she rode in a truck, perched like a bundle on top of the bench seat. Because of the speed, they said. I think of the first time we sat together by her fire when she grabbed my arm and told me she was afraid of me too and how she later relaxed enough to share simple jokes and baked *doidie* roots or *cuteperone* honeycombs with me. I especially remember her dignity and frank generosity and how she spent hours fanning one toddler while he slept through a hot afternoon. Like all the children of the Areguede'urasade, this boy was greatly beloved. He waddled around bedecked with necklaces and finery. Like the others, he was always whisper-quiet. I will not forget the scene the small children made, three or four of them playing in a tight circle, their bodies touching gently, smiling and shaking with laughter, but making absolutely no sounds other than the rasp of tiny feet on dusty earth.

Once settled with others in the village, these children were often beaten by the others. They were rarely comforted and frequently went hungry. Two years later, the same young boy, then about five years old, was periodically overcome with fits of hysterical rage when the other children would mistreat him or his sister. The adults of the other faction mocked him for being *sucio* and ignorant and bad, and it would drive him mad, until his small body locked up and his face turned purple and he would begin to howl. Once I came upon him alone in the middle of the afternoon outside the village, crying silently and stabbing a lizard with a stick. When he saw me, he walked away.

By 2007, the former members of the Areguede'urasade were split into two factions, one associated with the rising evangelical leader Achinguirai and one with Dejai. Each group had devised ways to communicate with one another in private, including trails that skirted any public space, and sight lines cleared between their houses. In public they remained silent, but in private they often summoned me to talk. Over the course of my fieldwork, I conducted numerous interviews with five of them.

In these stories, which often unfolded at night, they emphasized that they had decided to make contact but that they were ignorant of how things worked among the whites. I soon learned that it was considered off-limits for the New People to tell me that they preferred anything about life in the forest. They were told that if they did so, the *Cojñone* would force them to leave the village and return to the forest. There were metaphysical, as well as pragmatic, risks to violating these limits. The New People were taught that thinking about life before would produce an *ayipie deroco*, or a "bitter soul," which would offend Jesus. Such a bitter soul was vulnerable to weakening and amnesia, as in the expressions *chejna yayipie* ("my soul is finishing itself off") or *ayipie yui* ("my soul has been taken by another"). These thoughts and the corresponding feelings of sadness were likely to produce illness.

Becoming New was the same as becoming ashamed. Indeed, *ajengome* was considered to be the defining ontological state of *isagode*, or "touched ones," who had to be ashamed before their captors for giving up their own willpower. Yet becoming newly Indigenous meant that *ajengome* itself was a fractured and incoherent ideal. Should one be more ashamed before Dejai or his budding evangelical rival, Achinguirai? Should one *ajengome Dupade* as well as an imagined audience of *Cojñone* at the same time? What if some of these *Cojñone* were not missionaries but immoral culturalists and thus suspected of being sent by the Devil?

Despite or perhaps because of this incoherence, shame was the crucial medium by which the former Areguede'urasade were taught to interpret their position in the New World. They may have accepted multiple kinds of *ajengome* by their public silence, but they were keenly aware of the contradictions this implied. In the stories they told in private about their lives, they pointed to a fundamental deception. Siquei in particular offered a counter-narrative of contact:

I came back to camp. The women were crying. They were crying a lot. I said that Dejai and the others had been captured by the Ijnapuigosode [in 1986] but they were alive and living with them now. "They will kill us all, they will kill us all," they cried. "Quit crying," I said. "We will live with them too. Prepare your things. Don't cry and we will go with them now. . . . You all always say that you want to live with the *Cojñone*, so get ready and let's go." Ajua'nate was crying. She said that she was going to stay alone because Ajua doesn't want to talk to any *Cojñone*. I said, "Prepare your things and let's go. They won't kill us because they don't kill each other anymore. Asôre is there with them." The father of Asôre became angry. "Why didn't you stay with Asôre? They will kill him!" . . . Ebedai'date began to cry again, and she was crying loudly. Unoro'nate [his second wife] wanted to stay there too, but I said, "Prepare our things and let's go." Ebedai'date cried all the way to the camp. She was very afraid of the *Cojñone*. We thought they would kill us. Areguede said, "I want to stay one more night here." But my son stayed there so I must go too. He wanted to fight with them. So we all went. When we got there, Asôre said they believe in some kind of god that is in the sky. I asked if the Sun was their god, but everyone laughed. They said that no one had seen this god but that he had a lot of spirit.

By 2007, Siquei and the others said their decision to surrender was based on a mistaken idea about what life in *Cojñone-Gari* entailed. In such descriptions, the former Areguede'urasade members subtly positioned themselves against their condition as the lowest-ranking members of the Totobiegosode. Pointing out this fundamental mistake in their own judgment highlighted the divisions in the group when faced with the sudden prospect of living in *Cojñone-Gari*. Aasi's daughter, Uo, was part of the Areguede'urasade group. As she put it,

Then I stopped listening to the men. We all began to cry. Ebedai'date was crying too. My mother was crying too. I was crying. The women were crying. My mind went to what Areguede had told us before, that the *Cojñone* don't have only one disease but many. There I began to cry. Because I knew that they were *Cojñone*, that they were not my people. There I was, crying with all the others. That is how it happened that we now live with the *Cojñone*. There are many diseases here. They told us to stop crying. Areguede said, "We are going to stay here. Because Tie'de doesn't want to live with the *Cojñone*, that is why we'll stay here." If he had stayed in the forest, surely Areguede would not have died already. "Here we are going to stay," he said.

At least one former member of the Areguede'urasade pointed out further deceptions against them by Dejai in which he profited from the sale of their possessions before they were aware of the value of money:

The next day we came to the camp. We went there. It was very far. Dejai took the pots [traditional clay jars] that we now call trash. We didn't know that he would sell these things to the *Cojñone*. We said to him, "Are you going to use those?" He said, "Yes, we will use them to drink water with." But then he sold them. They told us it was trash, but the *Cojñone* wanted it. I didn't know any of the people there because I had never seen them.

Captive and captor, trash and tradition. The tensions of shame and rage erupted one morning in September 2007. The leaders from Arocojnadi and I had gathered in Chaidi for a meeting of the tribal organization. While we were waiting for the meeting to begin, Siquei walked by, as if to go work in his garden alone. The NGO had insisted shortly after contact that Siquei and Jotai be installed as "leaders" in the tribal organization, an action considered laughable by the other Totobiegosode who considered them the most subordinate and ignorant of all. It was another point of friction, as they, like the other officials of the tribal organization, received wages from the NGO for attending meetings. As Siquei passed by, Dejai began to berate him for skipping the meeting and for being lazy. This kind of insult, of course, was by no means exceptional. The major source of daily entertainment in Chaidi at that time was ridiculing the New People and Siquei's family in particular. Yet on that particular morning, Siquei turned and began to speak back.

Siquei accused Dejai of thinking he was more powerful but for no reason. He said that Dejai did not hunt but that he still ate enough food to be fat. Siquei said that his own status was superior to that of Dejai. He yelled in rapid-fire Ayoreo that Dejai had only killed a jaguar when he lived in the forest but that he had killed many enemies. This meant that Dejai was *cucha bisideque*, a worthless thing.

Dejai's authority had never been so directly challenged by anyone since 1986. He flushed with rage and stood up. He shouted that Siquei was worthless and that he should leave if he did not like life among the *Cojñone*. He said he should give back the tin roof on his house, that he was ignorant and stupid, that he knew nothing. Dejai screamed that he had power over Siquei as his *isasori* and that Siquei was afraid of him.

Siquei by this time was visibly upset. His muscles twitched and his digging stick trembled as his feet beat an irregular figure eight in the dust.

Dejai kept going. He grabbed his shotgun and said that if he hadn't contacted Siquei and the others, they would still be living like animals in the forest, naked, dirty, and afraid. He said that acting angry would not make anyone forget that Siquei was ashamed, would always be ashamed because he let Dejai contact his people. That he was no *dacasute* at all, but a worthless man.

Siquei tried to respond. But there in front of the encircled onlookers, the response died on his lips. It would not come out. He could only manage to stammer, spittle flying as he tried to form the words. The others began to laugh loudly and to mock his stutter, the sound of the jeers drowning out his voice. He retreated to his house and did not emerge again that day.

Refusing Shame

I spent Christmas 2006 in Arocojnadi. The other guest was Iodé, a fifteen-year-old girl who supported herself and a shifting cast of relatives by working as a *cuajajo*, or so-called little bird in town. At first, she seemed out of place among the smoky fires and gentle banter. It wasn't the single eyebrow painted across her face, white powder on her brown cheeks, or six-inch gold heels. It was that she would only talk in an exaggerated voice, pushy and tough, ridiculing the backwardness and remoteness of the villagers, nonchalant and unashamed when they called her a prostitute, replying, "*Puta'date uyu*, I am the mother of whores."

She told us her rates, how she charged Ayoreo men three dollars, Brazilians two dollars, Mennonites one dollar, and if someone was cute it didn't cost anything. Cheaper for the Strangers, she said, even though she often went hungry and old people were dying of it—the hollowed trunk of the death tree—all around her. More than the money she was in it for a husband, a *Cojñoi* husband—no lazy poor Ayoreo would do. She was sure it would happen soon, he would take her to the city, buy her a house, maybe even one with running water. She would be happy and if not, at least she would have a baby of her own, a pale one more beautiful than a brown one.

She told us about the men she knew, the good ones (who bought her things) and the bad (about whom she said little). My friendship with her began when she speculated aloud about all the perverse things I must have done in my country to be sent to that shitty village in the middle of nowhere, where there was no Coca-Cola or ice or tinned meat, and how mentally deficient I must be to stay there for so long. She didn't have a place to spend Christmas because her mother was gone, eight years in the city, a working woman too, first on the streets, then the rougher work on the trucking lines.

Iodé had met Jochade's daughter in Filadelfia, and they became friends. In 1931 Filadelfia was known by its Plattdeutsch name of Fernheim and it was a sleepy grid of thatched huts and pale Mennonite farmers from Prussia, wracked by cholera and clinging to faith, the end of the oxcart or railroad for these people who were looking for peace and sovereignty. In 2007, Filadelfia was a frontier boomtown, flush with cash from opening a raw wilderness with bulldozers and cattle and land prices that surged wildly. A town of six thousand people with a handful of churches and forty tiny clapboard stores, each one selling six hundred empanadas and eighty roasted chickens a day, they couldn't cook them fast enough. The city of fraternal love and the far away home looked like a town from rural Kansas in the 1960s, straight wide dirt roads, neat square brick houses hidden behind rows of dusty flowers blanched by the intense light. Fancy trucks and brick mansions and rusty bicycles and mud huts and blonde teenagers on Chinese motorcycles and slender girls and smooth muscled men up from Minas Gerais or the east, to start out or start over.

It may have been a Mennonite colony, but in those days Filadelfia was an Indian town, a modern day *pueblo de indios*. It was a place where no one was out of the stare of an Original Inhabitant for very long. Indians in the sun-baked dirt streets, stepping off the sidewalk in deference to pink-skinned, thick-bellied white men. Indians in dozens of businesses with names like El Indio, the Native, or Casa Adelia, dedicated exclusively to supplying Indians with the goods they liked or were expected to like—bright floral print skirts, oversized neon soccer jerseys, cheap cotton baseball caps, bland high-carbohydrate foods, alcohol. Everything bought secondhand and sold at top-shelf prices, inflated for the poor. Old Indian couples, their faces carved by the sun and disease and hard work and bad food, gathering empty beer cans early in the cool morning. Indians in the few patches of brush left on the margins of the town, stretched out in the shade. Indians blackout drunk in the afternoon sun. Indians bathing in the town's water reserves, crouched behind buildings, living behind metal shops, baking bread, shoveling gravel, stacking wood, trimming roses, driving tractors, waiting for work, asking for change. Invisible Indians murmuring in the brush. Indian families gathering *algarrobo* fruit for miles along the highways. Hundreds of Indians lined up quietly on an open flatbed trailer behind a tractor in front of the huge air-conditioned supermarket on payday. Young Indian men stacked into cattle trucks and hauled off to a ranch, two dozen on a single tractor, riding on the scoop or hood. Indians looking in through the windows. Indians watching. Like they were trying to figure something out.

The constant vigilance was palpable, especially, it seemed to the Mennonites. It was audible in the whispered fears commonly heard in Filadelfia, White man to White man, about an Indian uprising, lustful Indian desires for young white girls, their savagery. As one South African man told me, "We tried to do it here like we did it back home. But the Indians are worse than the blacks. They're so lazy, they don't want to work at all. They could improve themselves, but they just don't. Do you think they like to be dirty?" These anxieties were purged from time to time in bruises and semen and blood.

There were many different kinds of Indians in Filadelfia: Nivaclé, Guaraní Ñandeva, Enxet, Manjui. But Ayoreo-speaking people were the most conspicuous of all the watchers, dressed like pirates or refugees, prominent bones and flamboyant clothes, the poorest, the loudest, the boldest, the least furtive. Barefooted children with brazen eyes and gentle voices and shy smiles and hair dyed with peroxide or bleach, designs drawn with lipstick on smooth cheeks. Ayoreo, in general, asked you for what they wanted. They didn't fawn or playact or awkwardly grovel. As if they wanted you to know that they knew the score. In 2008, there was an Ayoreo camp in Filadelfia called Casa Pasajera that has since been bulldozed on the order of Mennonite officials and its replacements bulldozed in turn. Three hundred people crowded under plastic tarps in an abandoned lot of sandy grit with no access to running water or electricity, denied by the colony because they thought it would discourage them from staying. At night under crackling plastic around a fire listening to quiet jokes and stories, it could seem almost peaceful. A place where old warriors came to work, starve, and die—where young men fought with knives, children grew up fast, and everyone seemed to be waiting for something that never arrives. With its skeletal dogs, piles of trash, human excrement, and emaciated people, Casa Pasajera could make Ayoreo seem like refugees from some unnamable cataclysm.

But they were there to stay, and they made the town their own. They had names for its places and goods, like Their Stomachs, where Mennonite businessmen dumped slaughterhouse offal for them, or Grandmother (*Dacode*) for yerba mate, because it's something you never forget even if you have nothing else. Children were named not after *edopasade* clan ancestors but after objects of power and desire. There were at least two men named Rice (*Aocei*), one Tomato Paste (*Conservai*), two Noodles (*Fideo*), an Onion (*Cebollai*), a Fresh Meat (*Carnei*), a Dried Meat (*Cesina*), three Vegetable Oils (*Acei*), one Dried Biscuit (*Galleta*), several Sweets (*Dulcei*), a Sugar (*Asucai*), a Lettuce (*Lechuga*), two Milks (*Leche*), a Yoghurt (*Yogui*), and two Breads (*Pamaane*). There was a Train (*Maquina*), a Bulldozer (*Topadora*), two Tractors (*Tactoi*), a Chainsaw (*Motosierra*), a Flashlight (*Linternai*), a Battery (*Piladie*), a Mirror (*Espejoi*), a Motorcycle (*Motoi*), many Guns (*Poca*), and an Airplane (*Achorro*).

During their visits to Filadelfia, my Totobiegosode friends didn't stay with their ancestral Guidaigosode enemies in Casa Pasajera. Rather, the NGO charged with assisting them let them sleep in a wooden shed behind its brick office. There, they would sit for hours watching the pale figures walk by in gingham dresses or purr along on motorcycles. They had funny names in Ayoreo for these people, like Fish for one particularly pale man, Head of Bees for a guy with tightly curled hair, or Bulldozer for a spiteful fat lady. Those better known had names with a little more sting, things like Harelip for a woman who tightly compressed her lips when she was angry at them, Mean Spirit, the Mother of Whores, Fat Belly, and Old Lady. The Ayoreo watchers claimed to know which *cojñoi* was sleeping with which other *cojñoi*, who was cheating on their wife or husband, where they went, what kind of vehicles they owned, the personalities of their pets.

By February, Iodé's mother had come back and at the insistence of the NGO the Totobiegosode leaders had forbidden Iodé from sleeping in the shed behind the NGO because she was bringing a string of strange men there. They were worried the Totobiegosode girls would follow her example, and rightly so. She moved back to Casa Pasajera, but we'd run into each other from time to time at night in Filadelfia. She'd be there on the corner in some impossible outfit with one group of girls and holler to me out of the dark, "Lucas *ahaiquea*, where are you going?" followed by her raucous laugh. We'd chat about this or that, where her grandfather was, how many turtles I had killed this time, about the people we knew. Then she was gone.

Sometimes, a little teenage girl dressed for work would find me and relay a message from her. Iodé was working in the temporary camp of a group of woodcutters clearing pasture a hundred miles west or she was fine with her grandfather on a ranch to the north, or she was sick and needed medicine. The next time I saw her months later, her face was torn in a jagged raw streak from the right eye to below her ear, and she was thinner than normal. She was hugging a Nivaclé man and she let him go when she saw me, flashing a bright smile. When he'd gone, she told me she was pregnant. She had been worried because the working girls of a rival group were angry with her, jealous of the baby, she said. They had beaten her the week before, five or six of them in a group, their husbands too. They had tried to kick her in the belly, but she had rolled into a ball, like this, and it turns out that she had fooled them because her baby was okay. She said she couldn't keep it anyway, she didn't have any money and the men wouldn't like her anymore. For sure she wouldn't find a husband with a fatherless child. She wanted me to take it.

No matter how many times I was offered a child, it never failed to take my breath for a second, and I couldn't bear to look too long at the ones that no one wanted and everyone ignored. When I hesitated, she offered it to several of the Totobiegosode women, those wide-hipped matrons at the age of twenty-five, already so steeped in the details of making and sustaining life. They gently teased one another, saying it would be more beautiful than their own brown babies. "It will be so white," they laughed. "Like you!"

Most Ayoreo women work as little birds at some time during their life. I've seen an entire train car full of thirteen- and fourteen-year-old girls traveling between Santa Cruz and the Brazilian border towns, laughing and waving at me as they passed so small and brave and electric. Later I ran into them and their older female relatives who knew the ropes all fall-down drunk outside of the lowest, loudest bar at midnight in a backwoods dirt-road town where they still call Ayoreo *Bárbaros* and where they weren't served in restaurants. Where the dark brutality could well up and swallow you if you let it or if you weren't careful where you stepped.

The little birds played hard, lived fast, and often died young. They were routinely raped and murdered. Shame ran wild in darkened ditches, patches of thorny brush, plastic seat covers of broken-down trucks. Faced with several insoluble dilemmas, the little birds also rejected indigeneity and Ayoreo-ness. They offered their bodies to *Cojñone* but less willingly so to Indians. They positioned themselves as exceptionally close to *Cojñone* within Ayoreo communities, but they were often desirable to outsiders precisely to the degree of their supposed savagery. As one middle-aged client put it, man to man, on a dark corner along Filadelfia's main street, mistaking my familiarity with Iodé for complicity, "I prefer Indian girls. When they are older, forget about it. But you can find very young ones, new ones, especially these Ayoreo. They will do anything you want; they are just savages. You only have to give them a Coca-Cola and they are ready. But, you have to be firm with them, show them you are a man. They like it if you are tough on them, you know, give them a slap."

Somewhere at this extreme point of desire and disgust and despair, the incoherent force and tangled logics of shame split and doubled back yet again. The New People were violently taught to be ashamed of both the Christian God *Dupade* and *Cojñone* of all kinds. Yet these girls often described the selling of their bodies as precisely the opposite, a radical and unsettling kind of agency. They said they did not *ajengome* in front of the *Cojñone*. They didn't even call it work. They said *ore cana*, they were just playing. They refused to be ashamed by *Dupade* or the *cucha puyade* or *Cojñone* of any kind. In fact, they said, the *Cojñone* desired them. Even the Mennonites followed them like dogs, eager to "drink our bodies." With the money they earned, they supported extensive groups of people—parents, children, husbands. They had nicknames known only to their sisters. With their flamboyant fashion, unique slang, and brazen humor, they were something like role models for the young girls growing up hungry in the grit of unauthorized urban camps. The *Cojñone* bulldozed their camps by day, but at night, the same men returned.

Shame was a constant in the Place-Where-the-Black-Caiman-Walks, extending in all directions, the afterlife and texture of those points where colonial violence, bodily perspectives, and subjective immanence collided and collapsed under their own weight. Its incoherence filled in gaps of all kinds: the aporias of testimony, the impossibility of category, the failure to understand, the unassumable conditions of becoming, the limits of empathy. It mediated and inscribed the profound contradictions between opposed moral economies of Indigenous life in the Chaco. Shame was the remainder to constantly unraveling Ayoreo efforts to transform moral humanity; the testament to the fact that the conditions of life and being were never quite possible; the tension between the human and the inhuman; the impossibility of distinguishing between punishment and reward, innocence and guilt, right and wrong.[13]

It arose from the incapacity to manage the unmanageable, from the sense that self-objectification was always already elusive, from the onto-logical conditions of becoming Other, from the hunger of the Black Cai-man that it reflected and instantiated. Moreover, shame was contagious and it was no coincidence that shame began pursuing me just as I began to pursue it. Doing anthropology among Indians in the Chaco made you some kind of parasite; you just had to work out precisely what kind you wanted to be, or better yet not begin the journey in the first place. And I had even more difficulty identifying of what, precisely, I felt ashamed.[14] Like Primo Levi, I could find no obvious transgressions beyond the great-est transgression of them all: the very act of being there, of witnessing that to which it is impossible to reliably bear witness. It took me a long time to realize that shame was a riddle without an answer and that this was the only answer to discover. And I was not alone.

"Everything that was me was insufficient," wrote Miguel Bartolomé, who slept with a pistol under his pillow and had nightmares of malevo-lent spirits during his three month stay among Garaigosode in 1969.[15] At least four others I know of gave themselves over fully to the embrace of the Black Caiman, fits of skin rending insanity or depression followed with a shot in the head, a car driven off a hillside, the fetishes of tradi-tion and ethics feeble talismans against such shameful impotence. All the decent ones felt it to some degree, the darkness beckoning with rage, madness, despair.

One morning while I was staying in Filadelfia, Iodé returned to the garage carrying the groceries she had bought with the night's earnings. She offered me a little plastic box of yogurt. I thanked her and refused.

She was upset. "Why are you ashamed of being my friend?" she asked. "Eat this. It is not bad. I want to adopt you in the old way; you will be my older brother. I will be your sister, your little-bird-mother. Your whore-mother."

She handed me the yogurt again, and stretched out on the ground. My face burning, I opened the carton and began to eat.

SEVEN

Affliction and the Limits of Becoming

What sort of presence in our minds, what sort of whatness are they now to have? What sort of place in the world does an "ex-primitive" have? CLIFFORD GEERTZ

Every time we give up the will to know, we have the possibility of touching the world with a much greater intensity. GEORGES BATAILLE

Arocojnadi was a small settlement and anyone's absence was conspicuous. One evening, I noticed that Pejei—Jochade's nephew—was missing from his customary place around the communal fire. Pejei didn't appear the next night or the next or the one after that. It appeared that his entire household had left and that his shack was abandoned. No one chopping wood or hauling water, no fire inside. But someone was inside. I heard moans and mutters and a scream late at night.

Suddenly he was back. His laugh was strained, his smile too quick, he had nothing to say. The others were careful and gentle with him, making a point to share their food. When I asked him where he had been, Dasua interrupted.

"His *ayipie* left him," she said. "But now he's okay again."

Pejei nodded and smiled. I learned that he had spent the last three days tied to a post with coarse rope, thrashing and moaning and trying to run away to the forest.

Pejei was one of many Totobiegosode who were susceptible to a common form of madness called *urusori*. A handsome man in his mid-fifties with the build of a weightlifter,

172

he came from a long line of distinguished *dacasute* warriors and was said to resemble his father, the principal leader of the Totobiegosode whose band was hunted down and captured in December 1978.

By March 1979, Pejei's father had starved himself to death. The young Pejei was hired out to a local rancher and he returned several months later to learn that his entire family had died in his absence. He then married a young woman who five years later gave birth to twins, traditionally considered *puyaque*, or taboo.

Pejei's first bout of *urusori* came when missionaries pressured the couple to keep both infants, despite the fact that Pejei heard one of the children "speak" to him, thus predicting the death of his wife if they violated the prohibition. Back on the mission, they had little choice. True to Pejei's prophecy his wife soon died. Since then, attacks of *urusori* could strike him at any time and he was told to take antipyschotic medications.

Urusori could be caused by any profound fear. I was told that it was often triggered by frightening encounters with white men or things associated with them, such as the sudden appearance of a *Cojñoi* carrying a gun in the forest. The episode I witnessed with Pejei was attributed to an airplane that unexpectedly passed overhead at dusk. The sound of this airplane and the air it pushed down, "its breath," touched Pejei and the fright caused his *ayipie* to leave his body. Regardless of its cause, *urusori* is what happens when the *ayipie* newly reconstituted by contact and conversion leaves the body. When the *ayipie* leaves the body, the corporeal seat of moral sentiment and reason departs as well. In its absence, the immoral, irrational body is compelled to strip off all clothing, to flee from other Ayoreo, to return to the forest.

Ayoreo used the same word—*urusori*—to describe the state of being drunk or high.

The Ayoreo encampment of Barrio Bolivar was located some seven hundred miles away from the Totobiegosode settlements. It lay down a sandy road in a notorious *villa miseria* on the outskirts of Santa Cruz, Bolivia, a booming metropolis. There, you could buy ten cents worth of shoe glue, or *"ore ojare,"* spooned into a little plastic bag. It was an amount useless for actually gluing shoes, but it was precisely the minimum amount required to get high. For an increasingly large number of young Ayoreo-speaking people, this was the chosen means to escape an everyday life of crushing poverty, disfiguring disease, and routine violence. Young bodies joints askance unconscious on pavement in midday sun. They said that *ore ojare* makes you feel "like you can do anything,"

or "like you are Rambo." Like alcoholics or *coqueros*, Ayoreo people said those who become addicted to *ore ojare* have a *vicio*, a vice.

The vice of the glue sniffers was considered relatively benign compared with the affliction of those known as *Puyedie*. These were a fluid group of some two dozen Ayoreo who lived hidden in the tall grass of an abandoned lot behind the train station where they supported themselves with sex work of the most marginal kinds. The grass was their only shelter when it rained, the only source of privacy when they were with a client. It was a scene of winding paths, tattered blankets, sunken cheeks, swollen limbs, plastic bags. Their clients were from the poorest stratum of *cambas*, or non-Indigenous residents of Santa Cruz. Sometimes the *Puyedie* earned up to three dollars per client, but some paid in drugs. Others, including police officers, did not pay at all. The group included a rotating cast of men, who shared the earnings of the women and might intervene if someone was beaten or raped. "We earn well," one *Puye* said in a series of remarkable interviews with Irene Roca Ortiz, a Bolivian anthropologist who directed the first Ayoreo public health project in 2010–2012.[1] "But it is only enough for the coca paste."

Puyedie were largely defined by their addiction to smoking coca paste, an unrefined mash of coca leaves, sulphuric acid, and kerosene, gasoline, or benzol. Significantly, this paste was known in Ayoreo slang as *Puyai*, or "that which is taboo." Prior to contact, the word *Puyai* referred to the set of moral prohibitions established through the originary differentiations of humans and animals recounted in *adode* myth narratives. In the aftermath of world-ending violence and the upheavals of contact, the category of *Puyai* was evacuated. It was later applied to intoxicating drugs as well as to the domains of the past considered profane and immoral—all those practices considered to be offensive to the Christian God *Dupade*.

The mind-altering substance called *Puyai* delivered as much cocaine to the bloodstream as crack, along with additional toxic chemicals. It was just as addictive as crack, and one small envelope cost approximately fifty cents in 2010. The increasing traction of *Puyai* among urban Ayoreo youth was part of a wider regional trend, in which hundreds of thousands of street children and urban poor in Latin America were transformed into a consumer market for the waste by-products of cocaine manufacturing.[2] "The first time you try it, you never want to quit," as one Ayoreo woman put it. "You just want more, more, more."

The vice of the *Puyedie* was not restricted to the drugs. It also included eating dirt and the bricks that lined the bottom of open sewage ditches: "It is another of our vices so those bricks taste good to us." "They had piled a lot of pieces over there and each time someone passed by that place they grabbed some of those pieces with the black things, the sewage and they ate it."

One *Puye* woman said that it was the brick eating that led her to the coca paste smoking, rather than vice versa. She tried to hide it from her husband. "Then I would remember and try it again and I liked it again. . . . I ground the bricks into powder and I ate it, my mouth was full of it and one day he saw me and spoke to me but I couldn't reply because my mouth was full of dust. So he beat me." They said that some of the women forego any other kind of food and eat so many bricks that they turn yellow. "When that one defecates," they laughed, "it is pure brick that comes out." The life of the *Puyedie* is prohibited life.

The Limits of Transformation

The Totobiegosode *urusori* and the urban *Puyedie* bracketed the space of Ayoreo moral life, even as they made it newly eligible for extermination. For mainstream Ayoreo, such madness and vice marked the limit of transformation, the interruption of moral humanity, its failure to coalesce. They marked the end of the ideal self.

For those afflicted, it may well have been precisely the opposite. Madness and vice were radical forms of negative becoming. They turned the ontological murk of the colonial situation and its constitutive ties to death into the terms for self-transformation. For outsiders, these unravelings were merely the predictable evidence of the essential savagery of Ayoreo or what happens when a culture is destroyed—the stripped down, devalued core of biological life that remained, a dangerous and annoying and disgusting residue, begging for coins on the sidewalk and picking through trash and watching from the brush.

In these nonlinear entanglements of madness and culture, vice and becoming, life and death, the general Ayoreo project of becoming crashed against the instrumental incoherence by which colonial order was fused with the disorder of violence and then reduced not to economic rationality but to a space of death for Indigenous subjects, a process itself rationalized by the ways that difference snapped back into focus through this process only as a radical cosmological alterity that was external to modernity or the colonial situation itself and thus eligible for consumption. Contemporary politics, as well as colonial ones, were instantiated through the production and cannibalism of this difference. Yet through these extreme Ayoreo formations, such ontological murkiness was put at the service of contesting the limits of the space of death, of breaking and twisting them to their own purposes, and perhaps, even, of trying to transcend them.

Culture as a Politics of Life

Such Ayoreo life projects of madness and marginality provided stark counterpoints to the general empowerment of Indigenous peoples believed to characterize the era of "post-multicultural" citizenship in Latin America, and Bolivia especially. According to many scholars, the active legislation of cultural difference at state and international levels meant that the present was defined by the widespread empowerment of Indigenous peoples. Such empowering potentials, it was argued, arose from broad social movements through which Indigenous groups gave voice to a populist, grassroots "demand for a democratic government designed by the people themselves."[3] The recent anthropology of Latin American indigeneity was exciting precisely because it described how the political upheavals set into motion by 1990s multiculturalism—"an emancipatory development of world historical significance,"[4] "in which the very meanings of citizenship and the state were rethought"[5]—continue to evolve toward a long-anticipated end: a state unwillingly restructured or entirely overthrown by the creative protagonism of newly empowered "cultural citizens." Granting Indigenous peoples the right to culture was believed to reorient and revitalize democracy for all citizens. This horizontal imaginary allowed proponents to argue that the result was a reformed state whereby "the exclusions of the past will no longer go unnoticed."[6]

Yet increased political participation has not lessened socioeconomic inequalities for all Indigenous peoples in Latin America: it has unevenly redistributed them. In fact, disparities in health and wealth between Indigenous and non-Indigenous citizens in Latin America have either stagnated or increased over the last two decades of multicultural reforms.[7] This dialectic between exclusion and inclusion is characteristic of how contemporary politics are instantiated through competing moral economies of Indigenous life.

What has received less attention is how the obvious social apertures and logics of redeployed culturalism depend upon and create new regimes of what Didier Fassin calls biolegitimacy, or the unequal construction of the meaning and values of life.[8] For Fassin, contemporary biopolitics are best defined not so much as technologies for normalizing and controlling living beings but as the recent creation of sharp inequalities within "life as such." These unevenly felt bioinequalities instantiate how governance pivots upon a new politics of life, whereby the pursuit of its ideal definition also means deciding "who should live and in the name of what."[9] Who, exactly, is allowed to live in the name of culture? Who is allowed to die? What is at stake when guarantees of cultural citizenship become indistinguishable from the biological components of citizenship for certain historically oppressed peoples?[10]

It is obvious that some peoples and their descendants, especially those like Ayoreo long imagined to border the human/nonhuman, cannot or refuse to enact their physical and psychic alterity in ways that conform to the new criteria of difference valued within post-multicultural politics. They are those whom Clifford Geertz so memorably called "ex-primitives": ambiguous, unruly beings whose ties to legitimating origins are rendered impossible, unreliable, or newly suspect through active global investments in preserving primitive lifeways as a public good and radical imaginary.[11]

Such investments paradoxically authorize an amplified regime of violence against the supposedly deculturated ex-primitive. Those not eligible for the protections afforded to the cultural subject were also denied the rights of liberal citizenship.[12] From this vantage, it was no longer a mystery what sort of place in the world ex-primitives have: the zone of targeted marginality reserved for those populations who refuse or are denied the degree and liberally sanctioned kind of culture deemed necessary for becoming intelligible as fully human, Indigenous or otherwise. For officials, scientists, and citizens alike, the "whatness" of such groups increasingly dissolved not into performances of marginality aimed at securing rights but into the terms of affliction and death.

Genealogy of Madness and Vice

During my fieldwork, many Totobiegosode suffered from attacks of *uru-sori*. One man kept ripping off his clothes, trying to grab his spear, moaning that he had to kill the *Cojñone*. Another woman in Arocojnadi told me that it was the heat of a fever that made her want to cast off her clothes and run back to the forest. "My head was just too hot." In Bolivia, I heard similar stories from the early 1980s, such as the middle-aged man in Zapocó who woke up one day, stripped off all his clothing, grabbed his spear, and left. On his way out he told his wife that he was possessed and had to return to the forest. He wandered alone for several months before being shot and killed somewhere in Paraguay, or so the story goes.

I was told that *urusori* itself had undergone a fundamental change. Prior to contact, it was attributed to transgressing the *puyaque* moral prohibitions of ritually powerful spirit beings, particularly *Poji*, or Iguana. In the past, a person struck with *urusori* was thought to mimic and acquire the traits of *Poji*. Like an Iguana, the afflicted ones would run to the forest, sleep in the daytime in a hole they dug underground, and eat the raw, uncooked foods favored by Iguana. They acquired superhuman capacities to fly through the forest, appearing first one place then another far away. More than anything else, they were defined by fear, their animallike compulsion to run from or instinctively attack their own people, their terror of fire, their compromised language. In other words, those afflicted by *urusori* lost the defining core of their moral humanity. In the absence of their *ayipie*, they became nonhuman beings.

The story of one such possession goes like this. Once, *Poji* took away the *ayipie* of Cojnoquedi'de. He ran from a *Cojñoi*, then lay down on a piece of wood to sleep and had a bad dream about *Poji*, and that is when his sickness began. *Poji* came in his dream and took away his *ayipie*. He went alone to the forest with his wife and hunted at night. He could see at night because of his affliction. He said, "We cannot go back because the others are going to kill me." His wife said, "They aren't going to kill you." But he didn't pay attention to her. His people came looking for them and found his wife alone. They followed his tracks. Very far. But his tracks ended because he had flown through the air. Maybe with the help of some little birds, some doves. Then they heard a yell very far away.

That's when they knew he was no longer human but *uruso*. The next day they found him under the ground, buried in a hole, and they pulled him out. He grabbed his spear and tried to kill his own brother. They grabbed his spear but couldn't take it from him, even when they were all working together. Then he fell and he died. He didn't lay like a person but he lay bent like a *Poji* on the ground. They did many *ujñarone* but he was no longer breathing. For hours they left him there. When they looked again they saw one vein begin to move. Later his body began to move, and he started to breathe. They spoke to him but he did not answer. He was no longer a person but an animal, a *Poji*. His body was yellow. The people waited. Suddenly he jumped up and took off running. They followed him. His brother caught him and tied him to a tree and healed-by-blowing. He used strong *ujñarone* and wooden crosses. And finally he was cured. He was cured and then he became angry at the *Cojñone*, who *Poji* had used to take away his *ayipie*. When he was better, he went to kill the *Cojñone*. There is a place by Chaidi where he killed some of those *Cojñone* later.

In the upheavals of contact, cases of *urusori* proliferated. In 1963, the visiting scientist (and pedophile) D. C. Gajdusek noted—following an epidemic of deadly fever that killed one-third of the Ayoreo group—the cases of "two youths and one man who were in delirium and later with strange disoriented behavior ran off into the forest where they have presumably died. They had delusions and a strange manic fright behavior pattern."[13] Ayoreo-speaking people tell of many others.

Yet those afflicted with *urusori* in *Cojñone-Gari* experienced different symptoms than for the same affliction prior to contact. Post-contact *uruso* did not act so much like Iguanas or Nighthawks but like themselves and their ancestors prior to contact. Their humanity was not compromised by violating the taboos of a single spirit being but by residues of the peculiar humanity/nonhumanity instantiated within their incompletely transcended pasts and embodied in the flinches and memories of their flesh. The other Totobiegosode explained that Pejei's sickness was not caused by his *ayipie* vanishing or hovering in an indistinct state but by its returning to the past. If he ran out of medicine, his *ayipie* left his body and returned to the *cucha bajade*. It refused to stay focused on the present and the future, "*Chi ayipie echaiji cucha bajade, chajesa daquigade. Payipie que chicaji piquei.*"

Like Pejei, those afflicted by *urusori* turned against the trappings of moral life in *Cojñone-Gari*. Like Pejei, they attempted to destroy all traces of *Cojñone* around them, particularly the symbols of contact. They tore their clothes, they broke their dishes, they refused to eat the foods of the whites, they always tried to run to the forest. They were usually restrained by being tied to a tree or a post. Ayoreo people said that an *uruso* acted "like an animal," precisely because the symptoms bore a striking resemblance to past forms of human life that were later considered deeply profane.

Broken Lines of Flight

One thirty-three-year-old *Puye* woman, Rosy, began huffing gasoline in her early teens while living on an evangelical mission near Santa Cruz, a bustling metropolitan area of more than two million inhabitants. Like many Ayoreo people, she moved to an Ayoreo squatter camp on the outskirts of the city in her early teens, looking for a temporary wage-labor job. There, she switched from gasoline to shoe glue but quit after it caused her to abort a seven-month-old fetus. While in Santa Cruz, she often stayed at the place called the Casa Campesina, built by an advocacy NGO and intended as a temporary lodging place for Ayoreo-speaking people during their visits to the city.[14] At Casa Campesina, she learned how to smoke coca paste from other Ayoreo. At the time, her mother was away working on ranches. When she found out that Rosy had become a *viciosa* or one living "the life of vice," she began to cry. "She told me that she had heard that vices will kill me, that they are bad, that they are against God. That is why she began to cry, she began to feel bad all over." Her older sister tried to make her quit by tying her up with rope and beating her. "She beat me badly with a piece of wood but I am strong and I did not pay attention to her."

Rosy told of passing out and waking up in strange places, snatches of incoherent conversation, bribing police with money or sex, and a life defined by violent confrontations with Ayoreo and *Cojñone* alike. Rosy could not leave anything in their camp because the other *Puyedie* would steal it. Many of the *Puyedie* gave themselves the names of animals, like *Víbora* (Snake), *Caballo* (Horse), *Por-si-acaso* (Just in Case), *Salvaje* (Savage). Rosy said that she knew her vices were dangerous: "The vices kill us." Many of her friends had already died from their vices, more than she could keep track of: "More than ten, more than twenty, I don't know." She herself was frequently sick these days, from what she didn't know: "When you are a *viciosa*, any illness will grab you, you know." Even so, she said she wouldn't give up her vices: "They are sweet, *unejna*, to me." She said if she was taken somewhere else, away from her vices, she would return. Over the course of the interviews with Irene Roca Ortíz, the stories unraveled into disaggregated fragments: "She was jealous of her husband." "Snake." "I would fuck an animal or the Devil." "Who are you?" "Here comes Horse." "Let's go." Laughter.

Rosy emphasized that her deadly vices gave her life. She said that without her vices, she became more like an animal. She said she could not leave them: "You'd better tie me to a tree," she smiled, "or I won't stay." The same coarse rope is used to tie down all those afflicted with the *urusori* of madness or vice, to keep them from running off to an alterity at once inhuman and legitimating. And it is the same knot that always slips.

This slippage is comparable to what Gilles Deleuze, as part of his efforts to understand drug causality, referred to as the "lines of flight" constituting a social that is constantly escaping or leaking out in all directions: "The drug user creates active lines of flight. But these lines roll up, start to turn into black holes, with each drug user in a hole, as a group or individually, like a periwinkle. Dug in instead of spaced out."[15] João Biehl extends this insight to the analysis of pharmaceutical subjectivity in general, as a "continuous process of experimentation . . . an art of existence and as a material and means of sociality and governance" whose study "recasts totalizing assumptions of the workings of collectivities and institutions."[16] Such conceptualizations envision drug consumption as "a combined chemical / intimate / social / economic matter" that always bridges the psychic and the political-economic.[17]

Likewise, the "disordered subjectivities" of Ayoreo-speaking *Puyedie* or those possessed by the spirit of the past should not be understood as psychopathologies produced by the disintegration of culture, but rather, as radical forms of immanence that simultaneously instantiated and subverted the contradictory meanings and values attributed to Indigenous "life as such" within the contemporary. That is, they were the "dense transfer points" by which global political economies were fused with the most intimate forms of everyday experience.[18] Contemporary politics were instantiated through Ayoreo ontological formations and vice versa.

The stories told by Rosy about chemical lines of flight or by the *uruso* about the flight of his *ayipie* articulated precisely the fractured subjectivities of the ex-primitive, caught as they are between agency and erasure, human and nonhuman, vitality and public death. Yet it was also particularly telling that each kind of negative articulation was only legible to outsiders as the loss of culture and a return to an essential savagery. Considered the biological residues of culture death, ex-primitives were not eligible for the protections newly extended to cultural life. They were again treated as subhuman *Bárbaros* to be exterminated or consumed, as a social problem and a threat to civilized order.

On both sides of the border, Ayoreo faced systematic discrimination and were stigmatized as prostitutes and beggars, reduced to shadows of themselves by evangelical ethnocide.[19] In the Bolivian popular press, urban Ayoreo were described as evidence of what happens when "urban sprawl devours a culture." They were a group that had lost "the foundations for the reproduction of their culture" and whose "cultural fabric has been torn apart in the clash with mainstream society."[20] Or, as Sebastián Hurtado Rodríguez, the deputy governor of one province in the department of Santa Cruz, put it to journalists in 2011, Ayoreo in general were "a defect [*lacra*] of society," a group that "must become useful human material for society and not a visible defect, not a lamentable burden."[21] Few could articulate what kind of force emanated from these perceptions of Ayoreo as a subhuman blemish or burden, but there were times when everyone could feel it, when it built from thin air and unfurled in the atmosphere like a storm or a stench that settled on the tongue. If one walked with any young Ayoreo girl at night down the sandy roads and open sewers, the threat could be palpable. The unsmiling whistles, the mocking shouts, the ones who followed silently in the shadows.

The violence against Ayoreo-speaking people continues even while recent political developments in both Paraguay and Bolivia are often presumed to have ushered in a new era of social movements and Indigenous rights. But those tainted few not eligible for the protections afforded to "the Indigenous" are still considered *Bárbaros*. The local idioms of savagery and barbarism, long used to justify the slavery and genocidal extermination of Ayoreo-speaking people along the Chaco frontier, come back into focus on Ayoreo to the degree they are believed to embody an essential lack of social personhood. Being shorn of politically authorized culture does not only remove Ayoreo-speaking people from history; it strips them of humanity.

It is no coincidence that the "post-multicultural age" is marked by amplified violence against Ayoreo-speaking people, and young urban-dwelling women in particular. The teenager cut from vagina to chin in one long rip. The girl whose dismembered corpse was thrown into a vacant city lot in Santa Cruz. The young woman whose intestines were cut out and strung for ten yards along a fence line. The fifteen-year-old girl stripped naked and gang-raped by rich teenagers from the city who took pictures of it on their cell phones. The five-year-old girl sexually abused and beaten to death in a small Bolivian town. Two years later her fourteen-year-old sister, a part-time sex worker just elected as the 2012 "Queen of Carnaval" by the Ayoreo *barrio* in Santa Cruz, was found stripped, strangled, and mutilated in the trash behind a bar. Her corpse was only recognizable by the homemade tattoo of a heart on her arm.[22] (Indeed, some young urban Ayoreo say, with a laugh, that they get tattoos precisely so their corpses will be recognizable.)

Hypermarginality

The increasingly robust protections afforded to culture as a collective right in Latin America coincide with a legal definition restricted to something like a Herderian notion of bounded, stable difference or even cosmological alterity, as in recent legislation of "uses and customs," *usos y costumbres*, in Colombia; "particular forms of life," *formas peculiares de vida*, in Paraguay; or egalitarian administrative structures in Bolivia.[23] Together, such developments constitute a culturalization of legitimate Indigenous life—a neocolonial reduction whereby the state gains new moral authority to police Indigenous populations through granting the

ts to a symbolic cultural citizenship. Social movements aimed at se-
ng cultural citizenship while not contesting these limits paradoxi-
...y strengthen the denial of the humanity of those collective Native
subjects deemed "deculturated."

The result is that Latin American indigeneity today is a particularly
disjunctive field, comprised of minimally four, causally linked and simultaneous elements, including: a government of biolegitimacy realized
through a malleable category of culture that is increasingly robust but
applied with ever-narrrowing precision; the invigorated political agency
of those globally networked Indigenous populations able to successfully
claim culture by conforming in part to externally imposed definitions
thereof; the transnational hypervisibility of "isolated," "unconquered,"
or "traditional" primitive life; and the sociospatial relegation of supposedly "deculturated" ex-primitives to devastated hinterlands and the
margins of civic space, where the stigma of culture death is added to the
already trebled stigmas of race, place, and class.

This suggests that the widespread empowerment of authorized Indigenous subjects is now predicated on the *hypermarginality* of others:
those who do not fit within the increasingly policed matrix of cultural
life while also remaining at the very bottom of local socioeconomic class
hierarchies. Hypermarginality can be defined as a novel regime of social
depersonalization and structural violence deriving from the instrumental
conflation of politically authorized culture and Indigenous biolegitimacy
across distinct political domains. This amplification manifests the kinds
of politics emerging when a limited schema of cultural difference stands
in for the sanctity of life as a core moral value within secular democracy.
This redistributes bioinequality and the material techniques for dispossessing certain stigmatized groups, while co-opting those who benefit
into universalizing discourses of empowerment and putting all at the
service of refashioned regimes of governance.[24]

Indigenous hypermarginality occurs in the context of proliferating
regimes of sociospatial or scientific isolation and in the growing forces
of an exterritorializing containment around the world.[25] Such spreading forms of exclusionary closure contradict the common notion that
the present is best defined as the "Network Age," or as a rising epoch of
techno-political possibilities facilitated through horizontal attachments
and the democratizing breakdown of barriers of all kinds, from state to
species. Rather, we confront an increasingly polarized world. Here, the
idea that the present is defined by horizontal networks may well appear
as the fetish or emblem of the liberalism of a privileged few even as the
durable denial of association defines the lives of most. In this case, culture

reappears as a crucial medium for the stratifying craft of the neoliberal state.[26]

The hypermarginality of Ayoreo-speaking peoples in the Gran Chaco was maximally expressed among the *urusori* and the *Puyedie*. It was hard to imagine more thoroughly animalized beings, inserted as they were into contradictory moral, political, and economic orders through an unstable personhood of death. Like those stricken mad by a frightening encounter with *Cojñone* or dark spirits, the dehumanization of the *Puyedie* was a legible experience of negation for other Ayoreo. Both forms of *urusori* were lines of flight that organized vital contents in a recognizably nonhuman way along the axes of the ontological murk of the colonial situation, through rupturing rupture itself. Yet even these spiraling lines of flight were made brittle by the oppositional kind of negation reserved for the deculturated indigene within post-multiculturalist society.

Through their bouts of madness and their public death, profound contradictions between the "negative citizenship" of deculturated life, the resurgent moral economies of primitivism, and the figure of a savagery that must be sacrificed were uneasily reconciled.[27] The result for the hypermarginal subject was an amplified political erasure without—at that time—the possibility of magical powers or cultural revitalization or biomedical salvation or a healing reincorporation into market productivity. Rather, their routine animalization was a negative image by which neoliberal politics were instantiated through the fusion of culture and legitimate life.

The Neoliberal Fusion of Culture and Life

Colonial authorities, of course, have long deployed culture as an ideal through and against which valid Indigenous life is delimited.[28] As Michael Taussig describes, in Latin America this took the form of an "epistemic murk" through which colonial drives to create, extirpate, and ultimately enact savagery bent the magic of primitive alterity to the space of death itself erased by metanarratives of political-economic rationalities.[29] Early twentieth-century efforts by Latin American governments to solve the "Indian Problem" with forced acculturation programs arose simultaneously with the most grotesque forms of public extermination. If genocidal sacrifice failed, then stripping Indians of their less evolved cultures was believed to be a necessary first step for exposing them to the superior culture of rational modernity and integrating them into productive relationships with the nation and the market.[30] So was the

Indigenous rights movement birthed in blood and instrumental incoherence as well as the best of intentions.

It is perhaps not surprising that these foundational efforts to conceptualize Indigenous rights insisted on a tripartite scheme. They distinguished populations descended from ancient empires in the Andes and Mesoamerica from "aboriginal forest-dwellers" or primitives. They also proposed a third group: "marginals," or those "who find themselves placed halfway between two strongly different cultural *milieux* and whose main characteristic lies in an incomplete adhesion and participation both in the national and the aboriginal milieus."[31] That is, the most degraded Indians were not those with undeveloped cultures but those believed to have no culture at all. What is distinct about the contemporary moment is that the sociolegal logics of cultural diversity invert assimilationist schema while intensifying the violent marginalization of ex-primitives. Through the logics of culture, the terrifying murk of a colonial death space premised on consuming Indian flesh and souls is both disavowed and extended.

The marginal living conditions of Ayoreo-speaking people, of course, were not new either. Much like the disordered subjectivities of Ayoreo-speaking people, they reflected long histories of slavery, genocide, dispossession, and displacement. Yet recent realignments of governance, market, and citizenship amplified these preexisting inequalities to the point whereby those excluded from the matrix of culture were no longer deemed worthy of the same kind of life, if they were worthy of any life at all.

Disintegration and death defined the lives of supposedly deculturated ex-primitives—a process equally obvious in urban peripheries and former wilderness zones, now bulldozed, where state authority has long been precarious and arbitrary. This occurred against a backdrop of familiar and well-documented trends typical of late or neoliberalism, including increasing disparities in wealth and health between Ayoreo people and their non-Indigenous counterparts, the loss of viable ancestral territories to rampant agro-industrial expansion, internal social fragmentation, diminishing sources and opportunities for wage labor, a general disconnection of Ayoreo from macroeconomic trends, and social stigmatization. What was new at the time was how these trends were at once amplified, naturalized, and disguised through state guarantees of cultural rights. That is, the political governance of culture redistributed and fundamentally changed the nature of Indigenous marginality. This occurred through new linkages being made between authorized culture and legitimate Indigenous life.

As Loïc Wacquant has argued, the anthropology of neoliberalism is roughly divided into two approaches: "a hegemonic economic conception anchored by variants of market rule, on the one side, and an insurgent approach fuelled by loose derivations of the Foucaultian notion of governmentality, on the other."[32] The first approach to neoliberalism imagines a state retreating or withering in the face of an expanding market. The second approach defines neoliberalism as a fluid, transnational set of rationalities, calculations, or technologies that reorder the conduct of the governed themselves according to widespread appropriation of market logics of competition, efficiency, and use.

The notion of neoliberal (post-)multiculturalism, as developed by scholars of Latin American indigeneity, is appealing precisely because it promises to synthesize these two academic models. Thus, Nancy Postero argues that neoliberalism reveals "the state as an inefficient, often corrupt actor that only encumbers the market's neutral and unselfish actions."[33] At the same time, she suggests "the bottom line is that successful neoliberal subjects must govern themselves in accordance with the logic of global capitalism."[34] Here, the ascendancy of a neoliberal order is reflected in both the erosion of the state and the creation of subjects who govern themselves according to neoliberal rationalities.

Yet the neoliberal governance of Indigenous subjects is often presumed to subsequently follow an exceptional trajectory—post-multiculturalism—because of what is believed to be an inherent antagonism of "the Indigenous" and "the cultural" to "the neoliberal." Thus the present can be described as an era characterized by "a new form of protagonism that both incorporates and challenges the underlying philosophies of neoliberalism," through social mobilizations that "push to make [neoliberal state] institutions more inclusive."[35] In such models, the Indigenous appropriation of neoliberal logics explains the increased traction of broad social movements, insofar as this appropriation is imagined to create a "post-multicultural alternative" and "anti-neoliberal state" characterized by inclusion. Or, as the preamble to the 2009 Constitution of Bolivia puts it, "We have left the colonial, republican and neoliberal State in the past."

The Ayoreo figures of madness and vice suggest the opposite may also be the case. The exclusionary structures and logics of recent years are not always threatened by Indigenous mobilizations, but rather redistributed in response to them. This implies a dual reversal of some conventional conceptualizations of the relationships between the Indigenous and the neoliberal. First, it suggests that the figure of the post-multicultural Latin American state is characterized not by its erosion but by its recasting as the *indirect* manager of constantly shifting borders of life and death, human and nonhuman, embodied frontiers increasingly mediated by the polysemous category of culture. Second, it implies that imbuing a categorical feedback loop with the appearance of an intrinsic structural antagonism between the rational neoliberal and the cultural Indigenous is precisely what allows governance and violence to coalesce around a robust set of cultural rights and the increasingly thorough dispossession of certain stigmatized Indigenous populations at the same time. Reformed institutions do concede rights to a cultural citizenship, but in doing so, they gain new authority to enforce the boundaries of what does not count as legitimate or moral Indigenous life. Moreover, it calls attention to the fundamental incoherence and terror at the center of legal order.

The fact that structural inequalities and familiar exclusions are not disappearing but deepening for stigmatized populations is predictable if neoliberalism is approached not as the successful dismantling but as the strategic re-engineering of government, terror and violence and the forms of hierarchy and difference they require, in which the tenets of liberal citizenship remain the exclusive purview of a privileged few and become the terms of social death for many. In Latin America, this means that neoliberalism appears as a political reconfiguration of the moral value and practical limits of a kind of Indigenous life it both creates and consumes. What makes post-multicultural indigeneity a distinct formation of late liberalism is how this redistribution of the value and meaning of life does not primarily gain traction through the figure of market rationalities internalized by individuals but through their enshrinement in a politically retooled notion of culture. Culture here figures as a disjunctive matrix of subjection and dismemberment that is coproduced simultaneously by the state, nonstate political actors, and a transnational moral economy in which the cultural life of Indigenous subjects is indistinguishable from the legitimacy of that life.

This culturalization of legitimate Indigenous life occupies an entire global industry. It is consolidated through its outsourcing to what can be described as a global "culturalist humanitarianism," organized by NGO networks, funded by charity, and concerned with preserving the sanctity of cultural life. Cultural life, however, exists only as a collective and not as an individual life. Thus, cultural loss is given a greater moral weight than physical death; the sanctity of culture is privileged over the sanctity of bodily life. Structural violence against those considered to be insufficiently cultural is then glossed as an Indigenous failure to resist or an inability to fully comprehend their own origins. That is, the sociological conditions of violent dispossession are mistaken for ontological degradation and the "distorted characteristics of the victimizer" are imputed to the victims.[36] The instrumental incoherence upon which colonial violence depends is disguised and expanded.

According to these culturalist logics, the need for a moral remediation of ex-primitives spurs movements to safeguard a culture neglected or imperiled by its former possessors.[37] This is a colonizing operation that is crucial for the accumulations and expenditures of contemporary humanitarianisms and the emergence of culture as both a regime of biolegitimacy and a political theology animated by giving mass death the sanctity of life itself.[38] Culture is redeployed not as an empirical reality in need of a more precise catalogue of its contents or a more effective policing of its boundaries, but as a sustaining metanarrative of a governmental system whereby the tenets of democratic liberalism remain the exclusive purview of the privileged few, even while increasing inequality and amplified forms of dispossession define the lives of most.

Negative Immanence

Yet the peculiar play of order and disorder—of knowledge and nonknowledge—upon which culturalist violence depends is also generative of new relational ontologies among those it targets.

It is perhaps telling that these disordered Ayoreo ontological responses are not aimed against the logics of culturalism (yet) but at the epistemic murk of a revitalized colonial space of death. At this extreme convergence of exclusionary politics, madness and vice, the sensibilities of drug delirium and moral failure coalesce into something like an emergent Ayoreo formation of becoming. This process posits a new transformation of the human; it presumes that the Christian Ayoreo is not suitable for survival in the conditions of actually existing *Cojñone-Gari*. It takes the moral value of rupture and the capacity for self-transformation to the breaking point and doubles them back upon themselves. Whether it surges from the smoke of coca paste or the terrible exhaust of an airplane, *urusori* is the loss of the *ayipie*, the untethering of flesh and the spirit of moral humanity, the reflection of a colonial image of otherness and the shattering of a colonial mirror of production at the same time.

It is pure transgression, an inverted state of being associated with the savage, the animal, the contaminated, the contagious, the filthy, the abject, the abhorrent, the uncanny, the feared, the enemy, the Other, the past, the repressed, the agent, the spirits of the sun. It is a doubly or triply negative image, a form of negative historical consciousness rendered

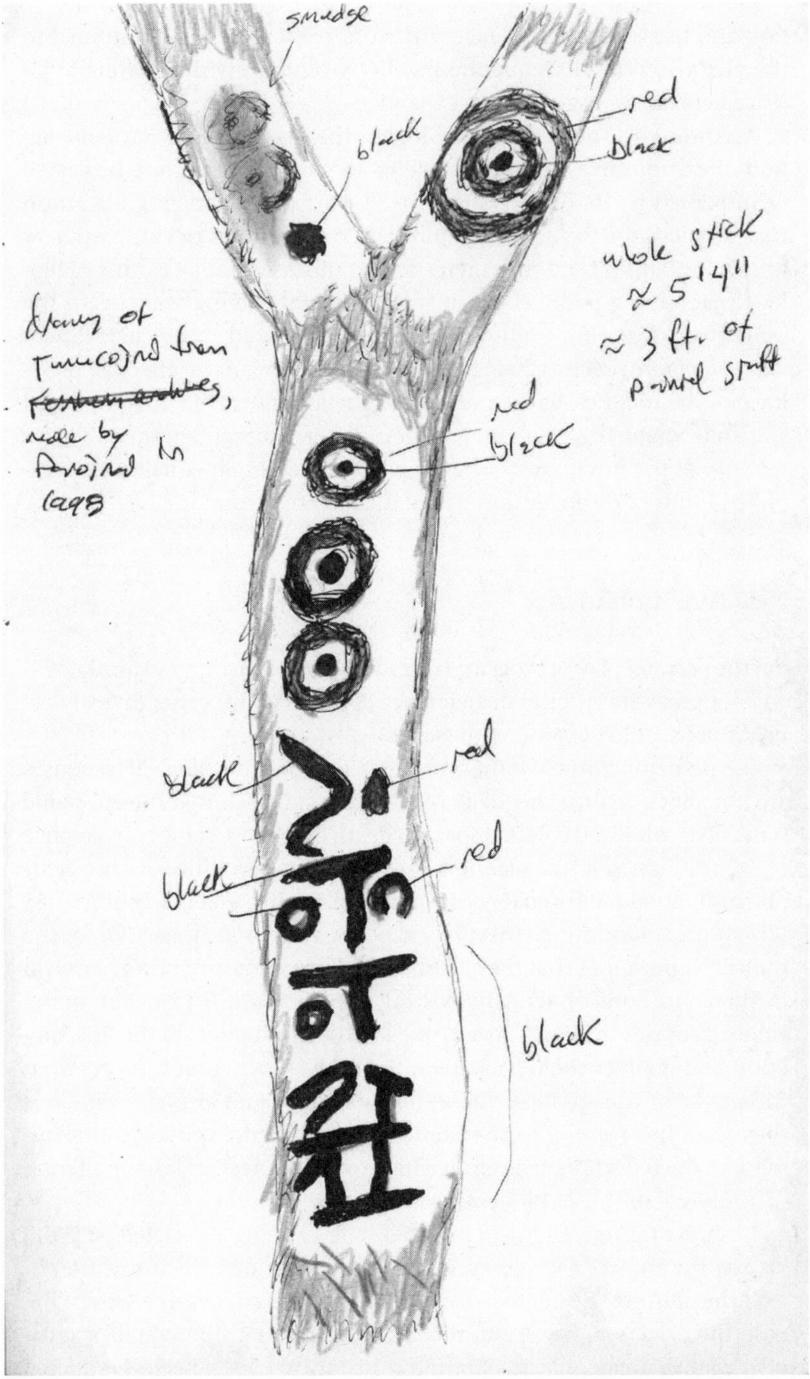
smudge
black
red
black
whole stick ≈ 5'4"
≈ 3 ft. of painted stuff
red
black
black
red
black
red
black

Listen to me
I was young
I was beautiful
I had gold teeth
They killed me
It doesn't matter to me
Here I am

EIGHT

The Politics of Isolation

Isolation: describes the situation of an indigenous people or part of one that occurs when this group has not developed sustained social relations with the other
members of national society, or that, having done so, has opted to discontinue
them. ARTICLE 2, PERUVIAN LAW NUMBER 28.736

Isolation may be the beginning of terror; it certainly is its most fertile ground; it
always is its result. HANNAH ARENDT

In northern Paraguay, the few Totobiegosode holdouts hidden in the forest are palpably present. Teenage soldiers warn
travelers to take care; the savages are everywhere. "They
aren't like you and me," a park ranger near the Bolivian
border told me in 2007. "They can be anywhere, we cannot
know." During my fieldwork, two ranch hands told me they
often went about their work armed. "You never know when
a savage Indian might attack you, no?" Pale men in SUVs
traveled the backroads inquiring after tracks and sightings.
One rancher told me he could always tell when the savages
were near, because the dogs acted up like they smelled a
wildcat or a storm. The concealed Ayoreo lurked just out of
sight, on the edges of wasted pastures, where the dust from
the heavy trucks drifted and rolled. The last wild Indians
of the Chaco, they assumed the same "nowhere-tangible,
all-pervasive, ghostly presence" of the sacrificial violence
destroying their homelands.[1]

They are just as real as they are fantastic. There are at least
two Ayoreo bands and several lone individuals roaming the
shrinking forests. One band, of unknown size and origins
(most likely a remnant Tunupegosode population), moves

in a great arc across the Bolivia-Paraguay borderlands around Echoi. The second is a Totobiegosode band led by Jotaine of the Picanerai clan, Jotai's father. In many ways, this band of eighteen people is a mirror image of the Areguede'urasade. They lived together as a single group until 2001. They are closely related. Among this small band are the Areguede'urasade's siblings and parents. The Jotaine'urasade also include children once imagined as future spouses of the children of the Areguede'urasade, to whom they were carefully matched by age, gender, and clan. As Siquei and the others once had, they pursue a life of nomadic concealment. They have structured their lives around the daily logistics of eluding starvation and death from the beings they think are pursuing them. Like the Areguede'urasade, they are keenly aware of the invaders pressing in from all directions. This begs the question: how do we make sense of these concealed people and their remarkable way of life? And how have our interpretations of their lives become inseparable from the fate that awaits them?

The Expedition

In November 2010, international attention was briefly focused on their plight. This attention took the form of a controversy over a scientific expedition to the northern Gran Chaco proposed by the London Natural History Museum. The expedition was to be comprised of sixty ecologists, biologists, and other experts on nature, and its aim was to document the biodiversity of the Chaco forest, a region described in a British newspaper article as "one of the most inhospitable, impenetrable and mysterious places on Earth."[2] The expedition would have been the first to quantify the biodiversity of this understudied area.

The British expedition was harshly denounced by Iniciativa Amotocodie, an NGO self-described as the "Isolated Peoples Protection Group" that claimed to be the legal representative of all the concealed Ayoreo-speaking groups in the area. The Cambridge-educated director of this NGO declared that the expedition was equivalent to an act of genocide against the so-called isolated Ayoreo. As he put it during a November 9, 2010, interview on BBC Radio 4, "It would be tantamount to genocide if an involuntary contact actually occurred, which would mean that there could be fatal consequences on both sides and the life-model of these people would break down, would collapse, and also the territory they belong to. This is tantamount to a genocide-like situation." Shaken by such accusations, government and museum officials suspended the expedition less than a week later.[3]

The relation this media event claimed—between the gathering of scientific facts and genocide—rested on the assumption that the concealed Totobiegosode were pure if fragile Others who "live in another world."[4] Their exceptional alterity, according to this widely circulated argument, was derived from two sources. First, their bodies and souls were believed to be inseparable from certain threatened domains of nature. As the NGO director put it in his BBC interview, the Ayoreo "live in complete interdependence with nature . . . in a great extension of completely virgin forest." Second, they were imagined to be the bearers of an uncontaminated culture that has not yet been "eroded" by contact.[5] The director described this conflation of pure culture and pure nature as "a principle of life."

Anthropologists and other theorists, of course, have long critiqued the primitivist trope of Indigenous populations that exist beyond "contact," history, or social relations as central to the logics of colonial domination.[6] Scholarship has shown that this trope of "uncontacted primitives" is an enabling principle for naturalized inequalities, structural violence, or imperialist nostalgia; a justification for the ongoing dispossession of Indigenous populations and the pathologization of local forms of knowledge or social memory; and a political field in which the positivist pretensions of anthropological expertise may be uncritically played out.[7] Difference, we now presume, is the result of longer *dureés* and wider relations.

Despite exhaustive ethnographic evidence to the contrary, the well-traveled fantasy about a form of cultural life conserved beyond the limits of modern society persists in rising again and again. From blockbuster films like *Avatar* and "first contact" tours in West Papua to recent UN human rights initiatives and best-selling books and the YouTube sensation created by aerial photos of remote Brazilian tribespeople, the figure of the isolated primitive is an increasingly powerful global imaginary. While the discipline of anthropology has moved on to concerns that it considers less problematic and more pressing, its trenchant critiques may register in different ways or not at all in the realm of popular politics.

Despite scholarly arguments, the notion of such "uncontacted," "unconquered," or "unreached" humankinds has been given new force within the political formations that ostensibly define the contemporary limits of legitimate life.[8] Today, political norms, moral arguments, and infrastructures of protection are being organized around the pressing imperative to police the boundaries of "voluntarily isolated" life. Those rallied to its defense are evoking subjective horizons, managerial logics, and human contents based on the urgency of preserving its imperiled form. Yet the moral defense of isolation as a principle of life articulates precisely the contradictions that occur when cultural preservation becomes

indistinguishable from the biopolitical governance of Ayoreo-speaking peoples. How is the category of isolated life brought to bear against Ayoreo humanity? What is at stake in this process for settled Totobiegosode and their concealed relatives?

Ambiguity of the Hidden

As part of my fieldwork, I spent two years working alongside the first tribal organization of the Totobiegosode, formed in 2005. The name chosen for this institution reflected its futuristic orientation: the Organización Payipie Ichadie Totobiegosode (OPIT), the Organization of New Totobiegosode *Ayipie*.[9] During this time, it was impossible to ignore the real social force that the fantasy of the isolated Indian exerted within and through Ayoreo life. It was a national legal category, a global framework for Indigenous rights, a form of moral reasoning about cause and effect, and a rallying cry for several competing NGOs. Yet it was difficult, if not impossible, for me to recognize this hypervisible "isolated" Ayoreo subject within the realities I encountered on the ground. I struggled to understand the tense gaps between the increasingly powerful category of isolation and the profoundly ambiguous Ayoreo sensibilities about the concealed groups.

To be sure, they were a matter of daily concern. Rumors about the forest people blew through the Ayoreo settlements like dust. They were cornered on this ranch or seen over there or shot at near here. "How much money," the leader of a Guidaigosode settlement once asked me, "do you think the NGO would pay if we captured them? Enough for a pickup?" Someone found their tracks and tried to follow. A man heard them whispering at dawn, invisible in the brush near his garden. They must have taken that lost bag of seeds or that one red shirt. They must be close, they must be coming back. People waited for them, but each time they slipped away.

The first time Ayoreo asked me to organize and lead an Indian hunt, I thought it was a joke. It was the summer of 2005 and I had stopped by for a short visit to my old Direquednejnaigosode friends in the urban camp on the outskirts of Santa Cruz, Bolivia. We sat in front of my adopted family's mud hut near the entrance, and anyone interested came over to talk. Two dozen people soon formed an irregular circle of makeshift chairs. Everyone knew I had spent time in Paraguay with the Areguede'urasade, and they were full of questions: how did they hunt? How did they look? Did they speak an archaic form of Ayoreo? Were they truly wild and mean? Were there others still in the forest?

Night fell too soon, music blared, and my hosts urged me to leave before several young men grew drunk and dangerous. As I made the round of handshaking, a strong-boned man in his early fifties named Beruide gripped my hand hard and stood up.

"Lucas," he began in a formal mode of address. "Take this." He reached up and hung a small wooden *potá* whistle around my neck.

"It is good that you are working with those Totobiegosode in Paraguay. They are bad and ignorant. They fear *Asojná* and they worship Satan." The group murmured assent.

"I am going to tell you my *ayipie*. I do not know what you think but I will say what I think anyways. Here it goes. I remember when we contacted Ichajnui in 1977 and it was good. I want you to get a project. I will help you and we will hunt down and capture those mean Totobiegosode in the forest. You will be the leader and I will be paid."

The gathered people looked at me expectantly. I did not know what to say. I tried to avoid a direct answer in the Ayoreo manner, but it came out poorly. When I departed, no one looked me in the eye.

Similar scenes were repeated on three other occasions, like the time when Bill Pencille's former slave boy, then a sixty-year-old man, sought me out and proposed that we hunt down and capture a forest band together. Invariably, such confrontations left me shaken. It was less the irony of being mistaken for an Indian hunter and more how these moments revealed a yawning gap between the reasons that I thought justified my presence there in the first place and the sensibilities of my closest companions.

Christian Ayoreo on both sides of the border exerted a constant pressure on the more settled Totobiegosode to contact their relatives remaining in the forest. It seemed that subjecting others was the only way their transformation into New People could be complete. Siquei and Jotai were especially targeted by Christian Ayoreo and missionaries like Bobby, until Jotai himself began to argue for the need to track down and contact his father's band.

For the Totobiegosode in Arocojnadi, this was a fraught proposition. In 1986, Jochade had given in to such pressure and guided the Campo Loro group to the forest camp. Although his role in that contact ultimately led to the move away from subordination on the missions, others like Dejai and Ducubaide saw it as a deep betrayal. It was a profoundly traumatic event for Jochade. Since then, he was the leader most strongly opposed to any attempt at hunting down the forest bands.

I remember sitting around the fire one cool evening in Arocojnadi when all at once out of nowhere a shiver seemed to run through the older people like an electric shock or a sudden wind that only they could feel. They flinched in synch as if by collective instinct. As if summoned by a terrible puppeteer that had been waiting all this time. Bodies strange and rigid, they tilted away from the firelight, lurched up, murmured half-words, hushed the children. Everyone grew silent, we listened intently to the darkness. I heard nothing, felt nothing but the nervous tension palpable and building. I looked at Dasua, her finger floating in the air, vaguely pointed northwest, where the edge of the forest loomed black and formless.

After ten or fifteen frozen minutes, she barked at the children to go inside, lie down, and be quiet. In the same curt tones, she told me to go sleep right then in the school building and not to come out until dawn. The next morning everything was back to normal. They later said the forest people were close by that night, although no one would say how exactly they knew.

Even during such moments, Jochade never permitted any talk of capturing them to gain momentum. He asked Yoteuoi, the former *daijnai* shaman, to place a carved message stick in the direction where the forest people were thought to be camping, so they would know there were

Totobiegosode and not *Cojñone* living in Arocojnadi. "They should not be bothered," said Jochade. And that was that.

For those who had profited so much from the 2004 subjection of the Areguede'urasade, however, another contact was a more enticing prospect. It meant another opportunity to consolidate authority by managing the flashflood of resources into the village. Once every couple of days in Chaidi, someone sitting around the evening fire would mention that he or she was thinking about the Jotaine'urasade. Others would chime in. Surely they are in such and such a place. Surely they are doing this or that. Surely they are scared. Surely they are close. Then someone would say, I really want them to arrive. They could stop running away. They could learn the Word of God. We should go find them and bring them back. Where did you say they were? We could be there soon.

In my stubbornness, I often tried to argue in favor of leaving them alone. Such tentative statements were usually ignored. Finally, Dejai had enough and one evening he turned to me with a steady stare. "Lucas, don't you want to see them too?" I did not respond. With a mocking laugh, he asked, "Wouldn't you like to come and take their pictures?" The others began to taunt me about my prior filmmaking and my concern for the Areguede'urasade. "When the Jotaine'urasade come out, won't you travel from your country to film them too?"

It seemed the concealed people elicited a sense of urgency in us all, but in profoundly ambiguous and different ways that I could never quite comprehend. Why indeed *was* I so concerned about their fate and why in that particular way? Alternately, what did they mean to their Totobiegosode relatives? Surely they were not only an avenue for status and power. Did they interrupt the new moral self, did they keep *Erami* and its spirits alive, did they reopen and salt the wounds of contact? Or did they make us all revisit past decisions and force us to acknowledge they were not as inexorable and as innocent as they once seemed?

In many ways, the forest bands had already become spirits. Once every two or three months, a story would arrive over the two-way radio that they had been murdered by a rancher. Their bodies buried in a pit, their camp dynamited by an airplane, their water source poisoned. An oil prospector found the corpses of six naked savages lined up head to toe under a tree in Bolivia. A Paraguayan peon confessed that his boss drove him to a massacre site and made him stack the bodies of ten savages in a pile and burn them and he could not forget the smell.

Totobiegosode in both communities took these rumors hard. Three times, I heard women sing the sobbing song of death for their relatives after hearing such news. Another time we made a trip to a place where someone had reportedly found bones. But all we found was dust. The details were always vague, the sources difficult to find. Usually no one pursued the stories; they lingered with others like heat waves.

The Politics of Isolation

There was little room for such contagious and haunting ambiguity within the political mobilizing around the protection of isolated people's rights. The more time I spent in the Chaco, the more blurred and murky the figure of the isolated primitive became. On one hand, it seemed to offer the potent and clear image of Ayoreo victimhood required for effective political advocacy. On the other hand, this was precisely the objectification of their humanity that actual Ayoreo so strenuously rejected as opposed to their project of moral transformation. Moreover, the global movement to protect Ayoreo isolation that gained momentum in the early 2000s did not seem to interrupt the dispossession of Ayoreo-speaking peoples at all. Rather, this fascination seemed to sustain and amplify colonial violence.

In November 2006, I participated in the United Nations Regional Seminar on "The Rights of Indigenous Peoples in Voluntary Isolation and Initial Contact in Amazonia and the Gran Chaco," where I witnessed an international political economy of cultural preservation coalescing around the image of isolation. Yet this organizing was premised on the sense that isolation was an empirical reality. It became an indexical sign for a kind of life imagined to exist independently of representational processes and relational ways of being.[10] This image was compelling, and its power was indelibly fused with the political visibility of the tribal organization OPIT and the Totobiegosode, whose proximity to "uncontact" was widely known and celebrated by the NGO. At the same time, the disconnect between Ayoreo sensibilities and the moral imperatives of this cultural advocacy made isolation a peculiar kind of hyperreality. That is, it was a model of reality that promised to create and impose the actualities it purported to describe, at least in the Gran Chaco.

Even more disturbing was how this disjuncture between an imposed hyperreality and Ayoreo self-understandings allowed the well-meaning interest in preserving "Ayoreo isolation" to slip into the disenfranchisement of actually existing Ayoreo people. Indeed, the global concern with isolation defined Ayoreo humanity in opposition. It diverted attention away from the political investments in denying the clear and vertical relationships that already bound the lives of the concealed groups so tightly to our own. At the same time, the isolation imaginary offered yet another justification for classifying Christian Ayoreo and settled Totobiegosode as degraded remnants of culture death and threats to the forest bands. Because the uncontacted were simultaneously everywhere and nowhere, their figurative isolation allowed the destruction of ancestral Ayoreo lands to continue unchecked. As I became more deeply involved in advocacy work throughout 2006–2007, I began to wonder if actual Ayoreo people were stripped of rights to the same degree that human rights were granted to the imaginary subject of isolation. Was valuing isolated life predicated on dehumanizing other Ayoreo?

The problem seemed to lie in the particular way that the politics of isolation redefined culture. By definition, the edges of isolated life were rigidly mapped onto the limits set around pure culture. That is, political investments in isolation assigned social force and human substance to the "serious fiction" of culture as a bounded, stable whole: a container for true difference. Many scholars have noted how such narrow definitions of authentic culture are a primary way in which the hard-fought gains of Indigenous rights movements may "boomerang" back into the very structures of oppression they aimed to disrupt.[11] What was distinct about the political and legal category of isolation is that, within it, social relation itself was a stark line of exclusion cutting through the category of culture. That is, it parsed Indigenous kinds of life into opposed regimes of legitimacy based on the degrees of relation between them and us. These two kinds of Indigenous life were mutually exclusive and demanded to be vertically ranked by politics. In such ways, the protection of isolated Ayoreo life created a new regime of Indigenous biolegitimacy. It rebordered culture and life and redefined the kind of life that the contemporary politics of indigeneity are interested in. In doing so, it extended a fundamental contradiction of contemporary indigeneity: contradictory limits of culture allowed for a kind of politics that may be set against the human lives these forms supposedly sheltered and protected. At stake was not merely a new technique of the self, but the uneven ascription of meaning and value to a kind of life imagined legitimate only to the degree it remained the precise opposite of the ideal subject of modern, network society.

This also made isolation a particularly fraught kind of knowledge. Although the category was politically effective to the degree it was perceived as a fixed or self-evident description of people like the forest Totobiegosode, the fact that it did not correspond to their realities meant it was also an incoherent form of intelligiblity whose contradictions mirrored the frictions between the diverse global projects that appeared to coalesce via their shared investments in isolated life. In such ways, the nonsensical logics of isolation echoed and reproduced the disordered logics of colonial violence. In order to understand how the politics of isolation was related to the ambiguous being of Ayoreo-speaking people, I set out to track this overlap between violence, culture, and life. It was a journey that took me from the grit of Ayoreo camps to the seemingly distinct domains in which their lives were objectified—international human rights law, multiculturalist state policies, humanitarian NGO practices, and genetic science—and back to the human tragedy unfolding in the Gran Chaco.

The Human Right to Isolation

The legislation of isolation as an international human rights issue began as a response to a particular problem—that is, through intense mobilizing by Indigenous organizations and NGOs in Peru and Brazil against the disastrous effects of multilateral development projects for certain Indigenous populations.[12] These largely successful mobilizations prompted a series of meetings funded by the Inter-American Development Bank (IDB) and the UN in 2005, 2006, 2007, 2009, and 2010 to establish an effective international lobby around isolation as a basic human right and domain of expert knowledge.[13] In the last six years, the UN, the OAS Human Rights Commission, and a series of precautionary measures issued by the Inter-American Court of Human Rights have begun to legislate isolation as an exceptional collective right, defined by a general state of vulnerability. The legislation of isolation, in other words, offers an apparently universal solution to a particular set of problems.

The clearest formulation of isolation as a shared existential state is found in the 2009 "Draft Guidelines on the Protection of Indigenous Peoples in Voluntary Isolation and Initial Contact," by the UN Human Rights Council. This document uses three criteria to define isolated peoples:

1. "They are highly integrated into the ecosystems which they inhabit and of which they are a part, maintaining a closely interdependent relationship with the environment in which they live their lives and develop their culture . . . ;
2. "They are unfamiliar with the ways in which mainstream society functions, and are thus defenseless and extremely vulnerable in relation to the various actors that attempt to approach them or to observe their process of developing relations with the rest of society, as in the case of peoples in initial contact;

3. "They are highly vulnerable and, in most cases, at high risk of extinction. Their extreme vulnerability is worsened by threats and encroachments on their territories, which directly jeopardize the preservation of their cultures and ways of life."

A similar definition of isolation has become national law in Peru, Ecuador, and Bolivia.[14] The legal government of isolation draws from two significant precedents. The first is the guarantee to self-determination as the fundamental Indigenous right. The condition of isolation, jurists have argued, is "the clearest and most unequivocal form in which they exercise their right to self-determination." Thus, legally protecting their human rights requires "a guarantee of respect for the no-contact principle . . . which represents the highest expression of their will." The second precedent is the right to culture. In the UN Human Rights Council document, the human rights of isolated peoples are linked to their status as "very vulnerable peoples whose cultures are at permanent risk of disappearing." Preserving these cultures both preserves "a valuable public good for humanity," and "protects the existence" of isolated peoples.[15] The legal case thus presumes that the condition of isolation is the maximum expression of Native desire or will, and that the validity of this life is entirely contained within the limits of uncontaminated culture.

In such definitions, isolated life is inseparable from pure nature, as well as pure culture. Today, isolated people legally resemble endangered nonhuman elements of nature in their rights to difference. Contrary to the life-forms that are viable within capitalist modernity, this regime of life/culture does not disrupt natural ecosystems, but is an integral part of them. As the NGO director who protested the expedition put it, "Without them something would be lacking in the forest, something related with their vitality and the validity of what we call biodiversity."[16] In such ways, isolation is not a human right at all, but a legal slot reserved for the latest reincarnation of natural man. Therefore, the respected Indigenous rights group IWGIA (International Work Group for Indigenous Affairs) can describe the concealed Ayoreo-speaking people in all seriousness as "a single, inseparable unit with their habitat . . . with which they live together in close communion."[17] Because it grants rights to a form of life that it cannot locate, isolation presumes a subject that is intelligible only in its sovereign absence. Paradoxically, it is a legal subject that must remain outside of law itself.

This means that isolation may allow for new forms of imperial guardianship, in which various state and transnational institutions compete to become the legitimate trustees of certain kinds of life. By claiming the power to authorize itself or third parties to act as guardians *ad litem* for isolated groups, the state prefigures them as wards or dependents. This also applies to international jurisprudence. The recent precautionary measures issued by the Inter-American Court to protect isolated groups, for instance, stem from a process called "third-party petitions," in which a third party can submit petitions on behalf of another if the actual injured party is deemed unable to submit a petition for itself.[18] The legal efficacy of isolation is predicated on such slippages between absence and agency. It presumes a kind of life that is only sovereign to the degree it reborders the human/nonhuman—that is, to the degree it is subsumed entirely into an external sign from which it is simultaneously excluded. This basic premise is not questioned by those actors and stakeholders, including Indigenous organizations, that are now competing to represent isolated groups and manage the resources marshaled on their behalf. This is the case for the handful of Ayoreo cultural brokers who are increasingly asserting their right to represent the interests of the concealed Totobiegosode in international forums.[19]

The legislation of isolation is in part derived from the ways in which the value of cultural diversity and biodiversity have become quantitatively the same within the logics of global capitalism.[20] Those working in international conservation, development, and human rights regard each as a global public good, an underprovisioned resource whose benefits ideally reach across borders, generations, and populations.[21] As Ismail Serageldin put it in a 1999 UNDP report, "Culture is an end in itself . . . it contributes to a society's ability to promote self-esteem and empowerment for everyone."[22] Yet whereas this cultural diversity is a value based on the recognition that "differences in human societies are parts of systems and relationships," and thus mutually constitutive effects of politics, history, and personhood, isolation presumes the inverse.[23] Isolation is predicated on a kind of cultural difference that exists in opposition to social relations. It is a radical form of difference that is inevitably contaminated by being entangled in wider networks. At the same time, its value is quantified in the capitalist terms of market exchange.

Isolation and Culturalism

The legislation of isolation presumes and creates contradictions within national multiculturalist policies as well as international law. The current mobilizing around isolation is only possible because of the well-documented juridical reforms across Latin America in the last two decades, particularly the rise of multiculturalism as an official state policy in Argentina, Bolivia, Brazil, Colombia, Ecuador, Guatemala, Mexico, Nicaragua, Paraguay, Peru, and Venezuela. Scholars have described how creating a pluralist and ethnically heterogenous state based on respect for human differences instead of their erasure promises "a radically new politico-legal order and conception of citizenship" that in effect redefines the national project in Latin America.[24] The emancipatory potentials of such reforms reside largely in their promise to decrease inequality by increasing political connectivity through new forms of citizenship and rights. Yet, as Jean Jackson and Kay Warren aptly point out, multicultural reforms remain unevenly spread and deeply contradictory for Indigenous peoples in Latin America.[25] This is even more complicated in places that have long been at the very margins of state rule, like the Paraguayan Gran Chaco.

As Terry Turner argued, multiculturalism transforms a Herderian concept of culture as a distinct worldview into a foundational human right.[26] Thus, article 98 of the 2009 Constitution of Bolivia argues that cultural diversity "constitutes the essential base of the Communal Plurinational State. . . . The fundamental responsibility of the state is to preserve, develop, protect and distribute the cultures that exist in the country." Multicultural citizenship, in other words, presumes a humanity that is defined by the universal capacity to intentionally produce cultural selves—a process that is a function of interdependence with others.[27] The category of isolation carries multiculturalist logics to such an extreme that they double back upon themselves. "Isolation as right" is premised on an appeal to a pluralist society built not around tolerance for diversity but around a state that polices the boundaries of culture as permanent borders that must be defended. The protection of this imperiled difference is not meant to insure a society in which everyone participates equally but one in which segregation is the only possible form of solidarity with isolated subjects—even while their territories are being actively transformed into the sites of hydrocarbon extraction or industrial agriculture.

While multiculturalist logics may make the state protection of isolation possible to imagine in the first place, the particular fusion of culture and life implied by the category of isolation stands the legitimating premises of such protections on end. The key difference lies in how isolation establishes a state policy around protecting a subjectivity that is incapable of change and self-representation.[28] The culture of the isolated is imagined to be a sui generis and stable outside to wider relations. Ironically, this seems to imply that the most valid Indigenous life can only exist outside of multicultural society! Thus, the state protection of isolation, or nonrelation, is seen within Peruvian Law 28.736 as the only way to "guarantee [isolated peoples] their rights to life and health, while safeguarding their existence and integrity." Likewise, the concealed Ayoreo groups are described as "living according to their ancestral cultural norms, in another world; their knowledge of the modern life culture of encompassing society is reduced to isolated fragments, which they are able to incorporate into their own worldview without altering its coherence."[29] Such familiar colonial conceits not only resonate with older representations of a racialized Ayoreo humanity as "an archaic culture," "mythical consciousness," or "wild and savage horde," but they also contradict the fundamental aims of multiculturalism by reducing Indigenous life to a static culture, and by denying Indigenous populations control over their own being.

Isolation can be considered a state of Indigenous exception, complete with its own norms and hierarchies.[30] Yet it always exceeds the state, especially in the Chaco, where the enforcement of law is already precarious and expeditionary. The question remains: How is this peculiar legal category translated into everyday practices and politics that Ayoreo cannot avoid?

Isolation and Humanitarianism

What is most important about the image of the isolated Ayoreo, of course, is what can be done with it. One month after the successful NGO protest against the planned Chaco expedition, the attorney general of ethnic and gender rights ordered a raid on its offices by national police.[31] Based on this raid, the director eventually faced criminal charges for allegedly embezzling a large sum of money.

This action was widely denounced by other NGOs in Paraguay. POJOAJU, an association of Paraguayan NGOs, promptly issued a statement in which they "energetically repudiate this abuse of power," and described the raid as a "disastrous precedent of state action against the organizations of civil society in the Chaco."[32] International organizations quickly followed suit, with Amnesty International condemning the state's actions against these "defenders of human rights" as a punitive reprisal for their denouncements against the expedition and for their advocacy against large landholders in the Chaco.[33]

In each case, the state's regulatory actions were described as attacks not against the NGO per se but as against the rights of the isolated Ayoreo the NGO supposedly defended. As the Amnesty International statement concluded, "this case demonstrates once again the void in the Paraguayan implementation of relevant international standards for indigenous peoples' rights." The fact that several Guidaigosode Ayoreo leaders themselves had requested the intervention against the NGO—for reasons unrelated to the expedition—was largely ignored.[34] Paradoxically, the presumed will of the isolated subject supplanted the voices of actual Ayoreo leaders and rendered them inaudible in the name of self-determination.

The unquestioned reading of NGO agendas as the human right of isolated Ayoreo subjects reveals the degree to which NGO labor has been crucial for translating the clear universals of "isolation as legal category" into the messy practice of everyday politics. Such institutions are the medium by which the divergent global values of isolation may become a single regime of authorized life.

In Paraguay, advocacy NGOs have replaced evangelical missionaries as the arbiters of "unreached" people.[35] One of the results of the post-dictatorship state reform project was the NGO-ization of Paraguay civil society, and the de facto privatization of cultural difference and its preservation.[36] Until the rise of the Federación para la Autodeterminación de los Pueblos Indígenas in the first decade of the 2000s, there had been no national indigenous movement in Paraguay. And during my fieldwork in 2006–2008, NGOs largely occupied the role filled by Indigenous federations elsewhere. It was rare to find an Indigenous community in the Chaco not affiliated with at least one NGO. While several provided critical services, many existed only as "briefcase NGOs" designed to capture aid money.[37] A common joke during my fieldwork was, "You got fired from your job and you're broke? Me too. I guess we have to open an NGO."

Two NGOs organized themselves around defending the rights of isolated Ayoreo-speaking people in Paraguay.[38] When international funders began to view Indigenous organizations as their ideal clients in the early 2000s, each of these NGOs supported the formation of a separate Ayoreo tribal organization that they funded, administered, and attempted to control. Not surprisingly, these institutions were also involved in a bitter and long-running conflict with one another. One institution—GAT—was affiliated with the most recently contacted bands of the Totobiegosode-Ayoreo subgroup. For GAT, the presence of isolated Ayoreo bands was used to justify their decades-old land claim on behalf of Totobiegosode people.

The second NGO—Iniciativa Amotocodie—successfully protested against the London museum expedition. Run by a Cambridge-educated European who was a former member of GAT, this NGO described how its work created an "Ayoreo policy of recuperation and revitalization . . .

[which] is bringing the cause of the modern Ayoreo (out of the forest) ever closer to that of the isolated groups, and the protection of them [is] becoming their own cause."[39] In such descriptions, the figure of the culturally pure isolated Ayoreo became a metaphor for the value of all contemporary Ayoreo people, even as the NGO's objective was glossed as the agenda of all Ayoreo-speaking people.

Despite their bitter and public disagreements, both NGOs produced strikingly similar imagery about isolated groups as a form of life that had "not yet had any contact whatsoever with modern civilization," and was in danger of imminent extinction.[40] They both invoked the sanctity of this imperiled life as justification for their intervention, and used it to connect to wider humanitarian narratives and global NGO networks. Survival International was particularly effective in raising international awareness around the plight of the concealed Ayoreo groups, beginning with its exposé of New Tribes Mission manhunts.[41] In recent years Survival has focused on the unchecked deforestation of Totobiegosode lands, and organized several direct actions around the issue, including popular demonstrations in 2010 at Paraguayan embassies across Europe by thousands of protestors waving signs that read "Save the Ayoreo." The narrative of saving this "Tribe that Hides from Man" from extinction was a predominant one; website visitors were urged to donate or support Survival's work by statements such as "The Ayoreo Need You," or "Their Future is in Your Hands."

Such imagery reinforced the notion that isolated life exists only as a state of emergency: the sovereignty of this life is contingent on the moral actions and financial charity of those in the Global North. This commonsense moralism prefigures NGOs as lifesaving institutions that "do good" by "giving voice to the voiceless" or taking anti-hegemonic positions against states and markets and empowering grassroots aims.[42] Thus, commentators have noted that "in Paraguay, civil society organized through NGOs plays a crucial role in promoting the protection of the territories and rights of the Ayoreo Indigenous families in the Gran Chaco. In this process, Iniciativa Amotocodie is distinguished as an NGO at the forefront of this protection, promoting as it does a unique participative model."[43] Such impressions are also necessary to insure the continued funding of these NGOs, mainly via charity groups and foreign aid offices of Norway, Holland, Germany, Spain, Switzerland, and the European Union.

These NGO actions were instrumental in raising awareness around the plight of the concealed Ayoreo, including the rampant destruction of their ancestral forest homelands. In practice, however, the imagined constituencies of such NGO politics also required erasing the unsettling voices of actual Totobiegosode people who were indifferent or opposed to the redemptive potential of the concealed groups. During my fieldwork, this was achieved by a variety of strategies that ranged from sincere attempts at dialogue and collaboration to blatant forms of domination, such as manipulating the information given to the leaders or threatening to withhold vital services from client communities if they disagreed with a particular NGO position. The common practice of paying leaders to participate in meetings blurred these lines between institutions even further. This process of bureaucratizing and domesticating valid Indigenous life was always reductive. As a means it never justified the ends of simply sustaining the NGO.[44] This inversion of roles was enabled by the emergence of the isolated subject as the fullest expression of what Alcida Ramos has called the hyperreal Indian "clones . . . [which] exist as if in a fourth dimension, a being with whom one enjoys having close encounters of whatever kind."[45]

Whereas Indigenous groups elsewhere have "turned to cultural forms of political struggle in direct defense of the reproduction . . . of their lives," both Paraguayan NGOs acted as if preserving pure culture required denying Indigenous peoples the capacity for self-objectification. While their funders assumed that defending cultural autonomy contributed to the "struggle to reassert the powers and values of human self-production," these NGOs produced the opposite effects in practice.[46] In this system, efforts to protect isolated life from capitalist pathologies actively reinforced the suppression of Ayoreo human rights and the denial of their capacities for self-objectification.

The Totobiegosode tribal organization was created by GAT in large part as a strategic response to the creation of another Ayoreo tribal coalition, Unión de Nativos Ayoreode del Paraguay (UNAP), by the rival NGO, Iniciativa Amotocodie. GAT perceived the new tribal organization as a potential threat to its institutional agenda, and was convinced that the solution was for the Totobiegosode to refuse to join UNAP and form their own organization. Despite these cynical calculations, I was hopeful that the new Totobiegosode organization by its very definition would open new spaces for Totobiegosode political agency.

At the time, the Totobiegosode leaders were kept in constant motion between meetings. They received wages from the NGO for attendance and in general they regarded leadership in the tribal organization as a form of wage labor. Such meetings were marked by a litany of empty promises and hollow gestures. They were key elements in a wider ritual of expenditure around the Totobiegosode land claim, which provided all participants with salaries under the pretense of the land claim. It was no coincidence that little legal progress had been made on the land claim since 1997. Regardless, the Totobiegosode right to land was indisputable and assured by both national and international law.

Assisted by an international law specialist, in 2007 Jochade, Dejai, Yakayabi and I traveled to the UN Permanent Forum on Indigenous Issues in New York to present the Totobiegosode land claim case. While at the meeting, we met up with Indigenous leaders from Ecuador and Peru. These leaders were in the process of forming an international coalition in defense of isolated people's rights with funds from the same groups I had met earlier at the UN Regional Seminar. At the time, Iniciativa Amotocodie was actively excluding the Totobiegosode from international venues where "isolation" was being discussed. Instead, this NGO was promoting UNAP as the sole representative of all concealed Ayoreo, a position the Totobiegosode stridently rejected. After all, among the leaders of UNAP were people who had actively hunted down Totobiegosode bands and captured them. At the Permanent Forum, we were able to make our case, and we invited the Ecuadorian and Peruvian leaders to include a visit to the Totobiegosode communities on the agenda for an impending trip to Paraguay. When the leaders followed through with their promise several months later, I thought it was a success.

The visitors arrived at Chaidi. Few of the former Areguede'urasade joined the meeting. Those that did sat quietly in the background. After the usual pleasantries, the visitors described their initiative. They said it was funded by the IDB and that it aimed to assert Indigenous protagonism in protecting the rights of isolated peoples. They invited OPIT and the Totobiegosode leaders to join their coalition. During the subsequent discussion, Dejai performed a dual discourse. He used the occasion to deliver a message in Ayoreo about what it meant to remain in the forest, even while he agreed in Spanish to sign the convention to protect them. While leaders like Dejai may interpret contact in distinct terms, they are also keenly aware it has become an effective way to be Indigenous.[47]

In this case, the legal category of isolation and its translation by NGOs often created the problems it presumed to solve. Totobiegosode people were forced to respond to this external image of themselves as one of the only possible resources for gaining financial or political leverage in the severely attenuated spaces of post-contact life or as a form of moral reasoning to which they stood in principled opposition or both at the same time. For many Totobiegosode managing isolation was part of the daily pragmatics of survival. This paradoxical situation was possible because, on a transnational scale, legitimate Ayoreo life was increasingly intelligible only as isolated life.

Isolating Life

A common conceit is that the boundaries of life itself are increasingly porous under the conditions of late capitalism, as biomedicine, organ transplantation, or genetic manipulation realigns life around an array of newly discrete deployments. Whereas scholarship has tended to focus on the relationships enabled by these shifts in science and technology, the rise of isolation over roughly the same time period suggests an alternative regime of biopolitics predicated on the inverse, or managing a form of life defined by the refusal of relation.

This process—in which the category of isolation evacuates and stands in for Ayoreo humanity to an ever greater degree—is mirrored in the techniques and concerns of genetic science. Outsiders like the Nobel laureate D. Carleton Gajdusek have long been fascinated by Ayoreo blood, not least because they imagined it to be the empirical evidence for a natural, human difference.[48] Ascriptions of isolation have oriented the study of Ayoreo biology since geneticist Francisco Salzano discovered a set of "unusual blood genetic characteristics" in the Ayoreo samples collected by Gajdusek in 1963.[49] As part of a well-documented research agenda first developed with James Neel in the 1950s, Salzano and his colleagues analyzed biological material from groups they believed most closely approximated prior stages of human evolution, targeting especially those they deemed most isolated, genetically diverse, and "pre-civilized," including the Yanomamo.[50] An attributed state of isolation was already generating its scientific validation ex post facto. Perez-Diez and Salzano thus describe their work on the Ayoreo as furnishing "data from one of the few remaining relatively unacculturated South American Indian tribes."[51] Much of this research was funded by the US Atomic Energy Commission, as elucidating the full range of pure genetic structures was considered necessary for understanding possible mutational damage caused by radiation.

Building on Gajdusek's dark legacy, isolation remains a technique by which scientists visualize, interpret, and evaluate the biological contents of Ayoreo being. In a recent set of papers based on Ayoreo blood samples collected in the 1960s and 1970s, scientists have concluded that the Ayoreo are "genetically peculiar" in two ways. First, they are defined by a relatively low rate of heterozygosity, or genetic variation within the group. Second, they are described as outliers that represent a maximum expression of genetic difference relative to other Indigenous groups in

South America, including neighboring Chacoan peoples. According to these studies, this exceptional biological difference is evident in blood proteins and gene expressions, as well as "an extremely reduced" number (2), kind (C/D), and distribution of mitochondrial DNA haplogroups.[52] These genetic traits, in turn, are interpreted as evidence of "founder effect" or isolation.[53] Today, Ayoreo DNA not only stands in for the political subject, but reduces its political subjectivity to the biological facts of isolation.[54] Through such techniques, the renewed scientific value of the isolated subject is segregated not only from politics, but also from actual bodies. Thus, the significant infrastructures dedicated to preserving Ayoreo blood in the form of plasma and glycerolized red cells are kept frozen for decades in labs across Brazil, Chile, and Argentina. Here, the desire to extract and preserve isolation produces it as an object for genomic management. And at this point, the particular disembodied objectifications of isolation—myth, tradition, culture, soul, blood, gene, law, victim, sacrifice, donation—become interchangeable.[55]

There is little place for humanity here. Materials that are purely biological or cultural are self-contained and self-evident signs, much like those painted figures from Brazil, shooting arrows at the circling planes over and over on YouTube. As the value of the isolated Ayoreo subject escalates, the value of the recently ex-primitive plummets. Humanizing the hyperreal isolated man presumes dehumanizing Ayoreo personhood in the present. Ayoreo people are once again prefigured as dying or already dead. The only hope for such degraded figures is a project of revitalization or being brought back to life. Thus, anthropologists like Miguel Bartolomé can argue:

In the moments that I am concluding these pages, March 2000, the great hope of Ayoreo cultural revitalization resides in the next exit from the forest of the "uncontacted" Totobiegosode, who have not undergone the deculturating impact of the evangelical ethnocide. As such, they are bearers of the ancient knowledge and cultural wisdom which their sedentarized countrymen have been obliged to renounce.[56]

In this schema, legitimate Ayoreo life is increasingly tethered to isolated life. And isolated life only comes into focus as petrified culture, and vice versa. Thus, the most legitimate form of Ayoreo humanity is that which is "still fully alive among the uncontacted groups."[57] Yet even these small groups that remain hidden in the forest are prefigured as fossils of themselves, ideally conserved by outsiders as a resource for a foreclosed future but already rejected into death by the immutable borders of relational impossibility. It is thus no surprise that the same NGO director can only imagine one option for their future, and not a practical one at that. Echoing the laboratory workers safeguarding Ayoreo blood samples, he has argued in print that "freezing the moment of contact" with isolated Ayoreo bands is the sole hope for preserving "the essence of their being."[58]

Yet neither life as such nor particular life projects can be entirely contained by such categories and the relational worlds they evoke. This is at the root of Ayoreo critiques of modernity and their refusal of such extant political definitions of culture as meaningful forms of self-objectification. It is, I believe, what Ayoreo-speaking people seem to imply by the concept of *pucuecaringuei*, a phrase that literally means something like "searching for what is emanated from oneself," but is translated as *vida*, or life. This is a concept that foregrounds becoming over being, and frames each as processes that are contingent on more than the forces contained within bodily limits. As my adopted father Yoteuoi once said to me as we walked north of Arocojnadi, "Pucuecaringuei is something that is outside but inside. Inside but outside. It is something you try to catch." He smiled. "But it is fast."

Isolation as Biolegitimacy

The politico-legal category of isolation, then, reiterates an extreme version of what Eric Wolf referred to as the "pool hall model of the world," in which certain domains of pure culture/nature/life are newly endowed with the qualities of billiard balls: disaggregated, bounded, and brightly colored objects colliding and spinning off one another.[59] What is most striking about this development is the degree to which outsiders interpret this figure of the isolated primitive as evidence of Indigenous self-determination and thus give it force to act through and against the life it ostensibly describes. The politics of isolation redefine culture and threaten to overwhelm the potential agency offered by other, more nuanced definitions of cultural difference. Indeed, isolation is politically effective precisely to the degree it erases or disavows the mediations (of history, empire, culture) by which indigeneity became a meaningful category in the first place.

Of the many contradictions articulated through the speculative politics of isolation, the redefinition of cultural difference is perhaps the most crucial. The category of isolation establishes a vertical hierarchy

of legitimacies in which bounded, ahistorical, and antirelational difference is privileged over and above the kinds of difference asserted within
local sensibilities or those taken as the product of unequal relations and
imperial histories. What is new about this framework for moral action,
uncritically assumed by a wide range of stakeholders (including a not
insignificant number of Indigenous rights activists), is that it extends
the ways in which such uneven cultural legitimacies become perceived
as biological legitimacies, and vice versa. Through mobilizing around
isolation, these distinct ways of interpreting cultural difference blur into
the ranking of Indigenous forms of life. Cultural legitimacy becomes indistinguishable from bioinequality.

This entails more than the simple naturalization of difference. Rather,
it also marks a shift in the kind of Indigenous life that global politics
is interested in. This shift—from a Native subject who is encapsulated
by imperial power but ideally self-determining to one who can only exist outside of social relations and representation in general—refracts the
similar shift from "the life of the refugee" to the "life of the sick" that
Didier Fassin uses to explain biolegitimacy as a mode of contemporary
governance. For Fassin, these changes are fundamentally about the fluid
stakes "with respect to the sort of life which is defended today and which
can enter at some point this state of humanitarian exception."[60] What is
distinct about isolated life is not its unique eligibility for inclusion into
already existing exceptional states, but rather, that it is a form of life created by such stagings of humanitarian exception and the transnational
government of emergency. Here, the state of exception produces the
only Indigenous subject that is capable of fully fitting within its pregiven
boundaries.

Questions remain: What relational worlds might these politics of isolation also imply for a critical public anthropology of indigeneity? Is it
enough to find relations where there appear to be none, or to put ethnography at the service of more effectively policing the borders of isolated
life from the many who would otherwise seek to dispossess or extinguish
it by brute violence? Does political anthropology require playing by already given rules? I don't believe so.

To be clear, what I am arguing is that there is an urgent need to imagine a subset of exceptional rights based on an ethical engagement with
the lived experiences of actual Ayoreo people instead of the imagined
dilemmas posed by a neocolonial fantasy with universalizing pretensions. Many failed attempts at cultural activism have convinced me that
such a framework can only begin by fostering broad Ayoreo protagonism

impossibility of political (solution

in mobilization around the future of the concealed groups; by effective regulation of para-statal organizations such as missionaries and advocacy NGOs; by the implementation of enforceable cross-border protocols for addressing the starkly asymmetrical relations that already exist with the concealed bands; and by the immediate titling and policing of an ancestral territorial base large enough to allow for their continued survival in the manner which they so choose. At the same time, it is acutely obvious to me that there is insufficient political will in Paraguay for any of this to occur.

After all these wanderings and years, I am left with my cocreated conceptual fantasies, in which culture is not a form of being restricted to the accuracy with which an a priori tradition may be replicated, but rather is a kind of becoming constituted through gaining and exercising the capacity to continually reset the terms of self-production and self-transformation. In these terms, "culture as becoming" *requires* meanings that are intrinsically unstable, fluid, and based on rotational social time. Humanity is thus not located in any biological essence or in reproducing any given formal practice but in the capacity to manage and objectify the terms of instability itself.

The Haunted Forest

Precisely one year after the controversy around the expedition, I heard that the concealed people had been contacted. Through messages pieced together over radios and cell-phones and Facebook, the Totobiegosode leaders told me to pack my bags. They wanted me to come. I wasn't sure why.

I pressed for details but there were no clear answers. Something had happened and no one seemed to know what. As when I first heard about the Areguede'urasade coming out, I stumbled around in a daze with ghostly visions of the Chaco always in my sight. I could barely sleep. When I did, I had terrible dreams of blood and dust and flies, and I woke covered in sweat. Was I really willing to return to a scene I had barely survived the first time? What could motivate such a trip? Given this terrible knowledge, could my presence be justified at all? As I began to sort through such questions, the story itself dissolved.

One person said that twelve naked Indians had been caught on a ranch near Chovoreca. Another said that a group had run away from the bulldozers working by Cerro León and laid down their spears, despondent, at the feet of a rancher. Someone surmised the NGO director had captured the Totobiegosode band. I heard the government sent a delegation. They returned two days late, having found nothing but the strange rectangular tracks of *parode* sandals and stories of naked brown bodies glimpsed at dusk. Within a week, everyone had dropped it completely.

The image never entirely dissipated, of course, not ever, not for any of us. In that land of linear pastures and bulldozers that never stop, concealed life presses against the senses. It carries the force of crisis and spirit wronged, its contents never safely subsumed within its sign. It is no coincidence that the increasingly fervent investment in the reality of isolated life coincides with the amplified dehumanization of actual Ayoreo-speaking people; the subhuman *Puye* and the *uruso* madmen find their mirror images in the flash of brown skin glimpsed at dusk. And so the Savages return to the forests from whence they emerged, where they wait and watch and will not forgive.

Yet the forest bands cannot be entirely subsumed into these images of them, at least not yet. One century after Ishi—the Yahi tribesman who gained notoriety as the "last wild Indian in North America"—stumbled into a California corral, the concealed Totobiegosode are facing a similar dilemma. This crisis is only intensified by the sedimented weight and numbing familiarity of the stories we think we know, the sense of an inevitable end foreclosing their futures as surely as the bulldozers.

Like the last Yahi, the forest people have retreated into a life of strenuously maintained concealment. The words once written about the Yahi in the 1870s can easily be applied to these Ayoreo today:

No human eye ever beholds them, except now and then some lonely hunter, perhaps, prowling and crouching for days over the . . . scraggy forests which they inhabit. Just at nightfall he may catch a glimpse of a faint camp-fire, with figures flitting about it; but before he can creep within rifle range of it the figures have disappeared, the flame wastes slowly out, and he arrives only to find that the objects of his search have indeed been there before him, but are gone. . . . For days and weeks together they never touch the earth. . . . They never leave a broken twig or a disturbed leaf behind them.[61]

We know this concealment is unprecedented in Ayoreo history. And we know that this concealment is due to their keen awareness of the beings that they carefully observe and that surround them more completely every passing season. Years have gone by with only the slightest traces of their presence: a glimpse of bare skin by a park ranger, a twig snapping at dusk, a half-hidden track of their *parode* sandals, a scrap of ash or bone.

We know they go to great lengths to conceal their existence from outsiders, waiting hours to cross a road, wiping out their tracks, hiding their fires. We know that they have developed a way to speak with whistles, that even their children are trained to silence. They read the tracks of bicycles and trucks and puzzle over the incredible speed and energy of their *Cojñone* enemies. They run far and fast if they see a bootprint out of place or unexpectedly hear a chainsaw or a tractor. Encounters with bulldozers are catastrophic events, putting the entire group at risk of death from starvation or thirst, each more devastating than the last. They often do not know where they can go or what they can do to escape. We know these people are often enraged and saddened at what they consider an invasion of their ancestral territory, but they rarely wish to risk a confrontation.

Still, they collect the aromatic wild honey and stalk the sharp-tusked peccaries and gather the ancient land tortoises. They bake the starchy roots of the *doidie* in ashes and eat the sweet fruits of *tokode* cacti and the *esode* trees. They make all of their clothing and bags from the leaf fibers of the *dajudie* plant. They meticulously craft bows and spears from *kaunange* and *aidie* hardwoods. They scavenge the roadsides or empty ranch houses for scraps of metal and plastic. They carry a piece of granite from Cerro León to sharpen the metal.

We know they tell *adode* myths and heal one another by sucking out or blowing away sickness with *ujñarone* curing chants. They smoke *sidi* tobacco and *canirojnai* roots to conjure visions of the future before their eyes. When necessary, they ask the spirits for help with *chugu'iji* performances of rage and with the wordless rhythms of the *perane*. They beat on hollow trunks of the *Cukoi* tree to call rain. They carve *tunucujnane* poles with magical designs and leave them as protection in their wake. They sing the same songs as their relatives in Chaidi and Arocojnadi. They are also convinced that death is coming for them, that their world is nearing its end.

Although there is much we do not know, it is clear that this long concealment is no primitive idyll. It is a way of life finely attuned to the daily logistics of concealment from the beings they believe are hunting them down. Their cosmology is partially a response to global political economies; it is a worldview already indistinguishable from the pragmatics of eluding starvation, capture, and death in the face of industrial agribusiness. This is the only kind of primitive society, the only kind of primitive life, that we have permitted to survive anywhere on the face of this earth and even so probably not for much longer. That these small groups of holdouts have been able to endure this long is a testament to their extreme resourcefulness and resolve.

They have managed to achieve a tenuous coexistence despite a permanent state of alarm. Faced with tragic and deteriorating prospects, they still properly care for their children, their sick, their dying, their elderly. When Areguede could no longer walk fast enough to keep up the group, Siquei carried him in a large bag on his back. Those in the forest, we can presume, do the same. They do not violate the *puyaque* restrictions, and each year they ritually create the world anew. Although they know they cannot escape invasion, war, famine, and intolerance, this has not been enough to make them surrender their compassion, their protagonism, their piety. The end of their world has not forced them to violate the essence of what makes them Ayoreode, Human Beings.

As for the rest of us? It is much less certain what the unfolding tragedy in the Chaco implies about our humanity. What have we surrendered to the feeble magic of isolation and to the allure of the vanishing? What price are we willing to pay for our cheap and unwitting denials, for brightly colored photographs?

Faced with the prospect of the next last first contact in a landscape being destroyed as rapidly as the forces of modern technology will allow, having already jettisoned the few real possibilities to envision any other outcome, we wait . . . for what? I do not know but it is highly likely that the day will dawn when they stumble terrified out of the scraps of forest or choose to take their own lives. Only one thing is certain: when they do, we will not be mourning merely the demise of the last wild Indians.

Behold the Black Caiman

Everyone that saw Him Fled! They fled at His horrible Form: they hid in caves and dens, they looked on one another and became what they beheld.   WILLIAM BLAKE

So where do these tracks lead, here at the end?

For me, they opened up into empty hands and away: out of synch like coming back from a war that only I had fought or maybe just dreamed. I did not escape unscathed from the Black Caiman—no one does—but those are stories for another time and even so I was not able to stay away.

I've been back every year since 2009, never for long enough to stay close. But often enough to know that some things have changed and others have not. Many of the people I knew have died: Jnupi, Ore Jno, Agá, Chicode, Simijáné, Codé, Ujñari, Emi, Ebedai'date, Juan, Dalila, Bajebia'date, Puasaquenejnamia, Aasi. Many new children have been born.

To be sure, there are other stories I could tell, other images I could conjure. More people are involved in political activism around cultural rights. Ayoreo have been elected to public office in Morales's Bolivia. Several young people are going to college.

I could evoke a scene from the 2008 Carnival procession in Santa Cruz, which featured a group of Ayoreo participants for the first time. Expected to perform their traditions in a public space, Ayoreo instead invented new costumes, donning fake *ayode* headdresses and painstakingly stitching Western-style bikini tops from *garabata* thread.

I could describe the ruined expanses of industrial agriculture in the Chaco, how the sun and the salt are creating areas of drifting sand in which little can grow—a white dream spreading across a landscape that is becoming a fossil of itself, fantasies of the Chaco as a hellish wilderness finally anchoring in hard fact. I could try to describe a new darkness, the confusion and rumors around an impending epidemic of HIV among the "little birds."

Or I could write of the protest in 2013, when the Totobiegosode tribal organization took direct action in defense of their land rights for the first time. They blocked the Trans-Chaco highway for several days in order to draw attention to the legal impunity that privileged *Cojñone* ranchers claiming land already titled in the name of Totobiegosode. The ongoing struggle to gain title to their ancestral territory merits a book of its own.

I could mention the North/South collaboration we arranged between Totobiegosode communities and an Indigenous-run NGO based in the United States, aimed at economic self-determination. I could write about the ways that my early audio and video recordings are now being taken up by Bolivian Ayoreo, or about the two young Ayoreo women who have begun recording interviews with elders on their own.

I could also conclude with portraits of the New People: Tié, who has sunken deeper into depression and silence; Sidabia, who has tuberculosis and is overrun with unkempt children; Cutai, who spends long stretches working on ranches and remains as cheerful and enigmatic as ever; or Siquei, who broke away from Chaidi and settled in Arocojnadi and then moved back. His eleven-year-old son found work on neighboring ranches, spraying pastures with herbicide.

Each of these dynamics might complement the interpretations I have offered in the preceding chapters and reveal the artifices that they too contain, unravel the hard edges of any conclusions. They might reiterate that I do not have the last word on any of this, that many would surely disagree with me, that writing is inseparable from becoming, that this is an overexposed snapshot of a time already past, a story we are all writing together, however we appear before one another—ready, set, go.

Yet this account emphasizes that synthesis is impossible. I have attempted to write close to contents in order to unsettle form, to show the unruly kinds of vitalism engendered by the social afterlife of ethnographic categories no less than world-ending violence, to suggest how the fevered pursuit of the primitive devolves into a contagious cascade of imperfect copies, to interrogate how the negation of the self may mutate into a

recognizable form of immanence, and to argue that the violent fixture of contradictory limits of legitimate life does not result in the end of human actuality but may be tripped up in such contradictions even as these limits harden into political fact and just as quickly begin to erode. These energies reverberate unequally within and against bodies, they linger hauntingly but not evenly in places and people. Together, they animate the shadowy figure that I have called the Black Caiman, the pursuit of which is the aim and method of this book.

I have chronicled the doomed efforts to pin down my quarry, to make ultimate sense of such nonsense. Every time I came close, the Caiman slipped away. If the forceful fictions of history and conversion continuously unraveled in the nonlinear weft and warp of daily life, so too did most hermeneutic aspirations. Ayoreo realities stretched my capacity for interpretation to its breaking point. If the New People offered no clear way out, Simijáné's schizoid chants or the misty fantasies of Echoi were just as perplexing. None were as profoundly disturbing as becoming an observing participant in the routine failure of meaning; in the tentative patchwork emergence through electronic media of a new project of transformation that was just as quickly undone; in the temporary optimism of apocalyptic futurism and the scant comfort of rotational time; in the deep fragmentation of subjectivity and the retying of sentimental chains; in the broken lines of flight to madness and intoxication as the deathly life best suited for New People in a New World. All the while the Black Caiman stalked closer to those few Totobiegosode hiding in the dwindling forest, and I powerless to give it pause.

In my decade-long pursuit of the Black Caiman, all of my initial explanatory ruses failed. Like my Totobiegosode teachers, I too was submerged under what Michael Taussig memorably described as "epistemic murk," the breakdown of knowledge through which colonial modes of production are inexorably fused with terror and the space of death. The key dialectical trick is that this epistemic murk not only unravels difference but continually reproduces alterity through fetishized images of wildness now imbued with magical force. Moreover, as Taussig wrote, this operation critically depends upon what seems to be its opposite, "the hermeneutic violence that creates feeble fictions in the guise of realism, objectivity and the like, flattening contradiction and systematizing chaos."[1] Appealing to the linear and the rational means being duped into lending essential momentum to the dialectics of domination. And appealing to form without recognizing how this engenders an opposite image does the same. This murkiness was acutely felt by all in the Chaco, where several distinct neocolonial projects were organized around apparently opposed images of moral Indian life and constantly interrupted one another. It was a system where the rational was the irrational, where similarity was difference, where expenditure was the point. Dysfunction was how it functioned.

Yet epistemic murk was only one medium through which the Black Caiman moved. This murkiness was simultaneously the stuff out of which Ayoreo actively redefined time and being, self and other. It was a sort of primordial soup for Ayoreo ontological allopoiesis. Pressurized by terror and dehumanizing violence, Ayoreo ontological theories and colonial images of alterity were dissolved into elements, constantly aligned into nonlinear constellations and then broken apart again. Embodied Ayoreo dispositions, durable and otherwise, emerged from this peculiar metaphysics in a way that both mirrored and exceeded the breakdown of representation. These dispositions looped colonial categories of Indigenous life back into somatic perceptions of the moral value of rupture and transformation in such a way that all were changed. The results, as Tom Abercrombie described in the Bolivian Andes, were creative Ayoreo heterodoxies and a newly bifurcated cosmos wherein colonial binaries were reproduced, reconciled, and unraveled.[2]

For Ayoreo-speaking people, this manifested in a tentative framework based on inscribing contact as a line of radical rupture. Precisely those elements that constituted the moral human within *Erami*, the forest/past, were evacuated and inverted in *Cojñone-Gari*, the space of the modern present. This inversion was imagined by many to be as profound as the difference between animals and humans, or between the world of the dead and the world of the living. It bracketed a form of moral reasoning about the contemporary that collapsed immanence and negation in a series of contradictory and nonlinear ways. While such disordered sensibilities modeled and reproduced the terms of the colonial situation, they could not simply be reduced to it. Rather, their Ayoreo proponents stridently reclaimed a capacity to transform self and world, in terms that were not autonomous but were also distinctly Ayoreo. The crucial divergence is that these fluid Ayoreo projects invariably protested a fundamental link in the colonialist and culturalist chain of cause and effect. They presumed that continuity is rupture and that being is always becoming.

The Black Caiman, of course, is also a trickster. And one of his trickiest tricks is to make such projects come back into focus only as a kind of ontological alterity that exists external to a colonizing violence. The newly politicized category of culture is a primary medium for this to occur in the present. Culture appears as the consummate logics of alterity, the soothing antidote to the turbulence caused when subjugated peoples assert their capacity to define their own subjectivities and the terms of their own ontological ruptures. All too often, culture now brackets an ineffective critique. It is precisely this quick step to locating a rationally ordered but antimodern Indigenous cosmology or ontology that makes perspectivist anthropology such an appealing and powerful rite of neocolonialism. Yet reiterating the myth of a stable ontological outside to modern rationality allows the fetish power of the primitive to join with redemptive desires and fully colonize its object. It poses as a protest of the reduction of the multiple into the one, in order to colonize multiplicity itself. This is the very logic that binds the tradition-seeking *Abujá* to the blood-taking scientist and to the soul-collecting missionary. This is the logic of the search for the primitive, the logic of the Indian hunt, the logic of isolation, the logic of death. Frantz Fanon was writing for indigeneity as well when he wrote that: "Ontology . . . does not permit us to understand the being of the black man. For not only must the black

man be black; he must be black in relation to the white man."[3] Is it any surprise that the myth of cosmological alterity is then celebrated as "the ontological self-determination of the Other"?[4]

Taking the tension between Ayoreo actualities and their epistemic trappings seriously means an unsettling confrontation with such masking myths. Signs of a domesticated cosmological alterity are nowhere to be found—at least not yet. Impossible to ignore, however, are the many ways that precisely these kinds of logics sustain, naturalize, and reproduce stark inequalities in the value and meaning of human life. Culturalism in its many reinvigorated guises now operates as an effective regime of what Didier Fassin calls "biolegitimacy." For Fassin, contemporary biopolitics are best defined not so much as coherent technologies for normalizing and controlling living beings but as the recent creation of sharp inequalities within "life as such." These bioinequalities manifest how authority flows through a new politics of life, in which the global pursuit of contradictory ideal definitions of moral life requires an active process of deciding "who should live and in the name of what."[5] It is a biopolitics predicated on rupture and discontinuity, on the vertical ranking of moral judgments, on blocked gradients and profound contradictions. It is precisely this kind of instrumental disorder that the contemporary politics of

Native culture disavows, instantiates, and reproduces. Through such tensions, contradictory forms of Ayoreo difference are produced, governed, and brought to bear within and against the always excessive substances of Ayoreo life.

In many ways, this account is just as guilty of cannibalism as those it has critiqued. It freezes an Ayoreo project of negative immanence that has surely already been negated. And I have implied throughout that this artificially fixed project contains a radical potential for rethinking the political and moral anthropology of indigeneity in South America. That is, the end ultimately washes me back to the beginning: anthropology's conceit to rescue the Native's point of view and effect a reflexive transformation through its faithful rendering.

Like Pierre Clastres and my other distinguished predecessors, I am not willing to surrender the hope that another world, another kind of being in it, is possible to conjure. And like Clastres and many others, I have turned to Indigenous worldviews in order to co-envision this possibility. Yet I never found a society against the state and I was compelled to renounce the search for the primitive. The kind of New World I encountered in the Chaco was a dystopic one. It was a terrifying reality we all share but one I am not certain that we—the New People, myself, or anyone else—can ultimately survive.

Ayoreo voices as recounted in this book deny the easy escapes of a stable outside, to be inhabited at our leisure. They let none of us off the hook. They mock and confound any attempt to find a new origin, to tell a linear story, to reanimate our logics of redemption. They inexorably pull us down into the vortical flows of rupture. And it is precisely in this downdraft of categorical implosion and inversion wherein something like their radical potential might lie.

Ayoreo sensibilities offer a sharp reminder that there are no cosmological outsides—no "unmodern ontologies"—to redeem our humanity and save us from the modern world we have made. If we have created this New World together, its seas of inequality, riptides of subjectivity, and islands of liberal ideals, then perhaps it is time to envision the next: a world not predicated on the essential difference of Indigenous peoples but on our shared capacity to transform ourselves and to objectify the common ways in which we do it.

It is ironic that this echoes recent calls by some progressive intellectuals, in which the task of the engaged scholar is to identify and create an "altermodern rationality." Disordered Ayoreo worldviews likewise hew closely to how these proponents define this redemptive rationality as a logics in which "becoming is prior to being and where the relation to alterity is not just a means of establishing identity but a constant process."[6] What makes the Ayoreo opening to altermodernity contain a radical potential, however, is that the generative murk of an Ayoreo project of "rupturing-becoming" is based on rupturing the very kinds of philosophical collapse between nature/culture, tradition/modernity, human/nonhuman that such theorists of altermodernity take for granted as constituting the Indigenous–European divide. That is, Ayoreo "worldviews" invert the very insights that scholars such as Hardt and Negri identify as the primary contribution that "an unmodern Indigenous ontology" makes to this wider project of envisioning altermodernity. Instead, Ayoreo call attention to the objectification of Indigenous cosmology as part and parcel of a wider regime of revisionary futurism, in which some vertically ranked world- and life-making projects count more than others.

The destabilizing Ayoreo project of negation protests modernity but to do so its protest is directed against all orders. It is especially directed against the figure of an Indigenous antimodernity through which modern and altermodern orders alike are oppositionally sustained. It emphatically asserts that what is insidious here is not the reduction of multiplicity to the singular but the ways that this metanarrative masks and requires the standardization of multiplicity itself—thereby undoing its radical potential, domesticating alterity, and making ontology available for governance. It suggests that this dynamic, intrinsic to late liberalism, requires more sustained ethnographic attention. And it reminds us that what is at stake here is not only the meaning and value of ex-primitive life but the meaning and value of our own.

This poses an unsettling question: How can we take seriously a kind of Indigenous worldview that is outside of the authorized outside precisely because it unmasks the forceful fictions of cosmological exteriority and in such ways reclaims the capacity for self-transformation according to terms that are not autonomous but are distinctly Ayoreo?

To do so requires a very different kind of political anthropology than that offered by Clastres and others. It begins by undoing the habits and categories by which indigeneity is an intelligible object, as suggested by the work of Terry Turner, Fred Myers, and many others. It begins by tracking the contradictory moral economies by which Indigenous life is unequally ranked, as Didier Fassin argues. It begins by inverting the presumed relationships between the form and content of indigeneity, as Tom Abercrombie shows. It begins by embracing the generative effects of negativity, as the work of Gaston Gordillo in the Argentine Chaco urges. It begins by pursuing not patterns and logics but the politics of the knowledge of what Georges Bataille called "nonknowledge."[7] It begins by moving away from culture-as-radical-content and toward culture-as-biolegitimacy. It begins by applying the heuristics of rupture and transformation to the arenas of public debate as much as to the social afterlives of prior analytic categories. It begins with the shift from an anthropology of Indigenous being to the anthropology of unauthorized becomings. And it begins with sustained reflection upon those murky zones where Indigenous projects of becoming and anthropological projects of knowledge come together, fall apart and, perhaps, meet up again.

Here, at the end of this lurching slog, I am brought up short. A ripple in the water, a sigh of fetid breath: the Black Caiman is finally near. My courage falters. Suddenly I am uncertain: Have I been stalking the Black Caiman or has the Black Caiman been stalking me? I take a deep breath, grip my flimsy spear tighter. I refuse to abandon the hunt after all these years, to seal myself into the form I have tried unsuccessfully to resist.

Instead, I remember white light slanting through dust and the warmth of a hand in the Place-Where-the-Black-Caiman-Walks. I have not invented the journey or the place, but have I loosed the terrifying form I set out to pursue and slay? Have I, too, become what I beheld? No matter. The story is written. The figure is carved.

And the Black Caiman moves ever closer. I feel him in the smoke, I smell him in the dust, I hear him in the forest, I see him in these words. There: a shadowy form, a menacing sway, merciless back, flat of eye, a quick rush, a silent sinking, dark water: the Black Caiman slips away.

I stand empty-handed in the mist. All I'm left with is a longing to find my way back, a yearning for those I'd like to consider my friends, a memory of an enviable piety and an unflinching strength, a haunting sense of what we once were and what we might yet become.

Acknowledgments

This project defined a decade of my life, and I have accumulated profound debts along the way. My greatest debt is to the Ayoreo people of Bolivia and Paraguay who fed me, tolerated my bad manners, corrected my predictable attitudes, reached out to me, and allowed something of their lives to get entangled with my own. I hope they do not regret it. The members of my seven adoptive families and the Totobiegosode people of Arocojnadi, especially, taught me a great deal about being human. In Chaidi and Arocojnadi, the following people deserve special thanks for the life-altering lessons they shared: Berui, Chamia, Chicode, Chicori, Chiri, Curiya, Curiya'nate, Cuteri'edo, Dajnidi, Dasujnai, Ducubaide, Edo'uejnai, Edua, Erui, Esoi, Estela, Gabidé, Gabi'date, Guiejna, Iboré, Idajaguabia, Joaine, Jororo, Mario, Mateo, Ñacore, Naka, Ñame, Ojnai, Orojoi, Parojnai, Porai, Pukoi'date, and Tojé. In Zapocó and Barrio Bolivar Bolivia: Ayahai, Chijñoi, Cuteri, Issac, Jose'date, Julia, Neke, Ñingomejei, Niño, Nojnaine, Ore Di, Ruben, Samané. Thanks are also due to Acani, Ajengome, Carlitos, Diyi, Ebe, Icaque, Nakale, Joini, Sobode, Tarobi, Umene. This research would not have been completed without the intellect and friendship of Chagabi Nevarino Etacorei, *yakayai*.

It could not have gotten off the ground without the support in 2001 of Jürgen Riester and the staff at the institution *Apoyo Para el Campesino del Oriente Boliviano* (APCOB), who hosted me as a Fulbright scholar in Bolivia. Similarly, Dra. Lida Acuña, then the Fiscala of Ethnic and Gender Rights, provided crucial support for my 2004 research in Paraguay, and Vanessa Jimenez of the Forest Peoples Programme was a key collaborator in 2006–2007. I gratefully acknowledge the financial support of Fulbright-Hays and the Wenner Gren Foundation for this dissertation fieldwork. Over the years, conversations with Hipólito Acevei, Manuela Alvarez, Amadeo and Ana Benz, José Braunstein, Volker von Bremen, Benno Glauser, Hannes Kalisch, Gundolf Niebuhr, Gustavo Paredes, Mirta Pereira, Charles Ramsey, Agustín Ribot, Honéssimo and Elodia Rojas, Elba Terceros, José Zanardini, and Chela Zolezzi offered key insights. Special thanks are also due to Bernd Fischermann for sharing unpublished oral histories of Totobiegosode Ayoreo people collected in the 1990s by the late Rosa Maria Quiroga, as well as to Gundolf Niebuhr of the Fernheim archive and to Jony Mazower of Survival International for permission to reprint images. Warm thanks are due to Vera and Verena Regehr for ongoing and stimulating exchanges. I could never have survived my initial time in the Chaco without the support of the Canova family of Asunción, whose home and humor provided a true refuge from 2004 to 2008. This account has benefited greatly from the experiences and generosity of anthropologists Paola Canova and Irene Roca Ortiz.

This project took preliminary shape through dialogues with teachers and colleagues in the Department of Anthropology and the Program for Culture and Media at New York University. Above all, deep thanks are due to my dissertation advisor Fred Myers and the members of my doctoral committee—Tom Abercrombie, Faye Ginsburg, Emily Martin and Bambi Schieffelin—for the insights and support they continue to share. For their advice, commentary, and conversation about this material over the years, thanks are also due to numerous people that made NYU and its environs such an exciting place to be in the early 2000s, especially T.O. Beidelman, Anya Bernstein, Imogen Bunting, Ernesto Ignacio de Carvalho, Emily Cohen, Ceridwen Dovey, Danny Fisher, Christopher Fraga, Aaron Glass, Rachel Lears, Deb Matzner, Laura Murray, Ram Natarajan, Todd Nicewonger, Lauren Paremoer, Rayna Rapp, Pilar Rau, Sandra Rozental, Stefanie Sadre-Orafai, Rafael Sanchez, Naomi Schiller, Emily Sogn, April Strickland, and Jennie Tichenor.

The project further developed while I was a 2010–2011 ACLS/Mellon early career fellow and visiting research associate at the School for Advanced Research in Santa Fe, New Mexico. By a wonderful equation that I have yet to work out, I found myself in a cohort of sharp scholars and extraordinary human beings. Gloria Bell, Cathy Cameron, Linda Cordell, Sarah Croucher, Doug Kiel, Steve Lekson, Teresa Montoya, and Melissa Nelson helped refine my thinking about this material while also gifting me an unforgettable year.

I was exceedingly fortunate to undertake the final revisions in 2012–13 as a Wenner-Gren Richard Carley Hunt Postdoctoral Fellow and a member at the Institute for Advanced Study. Thanks are due to Danielle Allen, Didier Fassin, and Joan Scott, faculty of the School of Social Science, for the opportunity. My conceptualization of this project was greatly sharpened through conversations with gifted scholars at the IAS and the 2012–13 Working Group on Ethnography and Theory. I am indebted to Vincent Dubois, Jens Meierhenrich, Nicola Perugini, and Michael Ralph for comments on different aspects of this manuscript. I am especially grateful to João Biehl and Didier Fassin for the indispensable feedback they so generously offered on extensive portions of the text, as well as to my friend Laurence Ralph, who read and insightfully commented on multiple versions of the chapters.

Special thanks are due to Gaston Gordillo for offering keen comments throughout the evolution of this project while also crafting a body of work that sets a high bar for anyone working on indigeneity in the Chaco. Thanks are also extended to the incomparable Mick Taussig for his wondrous support of this manuscript at a pivotal stage. The last iterations of the manuscript benefited greatly from close readings by Stephanie Malia Hom and Bunny McBride, in particular. David Bond has been a constant and crucial interlocutor over the years this project took shape, and his sharp comments have fundamentally improved its final form. I have been fortunate to work with David Brent at the University of Chicago Press, and I would like to thank him for his support of this project. A special acknowledgment is due to Priya Nelson, also at the press, for her faith in this project, editorial vision, and deft acumen in shepherding it to fruition. I am grateful to the book's anonymous reviewers for their comments, as well as Susan Karani and Linda Forman for editorial aid. Parts of chapter 5 originally appeared in *American Ethnologist*, parts of chapter 7 in *Current Anthropology*, and parts of chapter 8 in *Comparative Studies in Society and History*; thanks for the permissions to reprint.

Those closest to me have borne the brunt of my struggles with the Black Caiman. The steadfast support of my mother, father, and sister has been a constant anchor and compass during my itinerant adulthood, as has the friendship of a far-flung inner circle. My former undergraduate advisor and friend Harald Prins and his partner, Bunny McBride, deserve the final thanks. I first learned of the Ayoreo and anthropology from Harald, and I am lucky that this project has been a dialogue with him from start to protracted finish.

All too often, I have repaid each of these debts with absence. I fear these pages are but a pale shadow of the generosity, wisdom, and faith of so many. Any shortcoming, of course, is solely my own.

Notes

1. Kenneth Read, *The High Valley* (New York: Columbia University Press, 1965), xi.
2. *Asking Ayahai: An Ayoreo Story*, directed by Lucas Bessire, 41 min. (Documentary Educational Resources, 2004); *From Honey to Ashes*, directed by Lucas Bessire, 47 min. (Documentary Educational Resources, 2006).

INTRODUCTION

1. Pseudonyms are used throughout this account to protect the identities of people living and deceased.
2. My translation of Totobiegosode is not literal—*Totodie* (plural collared peccaries) + *gosode* (people group)—but rather refers to the origin story by which this name became applied to these people, formerly known as *Amotoco-gosode*, a phrase that combines the words for "forest that is clear underneath" and "people group."
3. Miguel Bartolomé, *El Encuentro de la Gente y los Insensatos: La Sedentarizacion de los Cazadores Ayoreo en el Chaco Paraguayo* (Asunción, Paraguay: CEADUC, 2000), 308.
4. Like all Indigenous peoples in Paraguay, the Totobiegosode's legal rights to their ancestral territories were constitutionally guaranteed. However, contradictory legal protocols meant that these rights were not implemented in practice. Since 1993, the Totobiegosode were mired in a pioneering claim for six hundred hectares of their ancestral lands. They had gained title to approximately one hundred hectares of this, in the form of four noncontiguous areas, in 1997. Since then, little progress had been made. The land they were

surveying when the contact occurred was in the process of being bought for them by a group of concerned Swiss citizens, a process coordinated by longtime ally Verena Regehr. For more on the Totobiegosode land claim, see Grupo de Apoyo a los Totobiegosode, *El ultimo canto del monte: Reclamo de tierra Ayoreo* (Asunción, Paraguay: Biblioteca Paraguaya de Antropología, 1998).

5. See Pierre Clastres, *Society Against the State: Essays in Political Anthropology* (New York: Zone Books, 1989); Philippe Descola, *In the Society of Nature* (Cambridge: Cambridge University Press, 1994); Philippe Descola, *Par-delá nature et culture* (Paris: Gallimard, 2005); Eduardo Viveiros de Castro, "Cosmological Deixis and Amerindian Perspectivism," *Journal of the Royal Anthropological Institute* 4 (1998): 469–88; Bruno Latour, "Perspectivism: Type or Bomb?," *Anthropology Today* 25, no. 2 (2009): 1–2.

6. See Eduardo Viveiros de Castro, "The Untimely, Again," introduction in *Archeology of Violence* (Los Angeles: Semiotext[e], 2010), 15.

7. Renato Rosaldo, *Culture and Truth: The Remaking of Social Analysis* (Boston: Beacon, 1989).

8. Quoted in Pierre Clastres, *Chronicle of the Guayaki Indians* (New York: Zone Books, 1998), 96.

9. This, of course, is a venerable tradition within critical theory. See esp. Stanley Diamond, *In Search of the Primitive: A Critique of Civilization* (New Brunswick: Transaction, 1981).

10. Ticio Escobar, *Misión etnocidio* (Asunción, Paraguay: Editora Litocolor, 1988), 37.

11. Allyn Maclean Stearman, *Yuquí: Forest Nomads in a Changing World* (Chicago: Holt, Rinehart and Winston: 1989), 148.

12. See Alvardo Bedoya Silva-Santisteban and Eduardo Bedoya Garlanda, "Servidumbre por Deudas y Marginación en el Chaco de Paraguay" (Geneva: International Labour Organization, 2005).

13. Michael Herr, *Dispatches* (New York: Knopf, 1977).

14. See Michael Steininger, et al., "Clearance and Fragmentation of Tropical Deciduous Forest in the Tierras Bajas, Santa Cruz, Bolivia." *Conservation Biology* 15 (2001): 856–66; see also Marcelo Zak, Marcelo Cabido, and John Hodgson, "Do Subtropical Seasonal Forests in the Gran Chaco, Argentina, Have a Future?," *Biological Conservation* 120, no. 4 (2004): 589–98.

15. International Work Group for Indigenous Affairs, "The Case of the Ayoreo: Report 4" (Copenhagen: International Work Group for Indigenous Affairs, 2010).

16. See Michael Taussig, *The Nervous System* (New York: Francis and Taylor, 1992).

17. See Terence Turner, "Indigenous and Culturalist Movements in the Contemporary Global Conjecture," in *Las identidades y las tensiones culturales de la modernidad*, ed. Francisco Del Riego et al. (Santiago de Compostela,

Spain: Federación de asociaciones de antropología del estado español, 1999).

18. See Viveiros de Castro, "Cosmological Deixis." See also Eduardo Viveiros de Castro, *Cosmological Perspectivism in Amazonia and Elsewhere* (Manchester: HAU Network of Ethnographic Theory, 2012).

19. Viveiros de Castro, "The Untimely, Again," 48.

20. Ghassan Hage, "Critical Anthropological Thought and the Radical Political Imaginary Today," *Critique of Anthropology* 32 (2012): 303.

21. Michael Hardt and Antonio Negri, *Commonwealth* (Cambridge, MA: Harvard University Press, 2009).

22. I develop this argument in conversation with David Bond elsewhere.

23. Michael Taussig, *Shamanism, Colonialism and the Wild Man: A Study in Terror and Healing* (Chicago: University of Chicago Press, 1987), 132.

24. For more on the generative effects of negativity, see esp. Gaston Gordillo, *Rubble: The Afterlife of Destruction* (Durham, NC: Duke University Press, 2014).

25. See Michael M. J. Fischer, "Double-Click: the Fables and Language Games of Latour and Descola; Or, From Humanity as Technological Detour to the Peopling of Technologies," unpublished paper, 2013.

26. See Didier Fassin, "Another Politics of Life Is Possible," *Theory, Culture, and Society* 26, no. 5 (2009):44–60; Didier Fassin, *Humanitarian Reason: A Moral History of the Present* (Berkeley: University of California Press, 2011).

27. See João Biehl and Peter Locke, "Deleuze and the Anthropology of Becoming," *Current Anthropology* 51 (2010): 317–51.

CHAPTER ONE

1. The term has two distinct meanings. It originated in Bolivia, where it refers to one particular anthropologist and is not derogatory. In Paraguay it has become a general category in dialogue with Ayoreo understandings of evangelical Christianity. It should be noted that this folk category is not evenly spread among all Ayoreo people in Paraguay. However, nearly every Ayoreo person in the central Chaco is familiar with the discourse of *Abujádie* as satanic.

2. See Michael Taussig, *The Devil and Commodity Fetishism in South America* (Chapel Hill: University of North Carolina Press, 1980); Taussig, *Shamanism, Colonialism and the Wild Man*; Gaston Gordillo, "The Breath of Devils: Memories and Places of an Experience of Terror," *American Ethnologist* 29, no. 1 (2002): 33–57; Gordillo, *Landscapes of Devils: Tensions of Place and Memory in the Argentinean Chaco* (Durham, NC: Duke University Press, 2004).

3. George Lukacs, *History and Class Consciousness: Studies in Marxist Dialectics*, trans. Rodney Livingstone (London: Merlin Press, 1971), 83. See also Emily

Apter and William Pietz, eds., *Fetishism as Cultural Discourse* (Ithaca, NY: Cornell University Press, 1993); Marc Edelman, "Landlords and the Devil: Class, Ethnic and Gender Dimensions of Central American Peasant Narratives," *Cultural Anthropology* 9, no. 1 (1994): 58–93; Thomas Keenan, "The Point Is to (Ex)Change It: Reading Capital Rhetorically," in *Fetishism as Cultural Discourse*, ed. Apter and Pietz, 152–185.

4. Whereas Taussig explores Devil imagery in relation to Indigenous alienation from capitalist means of production and the fetish of the primitive in situations of colonial terror, Gordillo examines how routine fears associated with sites and conditions of wage labor are crystallized in images of the Devil, and impinge on the social production of place. Both Taussig and Gordillo point to the ways that Devil imagery allows for a layering of fundamentally opposed forms of fetishism—capitalist and Indigenous—and the productive ambiguities between them. Such fetishism, however, does not reflect a misleading distortion of "real" or veiled conditions, but rather, instantiates social, racial, and economic hierarchies.

5. Primo Levi, *The Drowned and the Saved* (New York: Vintage, 1989).

6. According to post-contact narrations of prior Ayoreo cosmologies, the founding ancestors of each clan emerged into the world created by the sum total of the transformations of the Original Beings, and ran through it, claiming all they saw for themselves. What they claimed became their *edopasade*, beings with a special relationship to one clan. Those clans that came first therefore had more *edopasade* than those that came last. These "belongings" were particularly willing to help their human relatives and they included parts of or whole plants, animals, manufactured objects, minerals, colors, emotional states, space, environmental features, virtues, forms or abstract qualities, such as hard or soft, fresh or stale, beautiful or ugly, clean or dirty, straight or round, firm or loose, dull or shiny, spicy or bland, spiny or smooth, and even categories of elements, such as "old person thing," "that which is known but unspoken," "a limit that proceeds and ends repeatedly," "two parallel lines," or "the patterns of anteater tracks."

7. The ethnographic production of the accepted version of Ayoreo tradition dates from articles based on Heinz Kelm's visits to the missions of eastern Bolivia in the early 1950s. He was followed by Sebag who made brief visits to Tobité and Maria Auxiliadora in the early 1960s. By 1970, seven other young anthropologists were visiting Ayoreo missions: Ulf Lind and Walter Regehr with Guidaigosode at the New Tribes Mission near Cerro León, Marcelo Bórmida with short visits to the Jnupedogosode at the New Tribes Mission of Tobité, Kelm's student Bernd Fischermann with Erapeparigosode at the Franciscan mission of Santa Teresita, Bernardo Vallejo for his University of Texas master's thesis in Tobité, along with Branislava Susnik and Miguel Bartolomé with brief stays among the Garaigosode at the Sale-

sian mission of Maria Auxiliadora. Significant work was later done by José Braunstein, Paul Bugos, and Jurgen Riester.

8. This knowledge economy is contentious partly because so few of these experts have conducted substantial fieldwork. For various reasons, local ethnographic methodologies have, with few exceptions, emphasized paid interviews collected over short stays ranging from one day to six weeks. Thus, Miguel Bartolomé noted in 2000 that no one, including him, has "lived in a strict sense with the Ayoreo. Various anthropologists have visited them, but none has lived at length the experience of a family hut nor the exhausting hunts in the thorny forest." Even today, long-term participant observation and immersion in daily life are often discouraged as dangerous, naïve, and contaminating. As Bartolomé writes, his brief fieldwork in the Salesian mission of María Auxiliadora instilled in him "an intimate and profound comprehension of the irreducible difference between my world and that of the others," a recognition that led him to abandon the hope of "identifying" with Ayoreo people. See Bartolomé, *Encuentro de la gente*, 17, 171. Furthermore, many Ayoreo-speaking people may not have welcomed fieldworkers. David Maybury-Lewis and James Howe described their visit to Ayoreo people at El Faro Moro in 1977 in the following terms: "Their disinclination to talk to us was strikingly obvious, and when they did, our interpreter tended to answer for them or to paraphrase their answers as if they were homilies on the word of the Lord," in David Maybury-Lewis and James Howe, *The Indian Peoples of Paraguay: Their Plight and Their Prospects* (Cambridge: Cultural Survival, 1980), 70.

9. For example, see Maria Cristina Dasso, "La cosmogonía Ayoreo y su relación con el medio actual," *Moana estudios de antropología oceánica* 6, no. 1 (2001), n.p.; Dasso, "Alegría y coraje en la alabanza guerrera de los Ayoreo del Chaco Boreal," *Anthropos* 99, no. 1 (2004): 57–72; Anatilde Idoyaga Molina, "La muerte como cambio ontológico en los Ayoreo del Chaco Boreal," *Mitológicas* 5 (1990): 35–40; Idoyaga Molina, "Cosmología y sistema de representaciones entre los Ayoreo del Chaco Boreal," *Scripta ethnologica* 19 (1995); Idoyaga Molina, "Mito, temporalidad y estados alterados de conciencia: El surgimiento de relatos y las experiencias shamánicas entre los Ayoreo del Chaco Boreal," *Acta Americana: Journal of the Swedish Americanist Society*, 5 (1997): 61–76; Idoyaga Molina, "Cosmología y mito. La representación del mundo entre los Ayoreo," *Scripta ethnologica* 20 (1998): 31–72; Celia Mashnshnek, "Las nociones míticas en la economía de producción de los Ayoreos del Chaco Boreal," *Scripta ethnologica* 8 (1990): 119–40; Mashnshnek, "Expressions of Alterity among Some Ayoreo Mythical Beings," *Anthropos* 88, nos. 1–3 (1993): 201–5; Mashnshnek, "La fenomenología como propuesta metodológica para la comprensión de tres culturas etnográficas del Gran Chaco: Mataco, Pilagá y Ayoreo." *Scripta ethnologica* 17 (1995): 29–35.

10. For his critique, see Gaston Gordillo, *En el Gran Chaco: Anthropologias e historias* (Buenos Aires: Prometeo, 2006). Examples of this model are found in Marcel Bórmida, "Ergon y mito: Una hermenéutica de la cultura material de los Ayoreo del Chaco Boreal. Parte I," *Scripta ethnologica* 1 (1973): 9–69; and Bórmida, "Como una cultura arcaíca conoce la realidad de su mundo," *Scripta ethnologica* 8 (1984): 13–183. Consider the following quote from Bórmida: "Perhaps [contemplating Ayoreo myths] we experience then the nostalgia for something that we have lost forever, of a world made to human scale, that we have bartered for another in which we feel always more estranged, simply moving guests in a reality in which men and things seem to move without reason, like the gears of a giant machine whose end it is not given to us to know," in "Como una cultura arcaíca," 14.

11. Ibid., 2–18.

12. See Marcelo Bórmida and Mario Califano, *Los Indios Ayorea del Choco Boreal* (Buenos Aires: Fundación para la Educación la Ciencia y la Cultura, 1978); Fernando Pages Larraya, "El complejo cultural de la locura en los Moro-Ayoreo," *Acta Psiquiátrica y Psicológica de América Latina* 19, no. 4 (1973): 253–64.

13. Carmen Nuñez, "Asai, un personaje del horizonte mítico de los Ayoreo," *Scripta ethnologica* 5 (1979): 113.

14. See Anatilde Idoyaga Molina, "El daño mediante la palabra entre los Ayoreo del Chaco Boreal," Göteborgs Etnografiska Museum Arstryck (1980), 33. Also, Idoyaga Molina, "La significación de la mítica de Susmaningái (el coraje) en la cosmovisión Ayoreo," *Mitológicas* 13 (1989), 31.

15. Celia Mashnshnek, "La percepción del blanco en la cosmovisión de los Ayoreo (Chaco Boreal)," *Scripta ethnologica supplemento* 11 (1991), 68.

16. This not so surprising given that ethnographic knowledge was often filtered through the missionaries that early ethnographers relied on as translators. Thus, Marcelo Bórmida, John Renshaw, and Lucien Sebag worked with NTM missionary Paul Wyma in Tobité; and Puesto Paz, Ulf Lind, and Walter Regehr worked with Norman Keefe and Joyce Buschegger in Faro Moro and Campo Loro. Bernd Fischermann worked with Elmar Klingler in Santa Teresita; and Mario Bartolomé and Branislava Susnik worked with the Salesian priest at Maria Auxiliadora. The missions and missionaries themselves are fiercely defended in the foreword to Bórmida and Califano's 1978 book, *Los Indios Ayoreo del Chaco Boreal*, from "false" attacks by Fischermann and others of their roles in the "transculturation" of the Ayoreo: "Such twisting of the truth offends not only a group of dignified, sacrificing and honorable people, but also the Bolivian State that gave permission and support for their work" (see p. 9).

17. For effective critiques of this equation, see esp. Gordillo, *En el Gran Chaco*, and Mario Blaser, "Way of Life' or 'Who Decides': Development, Paraguayan Indigenism and the Yshiro People's Life Projects," in *In the Way of Development*, ed. Blaser, Feit, and McRae (London: Zed Books, 2004), 52–71.

18. Volker von Bremen, "Fuentes de caza y recolección modernas: Proyectos de ayuda al desarrollo destinados a los indígenas del Gran Chaco" (Stuttgart: Servicios de Desarrollo de las Iglesias, 1987), 7. See also von Bremen, "Dynamics of Adaptation to Market Economy among the Ayoreode of Northwest Paraguay," in *Hunters and Gatherers in the Modern World*, ed. Peter Shweitzer, Megan Bisele, and Robert Hitchcock (Oxford: Berghahn, 2000), 275–87.

19. Ibid., 13–15, 16, 18. A similar notion is advanced by John Renshaw's analyses of the "moral economy" of Chaco Indians, although in a slightly more nuanced formulation. For Renshaw, the Indians of the Paraguayan Chaco are a distinct cultural group because they share a moral order rooted in the social necessities of hunting economies. Their ethnic identity, he argues, arises from the unchanging egalitarian values associated with this mode of production. See Jonathan Renshaw, *The Indians of the Paraguayan Gran Chaco: Identity and Economy* (Lincoln: University of Nebraska Press, 2002), 259–61; see also Renshaw, "La propiedad, los recursos naturales y el concepto de la igualdad entre los indígenas del Chaco paraguayo," *Suplemento antropológico* 23, no. 1 (1988), 159–62, 168, 223–29.

20. Von Bremen, "Dynamics of Adaptation," 276–77.

21. Ibid., 278.

22. Ibid., 279.

23. This summary is based on the revised and translated version of 2001. The thesis is based on four stages of fieldwork from 1969 to 2001, conducted while employed by NGOs working with Ayoreo in a variety of capacities, and is divided into sections titled history, vital space, political organization, economy, material culture, games, life cycle, clans, kinship, language, mythology, clan possessions, taboos, division of the cosmos, concepts of the soul, dreams, nature and culture, rituals, blood, sickness, and shamanism. In it, Fischermann exhaustively catalogues a variety of myths and compiles a large amount of important original data. In many ways I hope the present work is an extension, rather than a critique, of Fischermann's lifelong research.

24. Bernd Fischermann, "La cosmovision del pueblo Ayoreode," trans. Benno Glausser (University of Bonn, unpublished thesis, 2001), 144.

25. Ibid., 76; see also 103–5.

26. Ibid., 104.

27. Ibid., 22.

28. Ibid., 70–73.

29. Ibid., 246–47.

30. Many of the Beings were reportedly transformed into animals at their own request, including Giant Armadillo (*Hochekai*), Nine-Banded Armadillo (*Ajaramei*), Three-Banded Armadillo (*Aruco*), Talking Parrot (*Suaiya*), White-Lipped Peccary (*Ñacore*), Collared Peccary (*Toto*), and Giant Anteater (*Yahogue*). Others were transformed because they refused to obey moral

norms, an act that some but not all elders attributed to the Christian God, *Dupade*. Most plants reportedly chose to transform themselves out of their desire to take on particular characteristics, such as strength, rootedness, or beauty. Some of these Original Beings chose to transform themselves out of rage or fear following conflicts with the larger group, as in the case of ritually powerful Nightjar (*Asojna*), White Butterfly (*Corabe*), and Shamanic Soul Matter (*Pujopie*).

31. Many contemporary Ayoreo-speaking people translate both *sarode* and *ujñarone* as "secret."

32. As Fischermann writes, "The results of the mythic happenings are reflected in the actual ordering of the world that for the Ayoreode is valid and immutable, and where certainly culture and nature enter into multiple mutual relationships although on the other hand they may be separated, especially in that which is concerned with the level of language. In respect to the relating and employment of myths among the actual Ayoreo there are two functions. On one hand they explain why nature and human society have been formed according to the actual modality, and on the other hand they argue why the forms of acting and rules are adequate and necessary for the relationships that exist as much within the society as well as between man and nature" (see Fischermann, "La cosmovision," 76).

33. Terence Turner, "The Crisis of Late Structuralism: Perspectivism and Animism: Rethinking Culture, Nature, Spirit, and Bodiliness," *Tipiti* 7, no. 1 (2009): 3–42.

34. As Viveiros de Castro puts it, "Such a critique, in the present case, implies a redistribution of the predicates subsumed within the two paradigmatic sets that traditionally oppose one another under the headings of 'Nature' and 'Culture': universal and particular, objective and subjective, physical and social, fact and value, the given and the instituted, necessity and spontaneity, immanence and transcendence, body and mind, animality and humanity, among many more. Such an ethnographically-based reshuffling of our conceptual schemes leads me to suggest the expression, 'multinaturalism,' to designate one of the contrastive features of Amerindian thought in relation to Western 'multiculturalist' cosmologies. Where the latter are founded on the mutual implication of the unity of nature and the plurality of cultures—the first guaranteed by the objective universality of body and substance, the second generated by the subjective particularity of spirit and meaning—the Amerindian conception would suppose a spiritual unity and a corporeal diversity. Here, culture or the subject would be the form of the universal, whilst nature or the object would be the form of the particular." Viveiros de Castro, "Cosmological Deixis," 469–70.

35. Eduardo Viveiros de Castro, "Perspectival Anthropology and the Method of Controlled Equivocation," *Tipiti* 2, no. 1 (2004), 4. For a more nuanced but related argument see Aparecida Vilaça, *Strange Enemies: Indigenous Agencies*

and Scenes of Encounter in Amazonia (Durham, NC: Duke University Press, 2010).

36. Viveiros de Castro, "Perspectival Anthropology," 474.
37. Lucien Sebag, "Le Chamanisme Ayoreo (I)," *L'Homme* 5, no. 1 (1965): 5–32; "Le Chamanisme Ayoreo (II)," *L'Homme* 5, no. 2 (1965): 92–122.
38. Viveiros de Castro, "Cosmological Deixis," 472.
39. Turner, "Crisis of Late Structuralism," 21.
40. For instance, dense, hardwood trees or fruit-bearing cacti were biological siblings whose chants were used together for the same illnesses. Like Ayoreo-speaking people, several of the patrilineal *cucherane* clans could be represented within a single *jogasui*, or extended family group, of the Original Beings. For example, three species of thorny, low-lying bromeliads used by Ayoreo for food and fibrous threads (*doria, doi'die,* and *doria ijnoi*) were sisters whose *ujñarone* were used to heal people from lung infections and to restore their appetite. Their *jogasui* included *cuya* (wild tree beans), *nahua* (tree cactus) and his sister *canirAsore* (kind of root), their maternal parallel cousins *abue* (species of tall cactus) and his brother *dujnangai* (another species of cactus), *tokoi* (small cactus), and *Jobé* (tarantula), as well as *ditai* (contamination by spilled blood) and *dikore* (trembling in the darkness). Every member of the *jogasui's ujñarone* was particularly strong for coughs and respiratory infections. *Nahua* and *CanirAsore* were believed to have been shamans and were part of the shaman's peer group *(uhode).* Thus, their *ujñarone* were reserved for treating shamans, and so on.
41. See Lucas Bessire, "Ujñarone Chosite: Ritual Poesis, Curing Chants and Becoming Ayoreo in the Gran Chaco," *Journal de la Société des Américanistes* 97, no. 1 (2011): 259–289.
42. The former leader of the 2004 band described an experience of *chugu'iji*: "Afterwards Jacore'date heard Asôre's whistle. I began to want my anger, because we were bad before in the forest. My anger arrives. My body was hot. It was so hot that it began to move, like it was boiling. Asôre's whistle said, *Ngoa, Ngoa, Ngoa* [which meant that something was coming]. I put on my *cobia* vulture feather collar. I grabbed my spear and went outside of our house, and there I made *chugu'iji*. I am angry, I said, 'These children will see what it is that I am thinking, *je je je je.* I am like the Black Caiman, *jeeee.* I will do like Ugaraihaine did before, *jeeeeee.*' I thought that the *Cojñone* would kill us, but I was very strong. That is why I acted as if I was the same as Ugaraihai'ne, who died right in his camp but was not afraid and killed some of his enemies."
43. Turner, "Crisis of Late Structuralism," 22.
44. These chants, it appears, never spread too far beyond the Jnupedogosode Ayoreo group and were only considered powerful for a short time.
45. This emphasis on vomiting has significant precedents in prior initiation techniques for aspiring *daijnane*, in which initiates were expected to drink

large quantities of green tobacco juice. If one vomited out this juice, then he was not able to continue in the initiation. If the initiate did not vomit the juice, he usually went into a trancelike sleeping state in which the tobacco activated his latent *pujopie* soul matter, a process that connected him to a variety of helper spirits, usually but not always belonging to his *edopasade*.

46. Indeed, the figure of the *Abujá* overflows with what Mick Taussig calls "mimetic excess," or "an excess creating reflexive awareness as to the mimetic faculty, that allows . . . the sympathetic absorption into wildness itself, as with the identification with the Indian." For Taussig, the colonial production of difference is inseparable from the capacity of imagery and its copies to harness such powers that were "once in the hands of seers and magicians who worked images to effect other images, who worked spirits to affect other spirits which in turn acted on the real they were the appearance of." In *Mimesis and Alterity: A Particular History of the Senses* (New York: Routledge, 1993), 254.

CHAPTER TWO

1. See Bartolomé, *Encuentro de la gente*; Bórmida and Califano, *Los indios Ayoreos*; Hein, *Los Ayoreo, nuestros vecinos*; Jean Dye Johnson, *God Planted Five Seeds* (New York: Harper and Row, 1966); Chuck Wagner, *Defeat of the Bird God: The Story of Missionary Bill Pencille, Who Risked His Life to Reach the Ayorés of Bolivia* (Grand Rapids, MI: Zondervan, 1967).

2. Fischermann, "La cosmovision," 33.

3. Ulf Lind, "Los Ayoreos, un pueblo guerrero," *Mennoblatt* 4 (February 16, 1971); reprinted in Hein, *Los Ayoreos, nuestros vecinos*, 16–17.

4. Here, I take it for granted that Ayoreo subjectivities are historically produced and have never been the function of a static "time of tradition," while they also attend to history as a product of subjective negotiations within nonlinear and multiple registers of social time. The past here refers to what Achille Mbembe has called, on p. 16 of his book *On the Postcolony* (Berkeley: University of California Press, 2001), "the [peculiar] time of entanglement," emerging from an unpredictable interlocking of presents, pasts, and futures, in which the present is defined as "experience of a time."

5. My understandings of place and negativity are informed by Gaston Gordillo's compelling ethnography of the Argentine Chaco, esp. his books *Landscape of Devils* (Durham, NC: Duke University Press, 2004) and *Rubble* (Durham, NC: Duke University Press, 2014). Such writings have helped me grasp how the place formerly called Echoi has an ambivalent social life in the present. It now exists in four contradictory ways: as a set of nostalgic memories of a lost social center; a haunting space of murder, contagion, and immorality; a zone for the state preservation of "pure culture"; and

a conspicuous lacuna within intergenerational social memory and official histories alike. Like the disordered subjectivities of Ayoreo-speaking people themselves, Echoi today is untimely and schizoid. It is equal parts ruin, fetish, and political fact. Such uncanny qualities, moreover, mirror precisely the existential dilemmas that define indigeneity for contemporary Ayoreo-speaking people. It is impossible to fully meet the contradictory requirements for "being Indigenous" in the Gran Chaco, where indigeneity is always a field of contestation between competing schema of the moral "Indian" rather than an a priori or standard identity. Here, I show how Ayoreo-speaking people inscribe these tensions into places, and translate their contradictions into boundaries that are spatial, temporal, and moral, at the same time.

6. All too often, those charged with morally saving, governing, or scientifically understanding "the Ayoreo" overlook the present significance of ruptured and irreconcilable senses of social time. In response to what they each perceive to be assaults on the empirical truth to which they subscribe, they labor instead at producing yet another seamless (ethno)historical or antiquarian narrative of the past that segregates and explains "the Indigenous" present. Such ethnohistorical projects commonly assume that the retrospective projection of culture as ethnographically defined in the present renders distinct domains of the past intelligible as perspectival narratives or politically authorizes them as empirically true. I take a different tack: here I describe the past as a domain against which the limits of modern moral life are created and evoked in contradictory ways. This contestatory tension is especially significant within the highly mediated fields of indigeneity, where contradictory external representations of the past are used to locate the value of Native humanity in the present. This chapter explores how such tensions are fused with the ambiguous sensibilities of Ayoreo being through the historical imaginaries of a single place defined by competing interests, a zone wherein the past itself is continually reordered. It argues against the notion that history exists primarily as a universal form legible by projecting its contemporary cultural content into the past, and is thus a coherent domain of shared experience in need of empirical validation by experts. Indeed, such notions of history are themselves under investigation in what follows.

7. See esp. Heinz Kelm, "Die Zamuco (Ostbolivien)," *Zeitschrift für Ethnologie* 88 (1963): 66–85; Heinz Kelm, "Das Jahrfest der Ayore (Ostbolivien)," *Baessler Archiv* 19 (1971): 97–140; Bernd Fischermann, "La fiesta de la Asojna o el cambio de las estaciones entre los ayoréode del Chaco Boreal," *Suplemento antropológico* 40, no. 1 (2005): 391–450.

8. Thus, the annual gatherings at Echoi—at least sometimes—were associated with shared participation in what is considered to have been the most important ritual of *Cojñone* precontact life. In doing so, the terms of this ritual—which reportedly reconstituted the moral boundaries of social life

and its relation to the cosmos—also could be seen as articulating a shared sense of pan-Ayoreo identification. By a common observance of this ritual, Ayoreo-speaking people minimized differences and fashioned themselves as part of a greater community based on a shared moral order.

9. See Walter Benjamin, *Theses on the Philosophy of History* (New York: Illuminations, 1969).

10. On the existence, historical emergence, and legal significance of Indigenous "joint-use areas," see Harald Prins, *The Mik'maq: Resistance, Accommodation and Cultural Survival* (Fort Worth, TX: Cengage, 1996).

11. See Heidi Scott, "Paradise in the New World: An Iberian Vision of Tropicality," *Cultural Geographies* 17 (2010): 77–101. See also Jorge Cañizares-Esguerra, *Puritan Conquistadors: Iberianizing the Atlantic 1550–1700* (Stanford, CA: Stanford University Press, 2006); Anthony Pagden, *The Fall of Natural Man: The American Indian and the Origins of Comparative Ethnology* (Cambridge: Cambridge University Press, 1982).

12. Juan Fernandez, *Relación historical de las misiones de indios chiquitos que en Paraguay tienen los padres de la Compania de Jesus* (Jujuy, Argentina: CEIC, 1994 [1726]), 167–68.

13. See Charles Clark, "Jesuit Letters to Hervas on American Languages and Customs," *Journal de la Société des Américanistes* 29 (1937): 97–145; Susanne Lussagnet, "Présentation de 'Arte de la lengua Zamuca,' por el Padre Ignace Chomé," *Journal de la Société des Américanistes* 47 (1958): 121–23; Susanne Lussagnet, "Vocabulaires Samuku, Morotoko, Poturero et Guarañoka, précédés d'une étude historique sur les anciens Samuku du Chaco bolivien et leurs voisins," *Journal de la Société des Américanistes* 50 (1961): 185–243. Similar Zamuco word lists were sent by the Jesuit priest Joachim Camaño in a series of correspondence to Hervas y Panduro for inclusion in his survey of the world's languages, in vols. 17–21 of the great "Idea dell'universo," published in the late 1780s. It is interesting to note that the word lists of both Chomé and Camaño replicate slight vocabulary and pronunciation differences typically associated today with the northernmost Ayoreo groups.

14. Fernandez, *Relación historical*, 167–68.

15. See esp. ibid., 188–99. These efforts were led by Father de Zea. In 1715, de Zea made contact to the south of San Jose with a group of Curacates. Fernandez, writing from de Zea's report, states that in the village there were also a number of visitors from the Zamuco tribe. His efforts to reach various Zamuco groups were so arduous that he named the forest south of San Jose "purgatory, so that whoever might come to this country in following years . . . will know what it will cost him." After several failed attempts, de Zea finally reached the northernmost Zamuco village on July 12, 1717. The Indians, who had long heard of the "fame of his arrival," "celebrated him with demonstrations of extraordinary happiness; they approached him all kneeling, and all of the men, one by one went kissing his hand;

the women wanted to do the same thing; but the holy man, undone by his tears of consolation, gave them to kiss the holy image of the Virgen that he carried in his hand." The next day, he gathered them all in the plaza of the village, and "asked them if they wanted the missionaries to come among them in order to teach them the faith of Jesus Christ, and to show them the road to heaven." According to de Zea, the chief replied that they had desired it for a long time, but that there was no one to teach them the faith or show them the necessary rules to follow. After erecting a wooden cross and praying for the blessings of San Ignacio, de Zea departed, promising to return in a year to live among them in a "great village" where all of their people would be united. He was never able to fulfill his promise, however, and died in 1719 at the age of sixty-five. Father Miguel Yegros and Brother Alberto Romero, an aged ex-merchant from a well-to-do Spanish family, returned to the site of de Zea's cross in October 1718, and sent out messengers bearing a bejeweled baton and a colored shirt to the Zamuco chief. Yegros writes that these "good people" greeted them with "incredible joy." The chief of the Zamucos told the Jesuits that he and his people had been waiting anxiously for their arrival for more than a year. Yegros asked for guides to help locate good lands and visited the other villages of the Zamucos in order to locate the best site for the mission. Particularly, he asked for permission to visit "those that were to the west near the Salinas, where the Father had been informed that there were very good sites for towns, waterholes, forests and palm savannah for cattle ranches." Romero was killed by the Zamuco the following year. Future attempts were made by Jaime Aguilar and Agustin Castañares.

16. Francois-Xavier Charlevoix, *Histoire de Paraguay* (Paris: Desaint, David and Duran, 1757); Jean Vaudry, "Relación histórica sobre la reducción de San Ignacio de Zamucos," *Boletín de la Sociedad Geográfica de Sucre* 30 (1933): 253.

17. See Fernandez, *Relación historial*, 228.

18. The forty-one-year-old polyglot Ignacio Chomé arrived in San Ignacio de Zamuco in late 1737 seeking a launching point for expeditions to "discover the Great Pilcomayo River and the barbarian nations that inhabit both of its banks." According to his description, the mission was located in the "center of infidelity," having "no communication" with the Chiquitos missions, and composed of Indians that "were ferocious in the extreme." In a May 17, 1738, letter written in San Ignacio for Father Pedro Vanthiennen, Chomé writes, "I had to stay with the Zamucos in order to learn their language, which is spoken in all of these lands. God gave me such grace that in five months of applying myself to studying this language, that I learned how to proclaim the truths of the Religion." Although his southerly expeditions in 1738 and 1741 with combined groups of Zamucos and Ugaronos were failures, Chomé's "Arte de la lengua Zamuca" provides the central link

between Zamuco groups and contemporary Ayoreo; see José Aguirre Acha, *La antigua provincia de Chiquitos limitrofe con la provincia del Paraguay* (Sucre, Bolivia: Imprenta y Litigrafia Salesiana, 1933).

19. See the *Status Missionum Chiquitensium Annum*, in the Jesuit Archives of Buenos Aires.

20. See Vaudry, "Relación histórica," 260–63. Chomé was named Father Superior of San Ignacio in 1740, and in a 1746 letter to Pedro Vanthiennen, he describes the final, calamitous years of the mission. Tensions among Zamuco groups and between the Indians and the priest increased. Chomé describes how the Ugaronos were divided into two groups: the Ugarono proper brought to the mission in 1732, and the Sapios brought to the mission in 1738. The second group was led by Santiago Dione, an elder whom Chomé describes as an "inveterate evil-doer." The two Ugarono groups soon began to "molest and threaten" the outnumbered Zamucos, who wanted to move to the Chiquitos missions. For their part, the Ugaronos, according to Chomé, by 1745 "thought to return to their old camps, abandon the Chrisitan faith and return to the primitive life of savages they had before the arrival of the Fathers." Dione developed a plan to attack the Zamucos and leave the mission with his group in August of that year, but a young Ugarono woman told the priest soon after they had left. Chomé "ordered the Ugaronos to follow the fugitives" and force them to return to San Ignacio. Once back on the mission, they formed a plot "to assassinate" the priest, which was abandoned when Dione suddenly died. Meanwhile, the Zamucos did not want to stay on the mission. Chomé was so afraid of attack by the Ugaronos that he requested an escort of two hundred Chiquitano Indians from the north, but the Father Superior, fearing that they would be killed, sent instead a single priest, Juan Esponella, with "sacred ornaments to celebrate the Mass."

21. This ethnohistorical terrain has been well traveled by scholars. See esp. Isabelle Combés, *Zamucos* (Cochabamba: Editoral Nomades–Universidad Catolica, 2009).

22. See letter titled "Descrubrimiento de la salina del pueblo de Santiago de Ordn del Señor Gobr. Don Melchor Rodrigues Barrioluengo hiso su Admtr. Don José Ramon Baca," Archivos Nacional de Bolivia. Incidentally, the priest locates the salt pans fifteen leagues to the south of the Jesuit corral and watering hole for cattle, which Chomé placed sixteen leagues to the north of San Ignacio.

23. See letter, "Melchor Rodriguez to Nicolas de Arrexondo," September 25, 1794, Archivos Nacional de Bolivia.

24. See Alcide D'Orbigny, *Voyage dans l'Amérique meridionale* (Paris: Bertrand, 1835), 689. Following the visit of the famed French naturalist, all post-independence sources relate Zamucos groups to the area of the Salinas. D'Orbigny describes seven hundred Guaranoca living in Santiago, fifty Zamucos in Santo Corazon, and five hundred Morotocas living in San Juan,

where they lived in a separate neighborhood and spoke a distinct language; its "smoothness and softness" and "agreeable euphony" led him to call Ayoreo "the Italian of the deserts." He writes, on p. 308, that "the Samucos have maintained, among the Chiquitos, a reputation for bravery and strength; the Morotocas are also the most feared of all the missions and those who command wherever they present themselves. Essentially good, social and the most hospitable and even affectionate with foreigners, their reception is open and happy. Everything indicates in them the affection for pleasures, especially for dancing, which the women practice with passion." Likewise, Bach wrote in 1842 that "the Nation of the Guaranocas, that is composed of around 200 families, occupies part of the Salinas of Santiago," and another group is found "around the margins of a great lake . . . at a distance of 20 leagues to the south of Salinas." The same information was used by José León Oliden to conclude that "a group of Samucos Indians exists in the old Jesuit mission of San Ignacio de Samucos . . . they are very close to the Salinas of Santiago and they know it well," as well as Belaieff in 1941 to describe the twentieth-century location of the Moros. See Mauricio Bach, *Descripción de la nueva provincia de Otuquis en Bolivia* (La Paz, Bolivia: Imp Unidas, 1842).

25. José Cardus, *Las misiones franciscanas entre los infieles de Bolivia* (Barcelona: Libreria la Inmaculada Concepción, 1886), 273–74.

26. Ibid., 275.

27. For example, of the last four traditional Totobiegosode political leaders— A'asi, Pejeide, Ugaguede, and Dejaide—three had actually been born as a Guidaigose. The lone exception, Pejeide, married a woman who had been Guidaigoto and was himself reportedly the son of a man who was first Ducodegose. This origin, however, did not prevent them from all becoming bitterly opposed to the "Guidaigosode" during their lifetimes. The same ethnic shift happened to the "pure" Totobiegosode group remaining in the forest. The leader of the Areguede-urasade group who made contact in 2004 was Areguede. He had been born as a Ducodegosi, and his wife had been a Guidaigoto. The reverse was also true, with many former Totobiegosode now found among the Guidaigosode. Even closer kinship ties reportedly linked the Totobiegosode to the Nyamocode and Iñojamuigosode, now-extinct groups that have been assimilated into the Garaigosode communities along the Rio Paraguay. This was in spite of the fact that both of these were usually allied with enemies of the Totobiegosode. According to elder informants, gage identity has always been something that one could choose or leave behind (*ore cha'a*). Such breaks usually consisted of one family or individual and the rest of the group. Because they implied social ruptures that could not be easily repaired, individuals rarely changed their gage identification more than once.

28. See Carlos Pastore, *El Gran Chaco en la formación territorial del Paraguay: Etapas de su incorporación* (Asunción, Paraguay: Criterio, 1989)

29. A number of pamphlets were circulated in Germany targeting those interested in "transplanting German thinking and attitudes to his new foreign homeland," describing fertile soil and opportunities to form German agricultural corporations and ranches in Paraguay. Among the many foreign colonies established in eastern Paraguay before the end of the nineteenth century was the Aryan settlement Elisabeth Forster-Nietzsche and her husband, Bernard Forster, founded in order to preserve a pure German ethnicity. See Peter Klassen, *The Mennonites in Paraguay*, vol. 2, *Encounter with Indians and Paraguayans* (Winnipeg, Canada: Mennonite Books, 2002), 8.
30. See Klassen, *Mennonites in Paraguay* 2: 32, 47, 61, 186–88.
31. See David Zook, *The Conduct of the Chaco War, 1932–1935* (New York: Bookman Associates, 1960).
32. See ibid., 61–62. See also Paul Bugos, "An Evolutionary Ecological Analysis of the Social Organization of the Ayoreo of the Northern Gran Chaco" (Evanston, IL: Northwestern University, unpublished dissertation, 1985).
33. See Paul Bugos, "An Evolutionary Ecological Analysis of the Social Organization of the Ayoreo of the Northern Gran Chaco" (Evanston, IL: Northwestern University, unpublished dissertation, 1985), 53–60; Estudios Rurales, *Apuntes para la colonización* (La Paz, Bolivia, internal document, 1981).
34. See Hein, *Los Ayoreo, nuestros vecinos*, 28.
35. Testimony of Gajnemeide, quoted in APCOB, *Entrevistas con CucAsoregosode* (Santa Cruz, unpublished document, 1993).
36. See Gonzalo Aguilar Dávalos, "El movimiento indígena en las tierras bajas en el siglo XX," in *Los Bolivianos en el tiempo*, ed. Alberto Crespo (La Paz, Bolivia: Editorial Indea, 1995).
37. See Johnson, *God Planted Five Seeds*; José Perasso, *Cronicas de cacerias humanas: La tragedia Ayoreo* (Asunción, Paraguay: El Lector, 1987).
38. In 1947, a Jnupedogosode group—which had been followed for years by missionaries—appeared at the Bolivian railroad camp of Ipias. The New Tribes Mission established Tobité with this group in 1948, the South American Indian Mission established Zapoco with Direquednejnaigosode in 1950, and Latvian Baptists established Rincon del Tigre with Cochocoigosode in 1952. Meanwhile, these northern Ayoreo obtained shotguns from the missionaries and waged devastating raids on their former enemies, the Guidaigosode, who soon also sought refuge on missions. By 1966, three additional missions had been established in Paraguay with the survivors, including the Salesian Catholic mission of Maria Auxiliadora inhabited by Garaigosode groups, the Franciscan Catholic mission of Santa Teresita formed with Guidaigosode groups, and the New Tribes Mission of Cerro León formed of Ducodegosode, Tiegosode, and Ijnapuigosode branches of the Guidaigosode confederation. Aided by enslaved Ayoreo whom they used as interpreters, missionaries from these often-feuding de-

nominations sent regular armed contact expeditions to bring nearby bands back to their missions. By the mid-1960s, New Tribes missionaries from Bolivia and Paraguay realized that the Moros and the Barbaros spoke the same language, and they began organizing international meetings of Ayoreo believers. En masse conversion to Christianity followed the first visit from Christian Ayoreo of Tobité to the Paraguayan mission of El Faro Moro in 1975. Families, bands, and confederations were divided in the upheavals of contact, and often people were resettled far from their ancestral territory. Many did not easily adapt to mission life. Some individuals chose to return to the forest, where they lived alone. Others walked for hundreds of miles between missions, or took long-term jobs on ranches. Ayoreo-speaking people independently established a continuous presence in the towns and cities of non-Ayoreo, including several groups who made first contact by appearing in downtown Filadelfia, Paraguay, in 1963 and 1964, much to the surprise of the town's Mennonite inhabitants.

39. For more on the complex affective ties between people and places that inform this argument, see esp. Keith Basso, *Wisdom Sits in Places: Landscape and Language among the Western Apache* (Albuquerque: University of New Mexico Press, 1996); Gordillo, *Landscape of Devils*; Tim Ingold, *The Perception of the Environment: Essays on Livelihood, Dwelling and Skill* (London: Routledge, 2000); Fred Myers, *Pintupi Country, Pintupi Self: Sentiment, Place and Politics among Western Desert Aborigines* (Berkeley: University of California Press, 1991); Kathleen Stewart, *A Space on the Side of the Road: Cultural Poetics in an "Other" America* (Princeton, NJ:Princeton University Press, 1996); Ann Stoler, "Imperial Debris: Reflections on Ruins and Ruination," *Cultural Anthropology* 23, no. 2 (2008): 191–219. The erasure of Echoi suggests how certain overlaps between the transnational politics of indigeneity and the residues of colonial violence might imply a distinct process of ruination. Echoi is not simply one more "corroded hollow," or "space on the side of the road" devastated by capitalist production or the accreted deposits of "imperial debris." As a ruin, it bears witness to a violence aimed at changing the spatialized limits of human life itself. But it is a process that is always partial and interrupted. The violent disordering of social time is inscribed into moral bodies at the divergent points at which they are carved into landscapes, giving form to haunting and incoherent new zones of repression, risk, and contagion.

40. Echoi today can be considered a form of "negative space," a vertiginous zone arising in opposition to a variety of contradictory social projects and everyday perceptions. As in the negative dialectics described by Theodor Adorno, the opposition of these sets of meanings does not lead to a transcendent synthesis but further contradictions. They decenter their objects and prevent closure. In such ways, an ethnography of Echoi is not a genealogy of redemption. Rather, Echoi is a spatial manifestation of the

negative dialectics at the core of colonial power and violence. The haunting memories of this place suggest an alternative genealogy of Ayoreo humanity that has been erased from scholarly accounts. Such memories could become important political resources for contemporary Ayoreo-speaking people and their frustrated nation-building projects. Yet much of the past and its potentials have become irretrievable. For more on negative dialectics, see Theodor Adorno, *Minima Moralia*, trans (London: EFN Jephcott, 1974); Gordillo, *Landscape of Devils*, 255–56.

CHAPTER THREE

1. See Richard Arens, ed., *Genocide in Paraguay* (Philadelphia: Temple University Press, 1976); Luke Holland, *Indians, Missionaries and the Promised Land: Photographs from Paraguay* (London: Survival International, 1980); Mark Munzel, *The Ache Indians: Genocide in Paraguay* (Copenhagen: International Work Group for Indigenous Affairs, 1973).
2. Rene Harder Horst, *The Stroessner Regime and Indigenous Resistance in Paraguay* (Gainesville: University Press of Florida, 2007), 110, 129.
3. Escobar, *Misión etnocidio*, 22
4. Ibid., 37.
5. See, for example, José Perasso, *Crónicas de cacerías humanas: La tragedia Ayoreo*, (Asunción, Paraguay: El Lector, 1987); Volker von Bremen, "Los Ayoreode cazados," *Suplemento antropológico* 22, no. 1 (1987): 75–94.
6. Pierre Clastres, "Of Ethnocide," reprinted in *Archaeology of Violence*, trans. Jeanine Herman (Los Angeles: Semiotext[e], 2010 [1974]), 103–13.
7. Quoted in Hein, "Los Ayoreos, nuestros vecinos," 216.
8. Escobar, *Misión etnocidio*, 40.
9. Ken Johnston, *The Story of the New Tribes Mission* (Sanford, FL: New Tribes Mission Press, 1985), 2–3.
10. Ibid., 38.
11. Ibid., 145.
12. *Brown Gold*, January 1944.
13. Johnston, *Story of the New Tribes Mission*, 22, 28. Also, consider the following quote from p. 254: "When we get to Heaven, all tongues, tribes, nations and peoples will be represented around the Throne, singing that new song to our wonderful Savior and Lord. One of these days, the last soul that is needed to complete the Body of Christ, the Church, will be saved. That last soul will very possibly be some tribesman out in the jungle somewhere. When the church is complete, the Lord will come back for his own with a shout, and 'so shall we ever be with the Lord.' Wouldn't it be thrilling to be the one to reach that last person with the Gospel message?"
14. See "Doors Opening to the World's Untouched Fields," *Brown Gold*, December 1945; "A Spiritual Iron Curtain," *Brown Gold*, November/December,

1946; "The Strategic Advance," *Brown Gold,* May 1944; "Work Starts at Boot Camp," *Brown Gold,* August 1944; "Looking Forward to VJ Day," *Brown Gold,* June 1945.

15. See "First Bolivian Conference," *Brown Gold,* June 1946; "Reaching Out in Bolivia," *Brown Gold,* February 1951.
16. Johnson, *God Planted Five Seeds,* 23.
17. Ibid., 20.
18. See also Jurgen Riester, *En busca de la loma santa* (Santa Cruz de la Sierra: Amigos del Libro, 1976).
19. Johnson, *God Planted Five Seeds,* 92.
20. Johnston, *Story of the New Tribes Mission,* 44–45.
21. "Brown Gold," *Brown Gold,* October 1943; *Brown Gold,* January 1944 (no title).
22. "Brown Gold?," *Brown Gold,* January 1997.
23. "Brown Gold," *Brown Gold,* October 1943, also quoted in Johnston, *Story of the New Tribes Mission,* 44.
24. Quoted in Johnston, *Story of the New Tribes Mission,* 45.
25. Michael Taussig, *My Cocaine Museum* (Chicago: University of Chicago Press, 2004), 5.
26. Karl Marx, *Capital,* vol. 1, *A Critique of Political Economy,* trans. Ben Fowkes (London: Penguin, 1990 [1867]), 183.
27. Ibid., 184–86.
28. Ibid., 187, 189–90.
29. Ibid., 228–31.
30. Ibid., 299.
31. Johnston, *Story of the New Tribes Mission,* 25.
32. "Go Ye Forth," *Brown Gold,* May 1944.
33. Consider the apocryphal story still told in eastern Bolivia about a hunter captured by the *Bárbaros.* One day, the story goes, a hunter from a back-woods town was captured by the *Bárbaros,* who tortured him, pulled out his eyebrows and eyelashes, chopped off his hair with a chisel, burned his body, and then carried him off to their camp. For some reason they did not kill him, and he learned to talk with them through sign language. Eventually, they made this man their king. After a year or more, the king took one of his friendly subjects and secretly returned to his wife, who no longer recognized him. Only after he told her the names of all the domestic animals in the yard did she let him in. The king soon left, carrying a heavy sack of sugar for his barbarian subjects. His panic-stricken wife informed the town police, who promptly gathered a posse of *vecinos.* The posse tracked the man back to the *Bárbaro* camp, and set up an ambush at dawn. After a fierce battle, the Christian *vecinos* captured more than fifty barbarians, including the king, who was recognized by one of the townspeople and promptly jailed. Eventually released after telling this tale, he was soon

employed by a New Tribes missionary. After eight months, however, he spent all of his wages on a load of sugar and returned to the jungle, from whence he has yet to emerge. In "A King in Captivity," *Brown Gold*, August 1945.

34. Johnson, *God Planted Five Seeds*, 112, 113, 148, 153; see also "Future Barbaro Work," *Brown Gold*, December 1947 / January 1948.

35. "A Friendly Contact with Savages," *Brown Gold*, August/September 1947.

36. Johnson, *God Planted Five Seeds*, 87–88.

37. See also, "Efforts toward the Unreached," *Brown Gold*, September/October 1946. Note that such relationships were often suddenly reversed when missionaries actually encountered Ayoreo groups. Johnson describes their first encounter with a Jnupedogosode band (including a young, flirtatious Simijáné) in September 1947: "Inez, too, had lost her tongue. She cowered against me, and did not want to leave my side; nor would I leave hers, for I was terrified of losing her. I remembered what Inez had once told me about the Ayores—that they might kill her because she had been too long with the *Cojñone*. I took my first good look at these strong dark bodies, whose long hair made them all the more ominous, and whose own fears kept them from smiling much. I breathed a prayer. I felt they were capable of anything. Several Ayores formed a knot around Inez and began talking to her. . . . I turned to Inez and asked, 'What are they saying, Inez? What are they saying?' Her face was an enigma, absolutely expressionless. Not even the shadow of a smile crossed her features. Was it indicative of rapt attention or cold fear? I hadn't a clue. A wave of remorse swept over me. I had brought her into this; if anything happened to her I would never forgive myself. Finally she seemed to understand how upset I was. 'Señora, they're only telling me about my brothers!' She said to me in a reproachful tone, with some impatience. With that I had to be satisfied." Ibid., 121.

38. Ibid., 88–104. Besides Guto'date, there were two other individuals who played similarly pivotal roles in enabling the first missionary contacts with Ayoreo-speaking bands. Both Comai and Ikebi were first enslaved, then acquired by missionaries. Comai was an orphan taken in 1949 from an Ayoreo band which had settled along the railroad line at San Jose, and eventually given to Bill Pencille. They worked together until Pencille returned to the U.S. definitively in the mid-1960s. Ikebi was the "first wild savage Moro" to be captured in Paraguay. The twelve-year old boy was roped near Bahia Negra Paraguay in 1956, before being publicly exhibited in a cage in Asunción, where for a small donation, onlookers could see an authentic savage. Several months later, he was placed in the care of the Salesian priests who were seeking a translator for their own Moro contact work. Ikebi worked with both the Salesians and the New Tribes missionaries facilitating first contacts throughout the 1960s and 1970s.

39. "Living with the Barbaros," *Brown Gold*, November 1948.

40. "Future Barbaro Work," *Brown Gold*, December 1947 / January 1948.

41. "Tobité," *Brown Gold,* February 1951.
42. See "From Tobité, Bolivia," *Brown Gold,* February/March 1952.
43. "What are the Savages Like?," *Brown Gold,* August/September, 1948.
44. "From Tobité, Bolivia," *Brown Gold,* February/March, 1952.
45. "Living with the Barbaros," *Brown Gold,* November 1948; also "Barbaros at Tobité," *Brown Gold,* January 1951.
46. In Wagner, *Defeat of the Bird God.*
47. "Barbaros at Tobité," *Brown Gold,* January 1950.
48. See "Chief Buried Alive," *Brown Gold,* August/September 1948: "This past week I have gone through some experiences and seen some things I hope I will never have to witness again. . . . I heard a cry, and could not believe my eyes. There was a hand pushed up through the ground. It was the hand of man; he was dead, but in with the corpse there was a sick baby crying. I went into camp and told them about it, but they just laughed and said the baby was sick. I went back the following day and the baby was still crying."
49. Ibid.: "I just sat back in amazement and horror to think of this baby in the hole with a corpse. I went back to the others and we pleaded to see if they would permit me to dig it up and try to bring it back to health. It was useless and for three days I went back to hear the same weakening cry. I didn't go back any more then, as it got so I couldn't sleep. When I hear a little one cry at night now, I just bawl thinking and trying to get that awful picture from my mind."
50. "Chief Buried Alive," *Brown Gold,* August/September 1948.
51. "Barbaros Plan Attack!," *Brown Gold,* January 1950.
52. Wagner, *Defeat of the Bird God,* 20.
53. Quoted in Holland, *Indians, Missionaries and the Promised Land,* 36.
54. See Munzel, *Ache Indians,* 1–34.
55. Ibid., 32.
56. Jehan Vellard, "Une mission scientifique au Paraguay," *Journal de la Société des Américanistes* 25 (1933): 316–17, 324. Quoted in Munzel, *Ache Indians,* 6.
57. See "Moro Encounter!" *Brown Gold,* May 1966; "Mission to the Moros," *Brown Gold,* February 1967; "Mission to the Moros," *Brown Gold,* May 1968; "Unrest in the Moro Camp," *Brown Gold,* January 1970.
58. New Tribes Mission, posted December 6, 2005, at http://www.ntm.org /train/news_details.php?news_id=2678.
59. Taussig, *Shamanism, Colonialism and the Wildman,* 288.

CHAPTER FOUR

1. Bartolomé, *Encuentro de la gente,* 289.
2. Ibid., 287.
3. Escobar, *Misión etnocidio,* 233.

4.   Jean Pierre Estival, "Os casadores e o radio: Sobre o novo uso dos meios de comunicaçao entre os Ayoreo do Chaco Boreal," *Revista Anthropologica* 17, no. 1 (2006), 104, 111.

CHAPTER FIVE

1.   See Joel Robbins, *Becoming Sinners: Christianity and Moral Torment in a Papua New Guinea Society* (Berkeley: University of California Press, 2004); Robbins, "Continuity Thinking and the Problem of Christian Culture: Belief, Time and the Anthropology of Christianity," *Current Anthropology* 48 (2007): 5–38.
2.   Thomas Abercrombie, *Pathways of Memory and Power: Ethnography and History Among an Andean People* (Madison: University of Wisconsin Press, 1998), 416.
3.   By "apocalyptic futurism," I refer to the ways in which apocalyptic schema organize meaning in the present from the vantage of an imagined future. For a reminder that writing about apocalypticism is never innocent, see Jacques Derrida, "Of an Apocalyptic Tone Newly Adopted in Philosophy," in *Derrida and Negative Theology*, ed. Harold Cowald and Toby Foshay (Albany, NY: SUNY Press, 1992). In Kathleen Stewart and Susan Harding's "Bad Endings: American Apocalypsis" (*Annual Review of Anthropology* 28 [1999]: 285–310), the authors describe how Indigenous apocalypticisms in Melanesia have been coproduced by ethnographic economies, becoming "artifact[s] of entwined practices of strategic mistranslation on the part of peoples undergoing some manner of colonialism and of strategic stigmatization on the part of the colonizing peoples" (287). Apocalypticism—as belief and descriptive economy—has often been preemptively attributed to Indigenous peoples thought to be inevitably doomed or vanishing, a "model for" Western eschatology and power posing as a "model of" Native futures. Indeed, cultural and literary critics have pointed to the many ways apocalypticism is a central trope within a vast array of post-Enlightenment political ideologies. See esp. Malcom Bull, ed., *Apocalypse Theory and the Ends of the World* (Oxford: Blackwell, 1995); Mike Davis, *Dead Cities and Other Tales* (New York: Zone, 2002); Richard Dellamora, ed., *Postmodern Apocalypse: Theory and Cultural Practice at the End* (Philadelphia: University of Pennsylvania Press, 1995); Frank Kermode, *The Sense of an Ending: Studies in the Theory of Fiction* (Oxford: Oxford University Press, 1967).
4.   See Norman Cohn, *The Pursuit of the Millennium: Revolutionary Millenarians and the Mystical Anarchists of the Middle Ages* (New York: Galaxy Books, 1970), for more on how apocalypticism is built around millennial precepts that "promise imminent collective salvation for the faithful in a paradise that will rise following an apocalyptic destruction ordained by the gods" (13).
5.   Vincent Crapanzano, *Imaginative Horizons: An Essay in Literary Philosophical Anthropology* (Chicago: University of Chicago Press, 2004), 2. Cf. Jane

Guyer, "Prophecy and the Near Future: Thoughts on Macroeconomic, Evangelical and Punctuated Time," *American Ethnologist* 34 (2007): 409–21.

6. This tannin-rich tree species was a valuable global commodity and highly sought from 1870 until the emergence of synthetic tannins in the 1920s. Old stumps bear witness to the teams of woodcutters who once traveled through the dense Chaco forest.

7. Branislava Susnik, "La lengua de los ayoweos-moros: Nociones generales," *Boletín de la Sociedad Científica del Paraguay y del Museo Etnográfico* 8 (1963): 133.

8. Ulf Lind, "Los Ayoreos, un pueblo guerrero," *Mennoblatt* 4 (1971): 4–8, quoted in Hein, *Los Ayoreos, nuestros vecinos*, 18.

9. Veena Das, *Life and Words: Violence and the Descent into the Ordinary* (Berkeley: University of California Press, 2006), 8.

10. For example, see Cathy Caruth, *Unclaimed Experience: Trauma, Narrative and History* (Baltimore: Johns Hopkins University Press, 1996).

11. See New Tribes Mission and Indian Settlement Board's "Proyecto Ayoreo: Economic Development at El Faro Moro, What Is Feasible? A Feasibility Study Conducted by a Joint Commission of New Tribes Mission and Indian Settlement Board" (unpublished document, January 1977), 5.

12. Maybury-Lewis and Howe, *Indian Peoples of Paraguay*, 70; Perasso, *Crónicas de cacerías humanas*, 40.

13. On this point, see also Volker von Bremen, "Los Ayoreode cazados."

14. Escobar, *Misión etnocidio*, 33.

15. It is important to note that *Dupade* is an Ayoreoization of the Guarani word *Tupa*, the name of the creator deity who resides in the sun. It was introduced to the Ayoreo language by Jesuit missionaries in the eighteenth century and taken up two centuries later by evangelical missionaries as denoting the closest approximation to "God" that existed in the Ayoreo language. For more on this, see Charles Clark, "Jesuit Letters to Hervas on American Languages and Customs," *Journal de la Société des Américanistes* 29 (1937): 97–145.

16. Kenelm Burridge, *Mambu: A Melanesian Millennium* (London: Meuthen, 1960).

17. Here, I am drawing on the notions of "rotational" and "lived" time developed in different ways by Henri Bergson, *Duration and Simultaneity* (Indianapolis: Bobbs-Merrill, 1965); Das, *Life and Words*; Achille Mbembe, *On the Postcolony* (Berkeley: University of California Press, 2001); Eugene Minkowski, *Lived Time: Phenomenological and Psychopathological Studies* (Evanston, IL: Northwestern University Press, 1970), as well as work on the disciplining of time through colonial evangelism, such as Tom Beidelman, *Colonial Evangelism: A Socio-Historical Study of an East African Mission at the Grassroots* (Bloomington: Indiana University Press, 1982); and Bambi Schieffelin, "Marking Time: The Dichotomizing Discourse of Multiple Temporalities," *Current Anthropology* 43 (2002): 5–17.

18. Abercrombie, *Pathways of Memory and Power*, 408–22.

CHAPTER SIX

1.   The anthropology of affect has historically focused on the cultural, spatial, and temporal variability in the emotional lives of "small-scale societies" and "non-western peoples." The most progressive of these projects aimed to reveal the universalizing pretensions of Enlightenment dichotomies of reason and passions or science and magic, and thus to undermine the imperialist pretensions whereby certain colonial Others—women, children, and primitives—were violently disenfranchised and denied equality within rational modernity insofar as their humanity or modal personality could be relegated to the sign of the irrational, the superstitious, and the emotional. At the risk of oversimplifying a rich and often contradictory body of scholarship, this progressive anthropological response focused on unsettling and diversifying precisely those categories of the person—rationality, interests, psyche, body, science—by which emotions had long been intelligible. To do so, anthropologists described the function, meaning and force of emotions in a wide range of social settings. For the interpretive anthropology of the cultural logics and meanings of emotion concepts, see esp. Jean Briggs, *Never in Anger: Portrait of an Eskimo Family* (Cambridge, MA: Harvard University Press, 1970); Catherine Lutz, *Unnatural Emotions: Everyday Sentiments on a Micronesian Atoll and Their Challenge to Western Theory* (Chicago: University of Chicago Press, 1988); Fred Myers, "The Meaning and Logic of Anger among Pintupi Aborigines," *Man* (1979), 589–610; Michelle Rosaldo, "Towards an Anthropology of Self and Feeling," in *Culture Theory: Essays on Mind, Self and Emotion*, ed. Richard A. Shweder and Robert A. LeVine (Cambridge: Cambridge University Press, 1984); Rosaldo, *Culture and Truth*; Edward Schieffelin, "Anger and Shame in the Tropical Forest: On Affect as a Cultural System in Papua New Guinea," *Ethos* 11 (1983). See also Sara Ahmed, *The Cultural Politics of Emotions* (New York: Routledge, 2004). For the political anthropology of social sentiments, see esp. Fassin, *Humanitarian Reason*; Ann Stoler, *Race and the Education of Desire: Foucault's History of Sexuality and the Colonial Order of Things* (Durham, NC: Duke University Press, 1995); Ann Stoler, "Affective States," in *A Companion to the Anthropology of Politics*, ed. David Nugent and Joan Vincent (New York: Blackwell, 2004), 4–20. For phenomenologically inflected projects on embodied affects, see Robert Desjarlais, *Body and Emotion: The Aesthetics of Illness and Healing in the Nepal Himalayas* (Philadelphia: University of Pennsylvania Press, 1992); Kathleen Stewart, *Ordinary Affects* (Durham, NC: Duke University Press, 2007); Jason Throop, *Suffering and Sentiment: Exploring the Vicissitudes of Experience and Pain in Yap* (Berkeley: University of California Press, 2010). My project is not to police the boundaries between these projects, but to draw insights from each and bring them collectively to bear on the case in question.

2. These diverse anthropologies of emotion, in turn, flow from divergent approaches to morality and ethics within social science more broadly: one, inspired by deontological ethics via Kantian sociology, which imagines morality as a given set of external norms and judgments imposed on individuals by the social; another consequentialist trend which assesses the practical outcome of ethical positions rather than their relationship to wider social systems; and a third, grounded in Foucault's genealogies of virtue, which imagines ethics as a technical process by which populations are managed and selves are governed. Didier Fassin, in distinguishing a moral anthropology from any of these singular projects, calls for a more fluid model that takes as its object the complex and multilayered "moral making of the world." See Didier Fassin, "Introduction," in *A Companion to Moral Anthropology*, ed. Didier Fassin (Oxford: Blackwell, 2012).

3. *Dipoi*, or explicitly offensive embarrassment, is produced when someone maliciously "hangs it out to dry," or "strikes someone with it," (as in the expression—*inguira dipoi*) and it refers to intentionally damaging someone's reputation or social standing by causing public shame. Such hanging out or hitting is related to publicizing certain kinds of words, such as the *pu'upi pogiedode* ("secret words"), *che'edo* ("talking behind one's back"), or *nipongeaque* ("publicizing something that you caught someone else doing").

4. See Myers, *Pintupi Country, Pintupi Self*, for a more developed analysis of the generative tensions between countervailing moral sentiments, normative personhood, and social structure among small-scale societies.

5. My own proximity to both the pollution of Ayoreo bodies and, presumably, their highly coveted "hidden knowledge" posed an uncomfortable dilemma to the professionals in the booming Ayoreo industry of the Chaco. Despite a certain macho cachet, this proximity was also regularly targeted in what aspired to be public insults.

6. William Ian Miller, *The Anatomy of Disgust* (Cambridge, MA: Harvard University Press, 1997), 8–9.

7. Ibid., 6.

8. Its many variants include *eteguedate*, one who is scorned or rejected strongly; *etepidi*, that which one holds against someone else; *etesoredate*, one who rejects something or someone very strongly; and *etesori*, or one who hates, one who despises. Both *etesori* and *eteto* can also refer to an enemy. In present contexts, the lines between sensory and violent oppositions become unstable and blurred.

9. This case is described in chilling detail in an unpublished 1987 exchange of letters between concerned Ayoreo leaders and public officials of Roboré.

10. A man named Juan told me the following story: "Several years ago, there was a young man working on an estancia near Tte Martinez, who had vanished after a while and no one knew where he was. The other Ayoreo

wanted to find him. They entered the estancia and found a place where something had been burned. A massive uproar ensued. The guy's name was Andres Picanerai. Some 480 Ayoreo from Campo Loro, Tunucujnai, Jesudi came with spears. Carlitos was afraid, and said, 'Let's go back.' The soldiers from the army came and blocked the path to enter the estancia with their trucks. The young people and Diyi wanted to go back because they were afraid of the soldiers. But the old Ayoreo said, 'What do I care about the soldiers. If they wound me, then I will die here on the land of my fathers.' He said that if the old people got wounded they would be even more angry. And so they went around the soldiers. The Mennonites ran away, and the soldiers cocked their guns. But the old people were not afraid and said, 'What do we care if you all shoot us?' The soldiers moved out of their way then. The old people found blood on the point of a stick. They had their lances and *cobiadie*. They thought the whites had killed the young Ayoreo. They found a place where something had been burned, maybe a person. They found blood on another stick. They were very upset and thought that the whites had killed Andres. They were preparing to kill the whites. But then someone found the young man. He appeared, and had been simply working on another estancia far away. The people felt foolish. But the whites had been afraid of them. There was another Indian missing, and surely it was he who the whites had killed."

11. Giorgio Agamben, *The Remnants of Auschwitz: The Witness and the Archive* (Cambridge, MA: Zone Books, 1999), 106.

12. "Unrest in the Moro Camp," *Brown Gold*, February 1970.

13. Becoming Indigenous in the Chaco meant becoming subject to several mutually incompatible versions of what Didier Fassin calls "moral economies." That is, it meant negotiating how the contemporary politics of indigeneity are instantiated through the production, circulation, and appropriation of profoundly contradictory norms, values, and emotions associated with moral life. The constant collisions between opposed regimes for bordering culture and Indigenous life—by ethnographers, historians, missionaries, and humanitarian NGOs—meant that indigeneity was a particularly fraught subject position. Fassin, *A Companion to Moral Anthropology*, 10–12.

14. "You examine yourself, you review your memories, hoping to find them all, and that none of them are masked or disguised," wrote Levi of his own doomed struggles with the shame of surviving Auschwitz. "No, you find no obvious transgressions, you did not usurp anyone's place, you did not beat anyone (but would you have had the strength to do so?), you did not accept positions (but none were offered to you . . .), you did not steal anyone's bread; nevertheless you cannot exclude it. It is no more than a supposition, indeed, the shadow of a suspicion: that each man is his brother's Cain." Levi, *Drowned and the Saved*, 81–82.

15. Bartolomé, *Encuentro de la gente*, 171; also 164–65.

CHAPTER SEVEN

1. Irene Roca Ortiz, Tania Cutamiño Dosape, and Rocío Picaneré Chiqueno, "Aspectos generales de la situación del derecho a la salud del pueblo Ayoreode," in *Pigasipiedie iji yoquijoningai*, ed. Irene Roca (Santa Cruz de la Sierra, Bolivia: APCOB, 2012), 101–278.

2. James Inciardi and Hilary Surratt, "Children in the Streets of Brazil: Drug Use, Crime, Violence and HIV Risks," *Substance Use and Misuse* 33 (1998): 1461–1480.

3. Nancy Postero, *Now We Are Citizens: Indigenous Politics in Post-Multicultural Bolivia* (Stanford, CA: Stanford University Press, 2007).

4. Justin Kenrick and Jerome Lewis, "Indigenous Peoples' Rights and the Politics of the Term 'Indigenous,'" *Anthropology Today* 20, no. 2 (2004): 4.

5. Jean Jackson, "Rights to Indigenous Culture in Colombia," in *The Practice of Human Rights*, ed. Mark Goodale and Sally Merry (Cambridge: Cambridge University Press, 2007), 205.

6. Postero, *Now We Are Citizens*, 225.

7. See Gillette Hall and Harry Anthony Patrinos, *Indigenous Peoples, Poverty and Human Development in Latin America: 1994–2004* (Washington, DC: World Bank, 2006); United Nations Development Programme, "Regional Human Development Report for Latin America and the Carribbean" (New York: United Nations Development Programme, 2010).

8. Fassin, "Another Politics of Life," 49.

9. Ibid., 52.

10. See Sarah Franklin and Margaret Lock, eds. *Remaking Life and Death: Toward an Anthropology of the Biosciences* (Santa Fe, NM: School of American Research Press, 2003); Adriana Petryna, *Life Exposed: Biological Citizens after Chernobyl* (Princeton, NJ:Princeton University Press, 2002).

11. See Clifford Geertz, "Life Among the Anthros," *New York Review of Books*, February 8, 2001, p. 22. Geertz borrowed and expanded this term originally coined by Dean MacCannell; see Dean MacCannell, *Empty Meeting Grounds: The Tourist Papers* (London: Routledge, 1992).

12. See also Wendy Brown, "Subjects of Tolerance: Why We Are Civilized and They Are the Barbarians," in *Political Theologies: Public Religions in a Post-Secular World*, ed. Hent de Vries and Lawrence Sullivan (New York: Fordham University Press, 2006), 298–317.

13. See D. Carleton Gajdusek, *Paraguayan Indian expeditions to the Guayaki and Chaco Indians, August 25, 1963 to September 28, 1963* (Bethesda, MD: National Institute of Neurological Diseases and Blindness, National Institutes of Health, 1963), 44.

14. See also Irene Roca Ortiz, "Del Chaco Boreal a la periferia urbana: Etnicidad Ayoreode en la ciudad de Santa Cruz de la Sierra." *Villa libre: Cuaderno de estudios sociales urbanos* 3 (2008): 73–102.

15. Quoted in João Biehl, "CATKINE . . . Asylum, Laboratory, Pharmacy, Pharmacist, I and the Cure: Pharmaceutical Subjectivity in the Global South," in *Pharmaceutical Self: The Global Shaping of Experience in an Age of Psychopharmacology*, ed. Janis Jenkins (Santa Fe, NM: School of American Research Press, 2011), 70.

16. Ibid., 72.

17. Ibid., 70.

18. Michel Foucault, *History of Sexuality*, vol. 1 (New York: Pantheon, 1978), 103. See also Mary-Jo Good et al., eds. *Postcolonial Disorders* (Berkeley: University of California Press, 2008).

19. For a comparative analysis of peri-urban migration in the context of Paraguayan Ayoreo-speaking people, see Paola Canova, "Del monte a la ciudad: La producción cultural de los Ayoreode en los espacios urbanos del Chaco central," *Suplemento antropológico* 46, no. 1 (2011): 275–316.

20. Anna Infantas, "Bolivia's Ayoreo Indians, Devoured by the City," *Inter Press News Service*, December 18, 2012, accessed January 12, 2013, http://www .ipsnews.net/2012/12/bolivias-ayoreo-indians-devoured-by-the-city/.

21. "Autoridad dice que Ayoreos son lacra," *Correo del sur*, June 26, 2011.

22. "La policía marca diciembra como un mes de extremo cuidado," *El día*, December 5, 2011.

23. See Turner, "Indigenous and Culturalist Movements."

24. In such ways, Indigenous hypermarginality instantiates and expands what Loïc Wacquant has described as "advanced marginality": (1) instantiates, in that Indigenous hypermarginality demonstrates many of the same definitive logics, properties, and causal ties to neoliberalism including an ethnic and class fragmentation through eroded sources of wage labor, a local disconnection from macroeconomic trends, sharp territorial dispossession, ecological dissolution, and increasingly targeted forms of dehumanizing stigmatization; and (2) expands, in that Indigenous hypermarginality is not confined to the urban ghettos of the post-Fordist metropolis nor their particular conflations of race, class, and place. Rather, it is a deterritorialized form of marginality specific to the oppositional modes of culture that define "the Indigenous" as an actionable political category and an inhabitable subject position. On advanced marginality, see Wacquant, "The Rise of Advanced Marginality: Notes on Its Nature and Implications," *Acta Sociologica* 39 (1996): 121–39; *Urban Outcasts: A Comparative Sociology of Advanced Marginality* (Cambridge: Polity Press, 2008); "Class, Race and Hyperincarceration in Revanchist America," *Daedalus* 139, no. 3 (2010): 74–90. On the oppositional modes of culture that define the category of the Indigenous, see esp. Thomas Abercrombie, "To Be Indian, To Be Bolivian: 'Ethnic' and 'National' Discourses" in *Nation-States and Indians in Latin America*, ed. G. Urbhan and J. Sherzer, 95–130 (Austin: University of Texas Press, 1991); James Clifford, "Indigenous Articulations," *Contemporary Pacific* 13, no. 2 (2001): 467–90; Marisol de la Cadena and Orin Starn, eds, *Indigenous Experi-*

ence Today (Oxford: Berg, 2010); Tania Li, "Indigeneity, Capitalism and the Management of Dispossession," *Current Anthropology* 51 (2010): 385–414; Kirk Dombrowski,"The Politics of Native Culture," in *A Companion to the Anthropology of American Indians,* ed. Thomas Biolsi (Oxford: Blackwell, 2008), 360–82; Fred Myers, *Painting Culture: The Making of an Aboriginal High Art* (Durham, NC: Duke University Press, 2002); Ronald Niezin, *The Origins of Indigenism: Human Rights and the Politics of Identity* (Berkeley: University of California Press, 2003); Alcida Ramos, *Indigenism: Ethnic Politics in Brazil* (Madison: University of Wisconsin Press, 1998); Anna Tsing, *Friction: An Ethnography of Global Connection* (Princeton, NJ:Princeton University Press, 2005).

25. Michel Agier, *On the Margins of the World: The Refugee Experience Today* (Cambridge: Polity Press, 2007); Zygmunt Bauman, *Collateral Damage: Social Inequalities in a Global Age* (Cambridge: Polity Press, 2011).

26. Loïc Wacquant, "Three Steps to a Historical Anthropology of Actually Existing Neoliberalism," *Social Anthropology* 20, no. 1 (2012):66–79.

27. For more on negative citizenship, see João Biehl, "Vita: Life in a Zone of Social Abandonment," *Social Text* 19, no. 3 (2001): 141.

28. See Michael Brown, *Who Owns Native Culture?* (Cambridge. MA: Harvard University Press, 2003); Jean Comaroff, *Body of Power, Spirit of Resistance* (Chicago: University of Chicago Press, 1985); Kirk Dombrowski, *Against Culture: Development, Politics and Religion in Indian Alaska* (Lincoln: University of Nebraska Press, 2001); Nicholas Dirks, *Colonialism and Culture* (Ann Arbor: University of Michigan Press, 1992); Ann Stoler, "Rethinking Colonial Categories: European Communities and the Boundaries of Rule," *Comparative Studies in Society and History* 31 (1989): 134–61.

29. Taussig, *Shamanism, Colonialism and the Wildman.*

30. See José Antonio Lucero, "Locating the Indian Problem: Community, Nationality and Contradiction in Ecuadorian Indigenous Politics," *Latin American Perspectives* 30 (2003): 23–48; Luis Rodríguez-Piñero, *Indigenous Peoples, Postcolonialism and International Law: The ILO Regime (1919–1989)* (Oxford: Oxford University Press, 2005), 55.

31. International Labor Organization, "Action to Give Effect to the Resolutions of the 1st Session of Committee of Experts on Indigenous Labour," ILO internal document CEIL.II.4.1954, quoted in Rodriguez-Piñero, *Indigenous Peoples,* 156. See also International Labor Organization, "The Second Session of the ILO Committee of Experts on Indigenous Labour," *International Labor Review* 70 (1954): 422–23.

32. Wacquant, "Three Steps to a Historical Anthropology," 68.

33. Postero, *Now We Are Citizens,* 15.

34. Ibid., 16.

35. Ibid., 17, 225.

36. Alejandro Portes, "Rationality in the Slum: An Essay on Interpretive Sociology," *Comparative Studies in Society and History* 14 (1972): 286.

37. See John Gledhill's, "Neoliberalism" in David Nugent and Joan Vincent, eds., *A Companion to the Anthropology of Politics* (New York: Blackwell, 2004), pp. 332–48, for a related argument.

38. See Fassin, *Humanitarian Reason*.

39. For more on the idea of possibilism, see Alfred Hirschman, *A Bias for Hope: Essays on Development and Latin America* (New Haven, CT:Yale University Press, 1971)

40. Sol Tax, "Action Anthropology," *Current Anthropology* 16, no. 4 (1975): 514.

CHAPTER EIGHT

1. Walter Benjamin, "Critique of Violence," in *Selected Writings*, vol. 1, ed. Marcus Bullock and Michael Jennings (Cambridge, MA: Harvard University Press, 1999 [1921]), 287.

2. Mike Swain, "The Lost Tribe of Green Hell: Cultures Clash in a Battle for the Future of the Planet," *Daily Mirror*, December 31, 2010.

3. Victoria Gill, "Conservation Expedition Poses Risk to Tribes," *BBC News*, November 9, 2010; John Vidal, "Natural History Museum Expedition Poses Genocide Threat to Paraguay Tribes," *Guardian*, November 8, 2010; John Vidal "Natural History Museum's Gran Expedition to Arid Chaco Halted," *Guardian*, November 15, 2010.

4. Benno Glauser, "Su presencia protege el corazon del Chaco seco," in *Pueblos indígenas en aislamiento voluntario y contacto inicial en la Amazonia y el Gran Chaco*, ed. Alejandro Parellada (Copenhagen: International Work Group for Indigenous Affairs, 2007), 220.

5. Benno Glauser, "Paraguay: Pueblos indígenas aislados," *Suplemento antropológico* 41, no. 2 (2006): 192.

6. My argument here is indebted to Fred Myers, "Locating Ethnographic Practice: Romance, Reality and Politics in the Outback," *American Ethnologist* 15, no. 4 (November 1988): 609–24.

7. For instance, see Johannes Fabian, *Time and the Other: How Anthropology Makes Its Object* (New York: Columbia University Press, 1983); Akhil Gupta and James Ferguson, eds., *Culture, Power, Place: Explorations in Critical Anthropology* (Durham, NC: Duke University Press, 1997); Marshall Sahlins, "Goodbye Tristes Tropes: Ethnography in the Context of the Modern World System," *Journal of Modern History* 65 (1993): 1–25; Michel Rolph Trouillot, *Silencing the Past: Power and the Production of History* (Boston: Beacon, 1995).

8. For more details on the range, estimated number, and distinct national politics around "voluntarily isolated" groups in Latin America, see esp. Parellada, *Pueblos indígenas*. Here, I am interested less in the precise empirical contents of the category of isolation, and more in the ways it acts as a generative imaginary and organizational schema for the politics of legitimate Indigenous life.

9. As the Totobiegosode leaders were quick to point out, this organization was not aimed at preserving the past but at consolidating the new life project of moral transformation.

10. Edwin Wilmsen, *Land Filled with Flies: A Political Economy of the Kalahari* (Chicago: University of Chicago Press, 1989), 4.

11. Margaret Keck and Kathryn Sikkink, *Activists Beyond Borders: Advocacy Networks in International Politics* (Ithaca, NY: Cornell University Press, 1998), 14. See also Robert Albró, "The Indigenous in the Plural in Bolivian Oppositional Politics," *Bulletin of Latin American Research* 24 (2006): 433–53; Alison Brysk, *From Tribal Village to Global Village: Indian Rights and International Relations in Latin America* (Stanford, CA: Stanford University Press, 2000); Charles Hale, *Mas que un indio: Racial Ambivalence and Neoliberal Multiculturalism in Guatemala* (Santa Fe, NM: School of American Research Press, 2006); Jean Jackson, "Culture, Genuine and Spurious: The Politics of Indianness in the Vaupes, Colombia," *American Ethnologist* 22 (1995): 3–27; Jean Jackson and Kay Warren, "Indigenous Movements in Latin America, 1992–2004: Controversies, Ironies and New Directions," *Annual Review of Anthropology* 34 (2005): 549–73; Kay Warren, *Indigenous Movements and Their Critics: Pan-Maya Activism in Guatemala* (Princeton, NJ: Princeton University Press, 1998); Deborah Yashar, *Contesting Citizenship in Latin America: The Rise of Indigenous Movements and the Postliberal Challenge* (Cambridge: Cambridge University Press, 2005).

12. See Beatriz Huertas-Castillo, *Los pueblos indígenas en aislamiento: Su lucha por la sobrevivencia y la libertad* (Copenhagen: International Work Group for Indigenous Affairs, 2002); Dora Napolitano and Aliya Ryan, "The Dilemma of Contact: Voluntary Isolation and the Impacts of Gas Exploitation on Health and Rights in the Kugapakori Nahua Reserve, Peruvian Amazon," *Environmental Research Letters* 2, no. 4 (2007): 1–12. In 1987, the Brazilian National Indian Service—FUNAI—created a Department of Isolated Indians. In Peru, the Indigenous organizations AIDESEP and FENAMAD have long been working to raise awareness around the human rights violations of isolated Indigenous groups in the upper Madre de Dios / Uyucali region by loggers, miners, and oil companies. The continuous advocacy by AIDESEP and FENAMAD led to the creation of five state reserves set aside for isolated groups between 1990 and 2002. Despite these advances, in 2003 the state allowed companies to begin drilling for natural gas and constructing pipelines in the heart of a state reserve inhabited by isolated peoples, the so-called Camisea fields discovered by Shell in the 1980s. This was financed, in part, by a seventy-five-million-dollar development grant from the Inter-American Development Bank. The Camisea project, and the anthropological report commissioned by the IDB on its effects, was widely criticized by AIDESEP and others. The disastrous effects of the project and intense lobbying by regional Indigenous organizations and international legal experts created a situation in which the IDB loan payments were

made contingent on concrete steps taken to protect isolated peoples. The result was the eventual creation of Peruvian Law 28736—the *Ley para la protección de pueblos indígenas u originarios en situación de aislamiento y en contacto inicial*—in 2006. Although this was the first national law of its kind, in 2007 the Peruvian President Alan Garcia issued a presidential decree stating that state reserves for isolated peoples can still be exploited for their natural resources.

13. See Vincent Brackelaire, *Diagnóstico de pueblos aislados* (Washington, DC: Inter-American Development Bank, 2006).

14. This includes the 2006 Peruvian Law 28736, Article 57.21 of the Ecuadorian Constitution, and Article 31 of the 2009 Bolivian Constitution.

15. United Nations Human Rights Council, *Draft Guidelines on the Protection of Indigenous Peoples in Voluntary Isolation and in Initial Contact of the Amazon Basin and El Chaco* (Geneva: United Nations document, 2009), 5–11, 14, 18.

16. Benno Glauser, "Paraguay: The Last Ayoreo in Voluntary Isolation," *World Rainforest Movement Bulletin* 87 (2004): 12.

17. International Work Group for Indigenous Affairs, "The Case of the Ayoreo: Report 4" (Copenhagen: International Work Group for Indigenous Affairs, 2010), 21

18. See Nickolas Boecher, "Third Party Petitions as a Means of Protecting Voluntarily Isolated Indigenous Peoples," *Sustainable Development Law and Policy* 10 (2009), 89.

19. There is a growing Ayoreo interest in taking protagonism in this issue, for various and complex reasons. See Aquino Picanerai, "De los dirigentes de la Union de Nativos Ayoreo de Paraguay," in *Pueblos indígenas en aislamiento voluntario y contacto inicial en la Amazonia y el Gran Chaco*, ed. Alejandro Parellada (Copenhagen: International Work Group for Indigenous Affairs, 2007), 234–36.

20. See Luisa Maffi, "Linguistic, Cultural and Biological Diversity," *Annual Review of Anthropology* 34 (2005): 599–617.

21. See Inge Kaul, Isabelle Grunberg, and Marc Stern, "Introduction," in *Global Public Goods: International Cooperation in the 21st Century* (Oxford: Oxford University Press, 1999), xix–xxxviii.

22. Ismail Serageldin, "Cultural Heritage as Public Good: Economic Analysis Applied to Historic Cities," in *Global Public Goods*, 240.

23. Arjun Appadurai, "Diversity and Sustainable Development," In *Cultural Diversity and Biodiversity for Sustainable Development* (Nairobi: United Nations Environment Programme, 2002), 16–20.

24. Rachel Sieder, ed., *Multiculturalism in Latin America: Indigenous Rights, Diversity and Democracy* (New York: Palgrave Macmillan, 2002), 5.

25. Jackson and Warren, "Indigenous Movements in Latin America," 549–73.

26. Turner, "Indigenous and Culturalist Movements."

27. Will Kymlicka, *Multicultural Citizenship* (Oxford: Clarendon, 1996).

28. The culture of the isolated subject is not envisioned as a set of shared capacities for self-production or reflexive metaobjectification, but a sui generis and immutable essence. Here, the instrumental polysemy of the culture concept that was previously the source of its effectiveness—as in venues of "strategic essentialism"—is impossible.

29. Glauser, "Su presencia protege el corazón," 220.

30. Isolation reveals the value reassigned to a perceived "outside" to capitalist modernities, even as it remains a floating signifier whose human contents are intrinsically unlocatable. It is defined not so much by the absence of positive law as the possibility of what Marc de Wilde describes as a "depersonalizing juridical violence" typical of contemporary states of exception. Indeed, it is hard to imagine a more completely depersonalized and profoundly theological legal subject than isolated man. See Marc de Wilde, "Violence in the State of Exception: Reflections on Theologico-Political Motifs in Benjamin and Schmitt," in *Political Theologies*, ed. Hent de Vries and Lawrence Sullivan (New York: Fordham University Press, 2006), 188–200. In such ways, the legal government of isolation always implies what Walter Benjamin called "law-making" or law-instating violence, a force capable of founding and reordering the relations that law is supposed to reflect and formalize. As Benjamin suggests, such dynamics reveal a fundamental lawlessness at the core of legalizing isolation, wherein laws are incapable of constraining the violent excess upon which their application seems to depend. See Judith Butler, "Critique, Coercion and Sacred Life in Benjamin's Critique of Violence," in *Political Theologies: Public Religions in a Post-Secular World*, ed. Hent de Vries and Lawrence Sullivan (New York: Fordham University Press, 2006), 201–19.

31. See Marvin Duerksen, "Allanamiento de ONG se basa en denuncias,"*ABC Color*, December 4, 2010; John Vidal, "Paraguayan Government Raids Offices of NGO Opposing Chaco Research Trip," *Guardian* December 15, 2010.

32. POJOAJU (Asociación de ONGs del Paraguay), "Caso de allanamiento de la Oficina de la ONG Iniciativa Amotocodie," press release, December 3, 2010.

33. Amnesty International, "Paraguay: Raid on NGO May Be in Reprisal for Its Public Criticism of Scientific Expedition," public pronouncement AMR 45/007/2010, December 6, 2010.

34. But see Marvin Duerksen, "Imputan a los directivos de Iniciativa Amotocodie por Lesion de Confianza," *ABC Color*, May 20, 2011; Marvin Duerksen, "Ayoreos exigen que una ONG devuelva mas de G. 2.000 milliones," *ABC Color*, June 7, 2011.

35. For a sophisticated genealogy of Paraguayan civil society see esp. Kregg Hetherington, *Guerilla Auditors: The Politics of Transparency in Neoliberal Paraguay* (Durham, NC: Duke University Press, 2011).

36. See esp. Blaser, " 'Way of Life' or 'Who Decides' "; Rene Harder Horst, "Consciousness and Contradiction: Indigenous People and Paraguay's Transition

to Democracy," in *Contemporary Indigenous Movements in Latin America*, ed. Erick D. Langer and Elena Muñoz (Wilmington, DE: Scholarly Resources, 2003), 103–34.

37. Alan Fowler, *Striking a Balance: A Guide to Enhancing the Effectiveness of Non-Governmental Organisations in International Development* (London: Earthscan, 1997).

38. Recent state reforms in Bolivia and Paraguay have created two very distinct political contexts for Ayoreo mobilizations around citizenship, claims to territory and culture, discourses of indigeneity, and human rights. See CEJIS, *El pueblo Ayoreo, entre el campo y la ciudad, el territorio y sus estrategias de sobrevivencia* (Santa Cruz, CA: CEJIS, 2005); Holland, *Indians, Missionaries and the Promised Land*; Stephen Kidd, "Los Totobiegosode reivindican su territorio tradicional," *Asuntos indígenas* 4 (Copenhagen: International Work Group for Indigenous Affairs, 1993); Postero, *Now We Are Citizens*; Elba Terceros Cuéllar, "Proceso de saneamineto de las tierras comunitarias de origen Ayoreo," in *Atlas de territorios indígenas en Bolivia: Situación de las tierras comunitarias de origen (TCOs) y proceso de titulación*, ed. José A. Martínez, 191–96 (Santa Cruz de la Sierra: Centro de Planificación Territorial Indígena de la CIDOB, 2000); Ultima Hora, "ONG reclama al gobierno de Paraguay proteger a los Ayoreos," August 9, 2007. Hans Heijdra offers a comprehensive and disturbing account of the complex relationships by which an Ayoreo tribal organization in Bolivia became entangled with state agencies and mega-development projects; its counterpart remains to be written for Paraguay. See Heijdra, *La nueva gente: Un estudio sociologico sobre las formas de produccion y las formas de organización en un contexto dinamico de una comunidad Ayorea del orienta Boliviana* (University of Agronomy Wageningen, unpublished MA thesis, 1987). The ways in which distinct formations of governmentality and colonialism have created divergent zones of political agency for Ayoreo people on either side of the Bolivia / Paraguay border is a topic that merits greater attention and one I hope to expand on elsewhere (see S. M. Hirsch, "Bilingualism, Pan-Indianism and Politics in Northern Argentina: The Guaraní's Struggle for Identity and Recognition," *Journal of Latin American Anthropology* 8 [2003]: 84–103; and Valerie Taliman, "Borders and Native Peoples: Divided, But Not Conquered," *Native Americas* 18, no. 1 [2001]: 10–16, on cross-border indigeneity). In this chapter, I focus on the Paraguayan context.

39. Glauser, "Su presencia protege el corazón," 230.

40. Glauser, "Paraguay: The Last Ayoreo," 12–13.

41. Holland, *Indians, Missionaries and the Promised Land*.

42. See William Fisher, "Doing Good? The Politics and Antipolitics of NGO Practices," *Annual Review of Anthropology* 26 (1997): 439–64; A. C. Hudock, "NGOs' Seat at the Donor Table: Enjoying the Food or Serving the Dinner?," *IDS Bulletin* 31 (2000): 14–18; David Hulme and Michael Edwards,

eds., *NGOs, States and Donors: Too Close for Comfort?* (New York: St. Martin's Press, 1997).

43. Alex Rivas, "Los pueblos indígenas en aislamiento: Emergencia, vulnerabilidad y necesidad de protección," *Cultura y representaciones sociales* 1, no. 2 (2007): 86.

44. NGO labor has produced isolation as a schema in which the global projects of accumulating profit, moral expenditure, romantic primitivism, and cultural activism are seemingly reconciled. When fixed through these arbitrary institutional biopolitics, isolation blurs the boundaries between the brute force of Benjamin's "law-preserving violence," and "iatrogenic violence," or violence as the inadvertent product of care that Laurence McFalls has located at the center of global humanitarianisms. See Laurence McFalls, "Benevolent Dictatorship: The Formal Logic of Humanitarian Government," in *Contemporary States of Emergency*, ed. Didier Fassin and Mariella Pandolfi (New York: Zone, 2010), 317–34. Such inversions are possible because of the particular conditions of indigeneity in the Paraguayan Chaco, a zone in which endemic corruption and contradictory legal protocols mean that enforcing any law in favor of Indigenous peoples is determined by wider political interests and unpredictable outbursts of state authority. Advocacy NGOs are funded as alternatives to this system, but in practice they rarely challenge its defining structures.

45. Ramos, *Indigenism*, 277.

46. See Turner, "Indigenous and Culturalist Movements," n.p.

47. See Myers, "Locating Ethnographic Practice," for a related point.

48. The earliest scientist to collect blood from Ayoreo-speaking people in Paraguay was Gajdusek, who visited the Chaco in 1963, shortly after the first Ayoreo mission was established near Teniente Montanía. At the time, the forty-year-old American pediatrician was deeply in the throes of a self-styled "quest" for "urgent opportunistic investigations of epidemiological problems in exotic and isolated populations," which he believed represented a new frontier for biomedicine and a unique opportunity for an ambitious young scientist to make himself a name. Accompanied by Lucien Sebag and Pierre Clastres, Gajdusek collected blood from seventy-two recently contacted Garaigosode. As he described on p. 44 of his diary, "We had a bit of difficulty getting blood specimens from the women and girls because they made a dreadful theatrical fuss, played coy and ran off every time I wanted to get a blood specimen from them. The boys however all had specimens drawn and the men then came up voluntarily and I bled them. Finally a few old women let me take specimens and slowly we managed to bleed the flirting, coy and jesting girls and younger women." "The men," Gajdusek wrote in his journal, "are husky, rather defiant and are often a bit sullen . . . [but they] seem to have preserved their pride, integrity and savagery." He was struck by their bodily presence: "The Moros

are filthy as are many desert dwellers . . . and in general manage to execute every conceivable unhygienic and rather revolting maneuver in the course of any given hour of the day. . . . They well illustrate the adage that every human is covered with a layer of shit—and also snot!—and that it is only a question of how thick the layer is!" Gajdusek noted that "our host considers the Moro immoral savages and has no sympathy for anything in their own culture . . . He seems to enjoy telling of their savagery and treachery and credits them with no modicum of human virtue at all." If Gajdusek dismissed Ayoreo reactions against his intrusions as a feminized "fuss," not so with the actual Indian blood he collected. Gajdusek's diary shows that he sent no less than eighteen letters and cables with detailed instructions for the storage, shipment and intact delivery of the samples between August 26 and September 25, 1963, despite traveling to and from remote places. While such attention to procedural details is of course simply good science, it is striking the degree to which Gajdusek's focus on the freshness of Indian blood contrasted with his detachment from the actual conditions of Indian life; the same conditions which his companions Sebag and Clastres found so darkly depressing. Blood in his account is indistinguishable from the value he placed on Indian life and his desire to consume its constitutive essence. Much like the souls collected by Bobby or the tradition sought by *Abujá*die, Indian blood in Gajdusek's diary figures as the material expression of a timeless truth that the scientist yearns to approximate and perhaps to incorporate.

49. Francisco Salzano et al., "Unusual Blood Genetic Characteristics among the Ayoreo Indians of Bolivia and Paraguay," *Human Biology* 50, no. 2 (1978): 121–36.

50. See Ricardo Ventura Santos, "Indigenous Peoples, Postcolonial Contexts and Genomic Research in the Late 20th Century: A View from Amazonia (1960–2000)," *Critique of Anthropology* 22, no. 1 (2002): 81–104; Ricardo Ventura Santos and Marcos Maio, "Race, Genomics, Identities and Politics in Contemporary Brazil," *Critique of Anthropology* 24, no. 4 (2004): 347–78.

51. A. A. Perez-Diaz and Francisco Salzano, "Evolutionary Implications of the Ethnography and Demography of Ayoreo Indians," *Journal of Human Evolution* 7 (1978): 254.

52. See C. Dornelles et al., "Mitochondrial DNA and Alu Insertions in a Genetically Peculiar Population: The Ayoreo Indians of Bolivia and Paraguay," *American Journal of Human Biology* 16 (2004): 479–88; A. S. Goicoechea et al., "New Genetic Data on Amerindians from the Paraguayan Chaco," *American Journal of Human Biology* 13 (2001): 660–67; S. Goulart et al., "Autosomal STR Genetic Variability in the Gran Chaco Native Population: Homogeneity or Heterogeneity?," *American Journal of Human Biology* 20 (2008): 704–11.

53. Sidney Dos Santos et al., "Autosomal STR Analyses in Native Amazonian Tribes Suggest a Population Structure Driven by Isolation by Distance," *Human Biology* 81 (2009): 71–88.

54. Linda Hogle, "Life/Time Warranty: Rechargeable Cells and Extendable Lives," In *Remaking Life and Death: Toward an Anthropology of the Biosciences*, ed. Sarah Franklin and Margaret Lock (Santa Fe, NM: School of American Research Press, 2003), 92.

55. The vision of Ayoreo humanity staged and objectified through genetic science is based on the same tense tautologies found in law, humanitarianism, and ethnographic expeditions in search of tradition. But the speculative promise of a geneticized isolation is retrospective as well as anticipatory. The value of the isolated genome may be premised on a redemptive expertise, but this depends on simultaneously reassigning value to a humankind that exists outside of society itself. See Mike Fortun, *Promising Genomics: Iceland and deCODE Genetics in a World of Speculation* (Berkeley: University of California Press, 2008), 12.

56. Bartolomé, *El encuentro de la gente*, 308.

57. International Work Group for Indigenous Affairs, *Case of the Ayoreo*, 14.

58. Glauser, "Paraguay: Pueblos indígenas aislados," 200; Glauser, "Su presencia protege el corazón," 232.

59. Eric Wolf, *Europe and the People without History* (Berkeley: University of California Press, 1982), 6–7.

60. Fassin, "Another Politics of Life," 52.

61. Quoted in Theodora Kroeber, *Ishi in Two Worlds: A Biography of the Last Wild Indian in North America* (Berkeley: University of California Press, 1961), 92.

CONCLUSION

1. Taussig, *Shamanism, Colonialism and the Wild Man*, 132.

2. Abercrombie, *Pathways of Memory and Power*, 110–13, 408–22.

3. Frantz Fanon, *Black Skin, White Masks* (New York: Grove, 2008 [1952]), 82.

4. Viveiros de Castro, "The Untimely, Again," 24.

5. Fassin, "Another Politics of Life," 52; see also Fassin, *Humanitarian Reason*.

6. Hardt and Negri, *Commonwealth*, 124.

7. See Georges Bataille, *The Unfinished System of Nonknowledge* (Minneapolis: University of Minnesota Press, 2004).

Bibliography

ABC Color. "Dudan de presencia de silvicolas en zona de Teniente Martinez." October 20, 2006.

———. "Fallecío ayer un Totobiegosode a raiz de problemas respiratorios." July 3, 2004.

———. "Hallan mas evidencias de posible presencia de silvicolas en el Chaco." September 14, 2007.

———. "Piden a Lugo que evite muerte de Totobiegosode." October 2, 2008.

———. "Ultimo grupo de Ayoreos silvicolas se siente acorralado por estancias." September 3, 2006.

Abercrombie, Thomas. *Pathways of Memory and Power: Ethnography and History among an Andean People*. Madison: University of Wisconsin Press, 1998.

———. "To Be Indian, To Be Bolivian: 'Ethnic' and 'National' Discourses." In *Nation-States and Indians in Latin America*, edited by G. Urban and J. Sherzer, 95–130. Austin: University of Texas Press, 1991.

Abu-Lughod, Lila. "The Romance of Resistance: Tracing Transformations of Power through Bedouin Women." *American Ethnologist* 17, no. 1 (1991): 41–55.

Adelson, Naomi. "Reimagining Aboriginality: An Indigenous People's Response to Social Suffering." In *Remaking a World*, edited by Das et al., 76–101, Berkeley: University of California Press, 2001.

Adorno, Theodor. *Minima Moralia*. Translated by E. F. N. Jephcott. London: Verso, 1974.

Agamben, Giorgio. *Homo Sacer: Sovereign Power and Bare Life*. Stanford, CA: Stanford University Press, 1998.

———. *Remnants of Auschwitz: The Witness and the Archive*. New York: Zone Books, 1999.

———. *State of Exception*. Chicago: University of Chicago Press, 2005.

Agier, Michel. *On the Margins of the World: The Refugee Experience Today*. Cambridge: Polity Press, 2007.

Aguilar Dávalos, Gonzalo. "El movimiento indígena en las tierras bajas en el siglo XX." In *Los Bolivianos en el tiempo*, edited by Alberto Crespo. La Paz, Bolivia: Editorial Indea, 1995.

Aguirre Acha, José. La antigua provincia de Chiquitos limitrofe con la provincia del Paraguay. Sucre: Imprenta y Litigrafia Salesiana, 1933.

Ahmed, Sara. *The Cultural Politics of Emotions*. New York: Routledge, 2004.

Albó, Xavier. "Bolivia: From Indian and Campesino Leaders to Councillors and Parliamentary Deputies." In *Multicultural Citizenship in Latin America*, edited by Rachel Sieder, 74–102. New York: Palgrave Macmillan, 2002.

———. "El retorno del indio." *Revista Andina* 9, no. 2 (1991): 299–366.

Albro, Robert. "The Culture of Democracy and Bolivia's Indigenous Movements." *Critique of Anthropology* 26, no. 4 (2006): 387–410.

———. "The Indigenous in the Plural in Bolivian Oppositional Politics." *Bulletin of Latin American Research* 24, no. 4 (2005): 433–54.

Amnesty International. "Allanamiento a ONG seria en represalia por sus denuncias publicas sobre expedicion cientifica," Public pronouncement AMR 45/007/2010, 2010.

———. "Paraguay: Raid on NGO May Be in Reprisal for Its Public Criticism of Scientific Expedition." Public pronouncement AMR 45/007/2010. December 6, 2010.

Anaya, James, and C. Grossman. "The Case of Awas Tingni v. Nicaragua: A New Step in the International Law of Indigenous Peoples." *Arizona Journal of International and Comparative Law* 19, no. 1 (2002).

APCOB. *Entrevistas con CucAsoregosode*. Unpublished document. Santa Cruz, 1993.

Appadurai, Arjun. "Diversity and Sustainable Development." In *Cultural Diversity and Biodiversity for Sustainable Development*, 16–20. Nairobi: United Nations Environment Programme, 2002.

———. *Fear of Small Numbers: An Essay on the Geography of Anger*. Durham, NC: Duke University Press, 2006.

———. *Modernity at Large: Cultural Dimensions of Globalization*. Minneapolis: University of Minnesota Press, 1996.

———. "Topographies of the Self: Praise and Emotion in Hindu India." In *Language and the Politics of Emotion*, edited by Catherine A. Lutz and Lila Abu-Lughod. New York: Cambridge University Press, 1990.

Apter, Andrew. "On Imperial Spectacle: The Dialectics of Seeing in Colonial Nigeria." *Comparative Studies in Society and History* 44 (2002): 564–96.

Arambiza, Evelio, and Michael Painter. "Biodiversity Conversation and Quality of Life of Indigenous Peoples in the Bolivian Chaco." *Human Organization* 65, no. 1 (2006): 20–34.

Arendt, Hannah. *The Origins of Totalitarianism*. San Diego: Harvest, 1973.

Arens, Richard, ed. *Genocide in Paraguay*. Philadelphia: Temple University Press, 1976.

ASCIM. *Report of Actual Conditions on the Mission of El Faro Moro*. Unpublished report. Filadelfia, 1977.

Askew, Kelly, and Richard Wilk, eds. *Anthropology of Media*. Malden, MA: Blackwell, 2002.

Assies, Willem. "Indigenous Peoples and the Reform of the State in Latin America." In *Challenge of Diversity*, edited by Assies, Willem, Gemma van der Haar, and Andre Hoekema, 3–22. Amsterdam: Thela Thesis, 2000.

Atkinson, Judy. *Trauma Trials, Recreating Song Lines: The Transgenerational Effects of Trauma in Indigenous Australia*. Melbourne: Spinifex Press, 2002.

Augaitis, Daina, and Dan Lander, eds. *Radio Rethink: Art, Sound, Transmission*. Banff: Walter Philips Gallery, 1994.

"Autoridad dice que Ayoreos son lacra." *Correo del sur*. June 26, 2011.

Auyero, Javier. "The Hyper-Shantytown: Neoliberal Violence(s) in the Argentine Slum" *Ethnography* 1, no. 1 (2000): 93–116.

Bach, Mauricio. *Descripción de la nueva provincia de Otuquis en Bolivia*. La Paz, Bolivia: Imp Unidas, 1842.

Badiou, Alain. *Deleuze: The Clamor of Being*. Translated by Louis Burchill. Minneapolis: University of Minnesota Press, 2000.

Barras, Bruno. "Life Projects: Development Our Way." In *In the Way of Development*, edited by Mario Blaser, Harvey A. Feit, and Glenn McRae, 47–51. London: Zed Books, 2004.

Bartolomé, Miguel. *El encuentro de la gente y los insensatos: La sedentarización de los cazadores Ayoreo en el Chaco Paraguayo*. Asunción, Paraguay: CEADUC, 2000.

Barume, Albert. *Heading toward Extinction? Indigenous Rights in Africa: The Case of the Twa of theKahuzi-Biega National Park, DCR*. Copenhagen: International Work Group for Indigenous Affairs, 2000.

Basso, Keith. *Wisdom Sits in Places: Landscape and Language among the Western Apache*. Albuquerque: University of New Mexico Press, 1996.

Bastien, Joseph. *Mountain of the Condor: Metaphor and Ritual in an Andean Ayllu*. Long Grove: Waveland Press, 1985.

Bataille, Georges. *The Accursed Share*. New York: Zone, 1991.

———. *The Unfinished System of Non-Knowledge*. Minneapolis: University of Minnesota Press, 2004.

Bauman, Zygmunt. *Collateral Damage: Social Inequalities in a Global Age*. Cambridge: Polity Press, 2011.

Bedoya Silva-Santisteban, Alvaro. *Servidumbre por deudas y marginación en el Chaco de Paraguay*. Geneva: International Labour Organization, 2005.

Beidelman, Tom. *Colonial Evangelism: A Socio-Historical Study of an East African Mission at the Grassroots*. Bloomington: Indiana University Press, 1982.

Bell, Catherine. *Ritual: Perspectives and Dimensions*. New York: Oxford University Press, 1997.

Benjamin, Walter. "Critique of Violence." In *Selected Writings*, vol. 1, edited by Marcus Bullock and Michael Jennings, 277–300. Cambridge, MA: Harvard University Press, 1999 [1921].

———. *Theses on the Philosophy of History*. New York: Illuminations, 1969.

Bergson, Henri. *Duration and Simultaneity*. Indianapolis, IN: Bobbs-Merrill, 1965.

Berlant, Lauren. *Compassion: The Culture and Politics of an Emotion*. New York: Routledge, 2004.

Bernand-Muñoz, Carmen. *Les Ayoré du Chaco septentrional: Étude critique à partir des notes de Lucien Sebag*. La Haya: Mouton, 1977.

Berraondo, Mikel. "Buscando protección: Pueblos en aislamiento frente al reto de los derechos." In *Pueblos indígenas*, edited by Parellada, 18–39.

Bessire, Lucas. "Apocalyptic Futures: The Violent Transformation of Moral Human Life among Ayoreo-Speaking People of the Paraguayan Gran Chaco." *American Ethnologist* 38, no. 4 (2011): 743–47.

———, director. *Asking Ayahai: An Ayoreo Story*. 41 min. Documentary Educational Resources, 2004.

———, director. *From Honey to Ashes*. 47 min. Documentary Educational Resources, 2006.

———. "Ujñarone Chosite: Ritual Poesis, Curing Chants and Becoming Ayoreo in the Gran Chaco," *Journal de la Société des Américanistes* 97, no. 1 (2011): 259–89.

Beteille, Andre. "The Idea of Indigenous Peoples." *Current Anthropology* 39, no. 2 (1998): 187–91.

Bettelheim, Bruno. "Individual and Mass Behavior in Extreme Situations." *Journal of Abnormal and Social Psychology* 38 (1943): 417–52.

Biehl, João. "CATKINE . . . Asylum, Laboratory, Pharmacy, Pharmacist, I and the Cure: Pharmaceutical Subjectivity in the Global South." In *Pharmaceutical Self: The Global Shaping of Experience in an Age of Psychopharmacology*, edited by Janis Jenkins. Santa Fe, NM: School of American Research Press, 2011.

———. *Vita: Life in a Zone of Social Abandonment*. Princeton, NJ: Princeton University Press, 2005.

———. "Vita: Life in a Zone of Social Abandonment." *Social Text* 19, no. 3 (2001): 131–49.

Biehl, João, Byron Good, and Arther Kleinman, eds. *Subjectivity: Ethnographic Investigations*. Berkeley: University of California Press, 2007.

Biehl, João, and Peter Locke. "Deleuze and the Anthropology of Becoming." *Current Anthropology* 51 (2010): 317–51.

Blaser, Mario. *Storytelling Globalization from the Chaco and Beyond*. Durham, NC: Duke University Press, 2010.

———. "The Threat of the Yrmo: The Political Ontology of a Sustainable Hunting Program." *American Anthropologist* 111, no. 1 (2009): 10–20.

———. " 'Way of Life' or 'Who Decides': Development, Paraguayan Indigenism and the Yshiro People's Life Projects." In *In the Way of Development*, edited by Mario Blaser, Harvey Feit, and Glenn McRae, 52–71. London: Zed Books, 2004.

Bloch, Ernst. *The Principle of Hope*. Translated by Plaice, Plaice, and Knight. Cambridge, MA: MIT Press, 1986 [1959].

Boecher, Nickolas. "Third Party Petitions as a Means of Protecting Voluntarily Isolated Indigenous Peoples." *Sustainable Development Law and Policy* 10 (2009): 89.

Bolivia. New Constitution of the Plurinational State of Bolivia, 2009.

Boltanski, Luc, and Laurent Thevenot. *On Justification: Economies of Worth*. Translated by Catherine Porter. Princeton, NJ: Princeton University Press, 2006.

Bórmida, Marcelo. "Como una cultura conoce la realidad de su mundo." *Scripta ethnologica* 8 (1984): 13–183.

———. "Ergon y mito: Una hermenéutica de la cultura material de los Ayoreo del Chaco Boreal. Parte I." *Scripta ethnologica* 1 (1973): 9–69.

———. "Ergon y mito: Una hermenéutica de la cultura material de los Ayoreo del Chaco Boreal. Parte II." *Scripta ethnologica* 2, no. 1 (1974): 41–107.

———. "Ergon y mito: Una hermenéutica de la cultura material de los Ayoreo del Chaco Boreal. Parte III." *Scripta ethnologica* 3, no. 1 (1975): 73–130.

Bórmida, Marcelo, and Mario Califano. *Los indios Ayoreo del Chaco Boreal*. Buenos Aires: Fundación para la Educación la Ciencia y la Cultura, 1978.

Borofsky, Robert. "Public Anthropology: Where To? What Next?" *Anthropology News* (2000).

Brackelaire, Vincent. *Diagnostico del pueblos aislados*. Washington, DC: Inter-American Development Bank, 2006.

Braunstein, José, and Elmer Miller. "Ethnohistorical Introduction." In *Peoples of the Gran Chaco*, edited by Elmer Miller, 1–22. Westport, CT: Bergin and Garvey, 1999.

Brecht, Bertolt. "Radio as a Means of Communication: A Talk on the Function of Radio." Translated by Stuart Hood. Screen 20.3–4.24 (1979–1980), 1930.

Bremen, Volker von. "Dynamics of Adaptation to Market Economy among the Ayoreode of Northwest Paraguay." In *Hunters and Gatherers in the Modern World*, edited by Peter Schweitzer, Megan Bisele, and Robert Hitchcock, 275–87. Oxford: Berghahn, 2000.

———. *Fuentes de caza y recolección modernas: Proyectos de ayuda al desarrollo destinados a los indígenas del Gran Chaco*. Stuttgart: Servicios de Desarrollo de las Iglesias, 1987.

———. "Los Ayoreode Cazados." *Suplemento antropológico* 22, no. 1 (1987): 75–94.

Briggs, Janet. "Ayore Narrative Analysis." *International Journal of American Linguistics* 39, no. 3 (1973): 155–63.

Briggs, Jean. *Never in Anger: Portrait of an Eskimo Family*. Cambridge, MA: Harvard University Press, 1970.

Brown, Michael. "Beyond Resistance: A Comparative Study of Utopian Renewal in Amazonia." *Ethnohistory* 38, no. 4 (1991): 388–413.

————. *Who Owns Native Culture?* Cambridge, MA: Harvard University Press, 2003.

Brown, Wendy. "Subjects of Tolerance: Why We Are Civilized and They Are the Barbarians." In *Political Theologies: Public Religions in a Post-Secular World*, edited by Hent de Vries and Lawrence Sullivan. New York: Fordham University Press, 2006.

Brown Gold. "Are the Heathen Lost?" January 1947.

————. "Are the Heathen Lost?" May 1981.

————. "Ayore Notes." August 1959.

————. "Barbaro Gleanings." October 1948.

————. "A Barbaro Miss." December 1948.

————. "Barbaros Acknowledge Aerial Contact." March 1949.

————. "Barbaros at Tobité." January 1950.

————. "Barbaros Plan Attack!" January 1950.

————. "The Barbaros Were Upon Us." July 1948.

————. "Bolivian Nomads." July 1945.

————. "Brown Gold." October 1943.

————. "Brown Gold." January 1997.

————. "A Captive Barbaro Woman in San Jose." August 1946.

————. "A Captured Barbaros Woman." August 1944.

————. "Chief Buried Alive." August–September 1948.

————. "Civilization a Curse to the Savage." August/September 1948.

————. "Defeat for the Bird God." February 1970.

————. "Doors Opening to the World's Untouched Fields." December 1945.

————. "Do You Really Believe the Heathen Are Lost?" December 1959.

————. "First Bolivian Conference." June 1946.

————. "A Friendly Contact with Savages." August/September 1947.

————. "From Tobité, Bolivia." February/March 1952.

————. "Future Barbaro Work." December/January 1947–48.

————. "A Glimpse of Tobité." May 1966.

————. "God Has an American Accent." January 1972.

————. "Go Ye Forth." May 1944.

————. January 1944.

————. "A King in Captivity." August 1945.

————. "Leave Those 'Happy' People Alone!" June 1980.

————. "Letter from Paul Fleming." November 1945.

————. "Living with the Barbaros." November 1948.

————. "Looking Back at Early Ayore Contacts." January 1956.

————. "Looking Forward to V-J Day." June 1945.

————. "Meet Mr. and Mrs. Ayore." October 1970.

————. "Mining Brown Gold." February/March 1953.

————. "Missionary Boot Camp." December 1944.

————. "Mission to the Moros: From Norman Keefe's Diary." September 1966.

————. "Mission to the Moros: From Norman Keefe's Diary." February 1967.

————. "Mission to the Moros: From Norman Keefe's Diary." May 1968.

————. "Mission to the Moros: From Norman Keefe's Diary." June 1968.

————. "Moro Contact: Trial and Triumph." November 1967.

————. "Moro Encounter!" May 1966.

————. "More Moros for Christ." December 1971.

————. "Move Toward the Moros." November 1965.

————. "Nuggets Aglow!" November 1945.

————. "Only an Ayore." January 1978.

————. "Paraguayan Nots." May 1965.

————. "The Pig People." July 1979.

————. "Reaching Out in Bolivia." February 1951.

————. "Satan Rules the Jungle." March 1950.

————. "Savages Come Out Again." May 1948.

————. "Second Barbaro Contact." October/November 1947.

————. "A Spiritual Iron Curtain." November/December 1946.

————. "The Strategic Advance." May 1944.

————. "Ten Years Reaching a Savage Tribe." May 1953.

————. "Testimony of Ecarai." January 1971.

————. "Tobité." February 1951.

————. "Tobité Report." July 1961.

————. "A Tribute to Those Who Have Gone Before." March 1972.

————. "Unrest in the Moro Camp." January 1970.

————. "What Are the Savages Like?" August/September 1948.

————. "What Color is Your Gold?" July 1954.

————. "Will There Be Any Heathen in Heaven?" November 1959.

————. "Work Starts at Boot Camp." August 1944.

Brysk, Alison. *From Tribal Village to Global Village: Indian Rights and International Relations in Latin America.* Stanford, CA: Stanford University Press, 2000.

Bugos, Paul. "An Evolutionary Ecological Analysis of the Social Organization of the Ayoreo of the Northern Gran Chaco." Northwestern University, unpublished dissertation, 1985.

Bull, Malcolm, ed. *Apocalypse Theory and the Ends of the World.* Oxford: Blackwell, 1995.

Burridge, Kenelm. *Mambu: A Melanesian Millennium.* London: Meuthen, 1960.

Butler, Judith. "Critique, Coercion and Sacred Life in Benjamin's Critique of Violence," In *Political Theologies: Public Religions in a Post-Secular World*, edited by H. de Vries and L Sullivan, 201–19. New York: Fordham University Press, 2006.

Caffrey, Patricia. "An Independent Environmental and Social Assessment of the Camisea Gas Project." Unpublished report. 2002.

Califano, Mario. "Enfermedad y terapia entre los Ayoreo (Zamuco) del Chaco Boreal." *Reunión anual de etnología* 1: 191–96. La Paz, Bolivia: MUSEF, 1987.

Canessa, Andrew. "Todos Somos Indigenas: Towards a New Language of National Political Identity." *Bulletin of Latin American Research* 25, no. 2 (2006): 241–63.

Cañizares-Esguerra. *Puritan Conquistadors: Iberianizing the Atlantic, 1550–1700*. Stanford, CA: Stanford University Press, 2006.

Canova, Paola. "Contested Spaces: Urban Ayoreo in the Mennonite Colonies of the Paraguayan Chaco." Department of Anthropology, University of Arizona, MA thesis, 2007.

———. "Del monte a la ciudad: La producción cultural de los Ayoreode en los espacios urbanos del Chaco central." *Suplemento antropológico* 46, no. 1 (2011).

Cardus, José. *Las misiones franciscanas entre los infieles de Bolivia*. Barcelona: Libreria la Inmaculada Concepción, 1886.

Caruth, Cathy. *Unclaimed Experience: Trauma, Narrative and History*. Baltimore, MD: Johns Hopkins University Press, 1996.

Casalegno, Ugo. "Les Ayore du Grand Chaco par leurs mythes: Essai de lecture et de classement des mythes ayore." Université Paris VII, 1985.

CEJIS. *El pueblo Ayoreo, entre el campo y la ciudad, el territorio y sus estrategias de sobrevivencia*. Santa Cruz, CA: CEJIS, 2005.

Charlevoix, Francois-Xavier. *Histoire de Paraguay*. Paris: Desaint, David and Duran, 1757.

Chase-Sardi, Miguel. La situación actual de los indigenas en el Paraguay. Asunción, Paraguay: BPA, 1972.

Chodoff, Paul. "The German Concentration Camp as a Psychological Stress." *Archives of General Psychiatry* 22, no. 1 (1970): 78–87.

Chomé, Ignacio. "Arte de la lengua Zamuco." *Journal de la Société des Américanistes de Paris* 47 (1958 [1745]): 121–78.

Clark, Charles. "Jesuit Letters to Hervas on American Languages and Customs." *Journal de la Société des Américanistes* 29 (1937): 97–145.

Clastres, Pierre. *Archeology of Violence*. Los Angeles: Semiotext(e), 2010.

———. *Chronicle of the Guayaki Indians*. New York: Zone Books, 1998.

———. "Of Ethnocide." Reprinted in *Archaeology of Violence*, translated by Jeanine Herman. Los Angeles: Semiotext(e), 2010 [1974].

———. *Society Against the State: Essays in Political Anthropology*. New York: Zone Books, 1989.

Clifford, James. "Indigenous Articulations." *Contemporary Pacific* 13, no. 2 (2001): 467–90.

Clough, Patricia. Introduction to *The Affective Turn*, 1–34. Durham, NC: Duke University Press, 2007.

Cobo, José R. Martínez. *The Study of the Problem of Discrimination against Indigenous Populations*. New York: United Nations Document E/CN.4/Sub.2/1986/7/Add.4, 1986.

Cohn, Bernard. *Colonialism and Its Forms of Knowledge: The British in India*. Princeton, NJ: Princeton University Press, 1996.

Cohn, Norman. *The Pursuit of the Millennium: Revolutionary Millenarians and the Mystical Anarchists of the Middle Ages*. New York: Galaxy Books, 1970.

Colchester, Marcus. *Salvaging Nature: Indigenous Peoples, Protected Areas and Bio-diversity Conservation.* Geneva: United Nations Research Institute for Social Development, 1994.

Collier, Steven, and Andrew Lakoff. "On Regimes of Living." In *Global Assemblages*, edited by Aihwa Ong and Stephen J. Collier, 22–40. Oxford: Blackwell, 2005.

Comaroff, Jean. *Body of Power, Spirit of Resistance.* Chicago: University of Chicago Press, 1985.

Comaroff, Jean, and John Comaroff. "Ethnography on an Awkward Scale: Post-colonial Anthropology and the Violence of Abstraction." *Ethnography* 4, no. 2 (2003): 147–79.

Combés, Isabelle. *Zamucos.* Cochabama: Editoral Nomades—Universidad Catolica, 2009.

Conklin, Beth, and Laura Graham. "The Shifting Middle Ground: Amazonian Indians and Ecopolitics." *American Anthropologist* 97, no. 4 (1995): 695–710.

Connerton, Paul. *How Societies Remember.* Cambridge: Cambridge University Press, 1989.

CORDECRUZ. Los Ayoreode: Diagnostico y problematica actual. Santa Cruz: CORDECRUZ, 1981.

Cordeu, Edgardo. *Transfiguraciones simbólicas. Ciclo ritual de los indios Tomaráxo del Chaco Boreal.* Asunción, Paraguay: Centro de Artes Visuales/Museo del Barro, 2003.

Cowan, Jane, Marie Dembour, and Richard Wilson, eds. *Culture and Rights.* Cambridge: Cambridge University Press, 2001.

Crapanzano, Vincent. *Imaginative Horizons: An Essay in Literary-Philosophical Anthropology.* Chicago: University of Chicago Press, 2004.

Das, Veena. *Life and Words: Violence and the Descent into the Ordinary.* Berkeley: University of California Press, 2006.

Das, Veena, et al., eds. *Remaking a World: Violence, Social Suffering and Recovery.* Berkeley: University of Califronia Press, 2001.

Dasso, Maria Cristina. "Alegría y coraje en la alabanza guerrera de los Ayoreo del Chaco Boreal." *Anthropos* 99, no. 1 (2004): 57–72.

———. "La cosmogonía Ayoreo y su relación con el medio actual." *Moana estudios de antropología oceánica* 6, no. 1 (2001).

Davis, Mike. *Dead Cities and Other Tales.* New York: New Press, 2002.

De la Cadena, Marisol, and Orin Starn, eds. *Indigenous Experience Today.* Oxford: Berg, 2010.

Deleuze, Gilles. *Pure Immanence.* New York: Zone, 2001.

Dellamora, Richard, ed. *Postmodern Apocalypse: Theory and Cultural Practice at the End.* Philadelphia: University of Pennsylvania Press, 1995.

Derrida, Jacques. "Of an Apocalyptic Tone Newly Adopted in Philosophy." In *Derrida and Negative Theology*, edited by Harold Cowald and Toby Foshay. Albany: SUNY Press, 1992.

Descola, Philippe. *In the Society of Nature*. Cambridge: Cambridge University Press, 1994.

———. *Par-delá nature et culture*. Paris: Gallimard, 2005.

Desjerlais, Robert. *Body and Emotion: The Aesthetics of Illness and Healing in the Nepal Himalayas*. Philadelphia: University of Pennsylvania Press, 1992.

De Wilde, Marc. "Violence in the State of Exception: Reflections on Theologico-Political Motifs in Benjamin and Schmitt." In *Political Theologies: Public Religions in a Post-Secular World*, edited by H. de Vries and L. Sullivan, 188–200. New York: Fordham University Press, 2006.

Diamond, Stanley. *In Search of the Primitive: A Critique of Civilization*. New Brunswick, NJ: Transaction, 1981.

Dirks, Nicholas. *Colonialism and Culture*. Ann Arbor: University of Michigan Press, 1992.

Dobrizhoffer, Martin. *An Account of the Abipones, an Equestrian People of Paraguay*. New York: Johnson Reprint Company, 1970 [1784].

Dombrowski, Kirk. *Against Culture: Development, Politics and Religion in Indian Alaska*. Lincoln: University of Nebraska Press, 2001.

———. "The Politics of Native Culture." In *A Companion to the Anthropology of American Indians*, edited by Thomas Biolsi. Oxford: Blackwell, 2008.

D'Orbigny, Alcide. *Voyage dans l'Amérique meridionale*. Paris: Bertrand, 1835.

Dornelles, C., et al. "Mitochondrial DNA and Alu Insertions in a Genetically Peculiar Population: The Ayoreo Indians of Bolivia and Paraguay." *American Journal of Human Biology* 16, no. 4 (2004): 479–88.

Dos Santos, Sidney, et al. "Autosomal STR Analyses in Native Amazonian Tribes Suggest a Population Structure Driven by Isolation by Distance," *Human Biology* 81, no. 1 (2009): 71–88.

Duerksen, Marvin. "Allanamiento de ONG se basa en denuncias." *ABC Color*, December 4, 2010.

———. "Ayoreos exigen que una ONG devuelva mas de G. 2.000 milliones." *ABC Color*, June 7, 2011.

———. "Imputan a los directivos de Iniciativa Amotocodie por lesion de confianza." *ABC Color*, May 20, 2011.

———. "Maniobran para retirar denuncia contra ONG Iniciativa Amotocodie." *ABC Color*, December 16, 2010

Duguid, Julian. *The Green Hell: Adventures in the Mysterious Jungles of Eastern Bolivia*. New York: Century Company, 1933.

Dyck, Noel. *Indigenous Peoples and the Nation-State*. Newfoundland: Memorial University of Newfoundland, 1985.

Edelman, Marc. "Landlords and the Devil: Class, Ethnic and Gender Dimensions of Central American Peasant Narratives." *Cultural Anthropology* 9, no. 1 (1994): 58–93.

Edwards, Michael, and Edward Hulme, eds. *Non-Governmental Organizations: Performance and Accountability Beyond the Magic Bullet*. London: Earthscan, 1995.

Eissler, KR. "On Isolation." *The Pyschoanlytic Study of the Child* 14 (1959): 29–60.

Enrica Pia, Gabriella. *La cultura material de los Ayoreo y el arte rupestre de Bolivia y Paraguay*. Asunción, Paraguay: CEADUC, 2006.

Escobar, Ticio. *Misión etnocidio*. Asunción, Paraguay: Editora Litocolor, 1988.

Estival, Jean-Pierre. "Os casadores e o radio: Sobre o novo uso dos meios de comunicacao entre os Ayoreo do Chaco Boreal." *Revista Anthropológica* 17, no. 1 (2006): 103–14.

Estudios Rurales. *Apuntes para la colonización*. Internal document. 1981.

Fabian, Johannes. *Time and the Other: How Anthropology Makes Its Object*. New York: Cambridge University Press, 1983.

Fanon, Frantz. *Black Skin, White Masks*. New York: Grove, 2008 [1957].

———. The Wretched of the Earth. New York: Grove, 1963.

Fassin, Didier. "Another Politics of Life Is Possible." *Theory, Culture and Society* 26, no. 5 (2009): 44–60.

———. *Humanitarian Reason: A Moral History of the Present*. Berkeley: Univeristy of California Press, 2011.

———. "Introduction." In *A Companion to a Moral Anthropology*. Oxford: Blackwell, 2012.

Feld, Steve. "Waterfalls of Song: An Acoustemology of Place Resounding in Bosavi, Papua New Guinea." In *Senses of Place,* edited by Steve Feld and K. H. Basso, 91–135. Santa Fe: School of American Research Press, 1996.

Ferguson, James. *Expectations of Modernity*. Berkeley: University of California Press, 1999.

———. "The Uses of Neoliberalism." *Antipode* 41, no. 1 (2010): 166–84.

Fernandez, Juan. *Relación historial de las misiones de indios chiquitos que en Paraguay tienen los padres de la Compania de Jesus*. Jujuy: CEIC, 1994 [1726].

Fischermann, Bernd. "Huida o entrega—vivir en aislamiento, el ejemplo de los Ayorei Totobiegosode." In *Pueblos indígenas*, edited by Parellada, 252–66.

———. "La cosmovision del pueblo Ayoreode." Translated by Benno Glausser. University of Bonn, unpublished thesis, 2001.

———. "La fiesta de la Asojna o el cambio de las estaciones entre los ayoréode del Chaco Boreal." *Suplemento antropológico* 40, no. 1 (2005): 391–450.

———. "Los Ayoreode." In *En busca de la Loma Santa*, edited by J. Riester, 65–118. Santa Cruz de la Sierra: Los Amigos del Libro, 1976.

———. "Una frontera frágil: Cultura y natura entre los Ayoreode." In *El último canto del monte*, edited by A. Cabrera et al. Biblioteca Paraguaya de Antropología, 29. Asunción, Paraguay: CEADUC, 1998.

———. "Viviendo con los Pai: Las experiencias Ayoreode con los jesuitas." In *Martin Schmid 1694–1772: Las misiones jesuíticas de Bolivia*, edited by Eckart Kühne, 47–54. Santa Cruz de la Sierra: Pro Helvetia, 1996.

Fisher, Michael M. J. "Double-Click: The Fables and Language Games of Latour and Descola; Or, From Humanity as Technological Detour to the Peopling of Technologies." Unpublished paper, 2013.

Fisher, William. "Doing Good? The Politics and Antipolitics of NGO Practices." *Annual Review of Anthropology* 26 (1997).

Fortun, Kim, Mike Fortun, and Steve Rubenstein, eds. "Introduction to Emergent Indigeneities." *Cultural Anthropology* 25, no. 2 (2010): 222–34.

Fortun, Mike. *Promising Genomics: Iceland and deCODE Genetics in a World of Speculation*. Berkeley: University of California Press, 2008.

Foucault, Michel. *The History of Sexuality*. Vol. 1. New York: Pantheon, 1978.

Fowler, Alan. *Striking a Balance: A Guide to Enhancing the Effectiveness of Non-Governmental Organisations in International Development*. London: Earthscan, 1997.

Franklin, Sarah, and Margaret Lock, eds. *Remaking Life and Death: Toward an Anthropology of the Biosciences*. Santa Fe, NM: School of American Research Press, 2003.

Fraser, Nancy. "Rethinking the Public Sphere: A Contribution to the Critique of Actually Existing Democracy." In *Habermas and the Public Sphere*, edited by C. Calhoun, 109–42. Cambridge, MA: MIT Press, 1991.

Freud, Sigmund. "Inhibitions, Symptoms and Anxiety." *SE* 20 (1926): 77–172.

Furlong, Guillermo. *Joaquin Camaño y su noticia del Gran Chaco*. Buenos Aires: Librería del Plata, 1955 [1778].

Gajdusek, D. Carleton. *Paraguayan Indian Expeditions to the Guayaki and Chaco Indians, August 25, 1963 to September 28, 1963*. Bethesda, MD: National Institute of Neurological Diseases and Blindness, National Institutes of Health, 1963.

Geertz, Clifford. "Life among the Anthros." *New York Review of Books*, February 8, 2001, p. 18–22.

Gell, Alfred. *The Anthropology of Time: Cultural Constructions of Temporal Maps and Images*. Oxford: Berg, 1992.

Gill, Victoria. "Conservation Expedition Poses Risk to Tribes," *BBC News*, November 9, 2011.

Gilmore, Harlan. "Cultural Diffusion Via Salt." *American Anthropologist* 57, no. 5 (1955): 1011–15.

Gimlette, John. *At the Tomb of the Inflatable Pig: Travels through Paraguay*. London: Knopf, 2004.

Ginsburg, Faye. "Embedded Aesthetics: Creating a Discursive Space for Indigenous Media." *Cultural Anthropology* 9 (1994): 365–82.

———. "Indigenous Media: Faustian Contract or Global Village?" *Cultural Anthropology* 6 (1991): 92–112.

———. "The Parallax Effect: The Impact of Indigenous Media on Ethnographic Film." *VAR* 11, no. 2 (1995): 64–76.

———. "Screen Memories: Resignifying the Traditional in Indigenous Media." In *Media Worlds: Anthropology on New Terrain*, edited by Faye Ginsburg, Lila Abu-Lughod, and Brian Larkin. Berkeley: University of California Press, 2002.

Ginsburg, Faye, and Fred Myers. "A History of Aboriginal Futures." *Critique of Anthropology* 26, no. 1 (2006): 27–45.

Glauser, Benno. "Paraguay: Pueblos indígenas aislados." *Suplemento antropológico* 41, no. 2 (2006): 187–200.

———. "Paraguay: The Last Ayoreo in Voluntary Isolation." *World Rainforest Movement Bulletin* 87 (2004): 12–13.

———. "Su presencia protege el corazón del Chaco seco." In *Pueblos indígenas*, edited by Parellada, 220–33.

Gledhill, John. "Neoliberalism." In *A Companion to the Anthropology of Politics*, edited by David Nugent and Joan Vincent. New York: Blackwell, 2004.

Goicoechea, A. S., et al.. "New Genetic Data on Amerindians from the Paraguayan Chaco." *American Journal of Human Biology* 13 (2001): 660–67.

Good, Mary-Jo, et al., eds. *Postcolonial Disorders*. Berkeley: University of California Press, 2008.

Gordillo, Gastón. "The Breath of Devils: Memories and Places of an Experience of Terror." *American Ethnologist* 29, no. 1 (2002): 33–57.

———. *En el Gran Chaco: Anthropologias e historias*. Buenos Aires: Prometeo, 2006.

———. "Hermeneutica de la illusion: La etnologia fenomenologica de Marcelo Bórmida y su construccion de los indigenas del Gran Chaco." *Cuadernos de Antr. Social* **(**1996).

———. *Landscape of Devils: Tensions of Place and Memory in the Argentine Chaco*. Durham, NC: Duke University Press, 2004.

———. *Rubble: The Afterlife of Destruction*. Durham, NC: Duke University Press, 2014.

Gordillo, Gaston, and Silvia Hirsh. "Indigenous Struggles and Contested Identities in Argentina: Histories of Invisibilization and Reemergence." *Journal of Latin American Anthropology* 8, no. 3 (2003): 4–30.

Gordon, Avery. *Ghostly Matters: Haunting and the Sociological Imagination*. Minneapolis: University of Minnesota, 1997.

Gordon, Linda. "Internal Colonialism and Gender." In *Haunted by Empire*, edited by Ann Stoler, 427–52. Durham, NC: Duke University Press, 2008.

Goulart, S., et al. "Autosomal STR Genetic Variability in the Gran Chaco Native Population: Homogeneity or Heterogeneity?" *American Journal of Human Biology* 20 (2008): 704–11.

Gow, Peter. *Countering Development: Indigenous Modernity and the Moral Imagination*. Durham, NC: Duke University Press, 2008.

Graber, Christian. *The Coming of the Moros*. Scottdale, PA: Herald Press, 1964.

Graham Kerr, John. *A Naturalist in the Gran Chaco*. Cambridge: Cambridge University Press, 1950.

Grassian, Stuart. "Psychopathological Effects of Solitary Confinement." *American Journal of Psychiatry Online* 140 (1983): 1450–54.

Gray, Frank, and Elia Murphy. "The Unlikely Missionary: Radio Rises to the Challenge of the Unreached People Groups." Posted on http://www.missionfrontiers.org/2000/05/unlikely.htm, 2000.

Greenblatt, Stephen. *Marvelous Possessions: The Wonder of the New World*. Chicago: University of Chicago Press, 1991.

Grubb, W. Barbrooke. *An Unknown People in an Unknown Land : An Account of the Life and Customs of the Lengua Indians of the Paraguayan Chaco, with Adventures and Experiences Met with during Twenty Years' Pioneering and Exploration amongst Them*. London: Seeley, 1911.

Grupo de Apoyo a los Totobiegosode, *El ultimo canto del monte: Reclamo de tierra Ayoreo*. Asunción, Paraguay: Biblioteca Paraguaya de Antropología, 1998.

Guillermoprieto, Alm. "Bolivia's New Order." *National Geographic* 23, no. 1 (2008): 56–71.

Gupta, Akhil, and James Ferguson, eds. "Beyond Culture: Space, Identity and the Politics of Difference." In *Culture, Power and Place: Explorations in Critical Anthropology*, 33–52. Durham, NC: Duke University Press, 1997.

Gustafson, Bret. *New Languages of the State: Indigenous Resistance the Politics of Knowledge in Bolivia*. Durham, NC: Duke University Press, 2009.

Guyer, Jane. "Prophecy and the Near Future: Thoughts on Macroeconomic, Evangelical and Punctuated Time." *American Ethnologist* 34, no. 3 (2007): 409–21.

Hacking, Ian. "The Looping Effect of Human Kinds." In *Causal Cognition*, edited Dan Sperber, David Premack, and Ann Premack, 351–94. Oxford: Clarendon Press, 1995.

Hage, Ghassan. "Critical Anthropological Thought and the Radical Political Imaginary Today," *Critique of Anthropology* 32, no. 3 (2012): 285–308.

Hale, Charles. *Mas que un indio: Racial Ambivalence and Neoliberal Multiculturalism in Guatemala*. Santa Fe, NM: School of American Research Press, 2006.

———. "Rethinking Indigenous Politics in the Era of the Indio Permitido." NACLA Report on the Americas (2004): 16–21.

Hall, Gillette, and Harry Anthony Patrinos. *Indigenous Peoples, Poverty and Human Development in Latin America: 1994–2004*. Washington, DC: World Bank, 2006.

Hallowell, Alfred Irving. "Acculturation Processes and Personality Changes as Indicated by the Rorschach Technique." *Rorschach Research Exchange* 6 (1942): 42–50.

Hardt, Michael, and Anotonio Negri. *Commonwealth*. Cambridge, MA: Belknap, 2009.

Heijdra, Hans. "La nueva gente: Un estudio sociologico sobre las formas de produccion y las formas de organización en un contexto dinamico de una comunidad Ayorea del oriente Boliviana." University of Agronomy Wageningen, unpublished MA thesis, 1987.

———. *Participación y exclusión indígena en el desarrollo: Banco mundial, CIDOB y el pueblo Ayoreode en el proyecto tierras bajas del este de Bolivia*. Santa Cruz: APCOB, 1997.

Hein, David. *Los Ayoreos, nuestros vecinos: La misión del norte chaqueño*. Asunción, Paraguay: El Lector, 1990.

Heller, Agnes. *The Power of Shame: A Rational Perspective*. Boston: Routledge and K. Paul, 1985.

Helmreich, Stefan. "An Anthropologist Underwater: Immersive Soundscapes, Submarine Cyborgs, and Transductive Ethnography." *American Ethnologist* 34, no. 4 (2007): 621–41.

Herr, Michael. *Dispatches*. New York: Knopf, 1977.

Herzfeld, Michael. *Cultural Intimacy: Social Poetics in the Nation-State*. New York: Routledge, 1997.

Hetherington, Kregg. *Guerilla Auditors: The Politics of Transparency in Neoliberal Paraguay*. Durham, NC: Duke University Press, 2011.

Hirsch, S. M. "Bilingualism, Pan-Indianism and Politics in Northern Argentina: The Guaraní's Struggle for Identity and Recognition." *Journal of Latin American Anthropology* 8 (2003): 84–103.

Hirschman, Alfred. *A Bias for Hope: Essays on Development and Latin America*. New Haven, CT: Yale University Press, 1971.

Hodgson, Dorothy L. "Precarious Alliances: the Cultural Politics and Structural Predicaments of the Indigenous Rights Movement in Tanzania." *American Anthropologist* 104, no. 4 (2002): 1086–97.

Hogle, Linda. "Life/Time Warranty: Rechargeable Cells and Extendable Lives." In *Remaking Life and Death*, 61–97. Santa Fe, NM: School of American Research Press, 2003.

Holland, Luke. *Indians, Missionaries and the Promised Land: Photographs from Paraguay*. London: Survival International, 1980.

Horst, Rene Harder. "Consciousness and Contradiction: Indigenous People and Paraguay's Transition to Democracy." In *Contemporary Indigenous Movements in Latin America*, edited by Erick D. Langer and Elena Muñoz. Wilmington, DE: Scholarly Resources, 2003.

———. *The Stroessner Regime and Indigenous Resistance in Paraguay*. Gainesville: University Press of Florida, 2007.

Hudock, A. C. *NGOs and Civil Society: Development by Proxy*. Cambridge: Polity, 1999.

———. "NGOs' Seat at the Donor Table: Enjoying the Food or Serving the Dinner?" *IDS Bulletin* 31 (2000).

Huertas-Castillo, Beatriz. *Los pueblos indigenas en Aislamiento: Su lucha por la sobrevivencia y la libertad*. Copenhagen: International Work Group for Indigenous Affairs, 2002.

Hulme, David, and Michael Edwards, eds. *NGOs, States and Donors: Too Close for Comfort?* New York: St. Martin's Press, 1997.

Hurtado, Magdalena, et al. "The Epidemiology of Infectious Diseases among South American Indians: A Call for Ethical Research Guidelines." *Current Anthropology* 42 (2001): 425–32.

Idoyaga Molina, Anatilde. "Cosmología y mito: La representación del mundo entre los Ayoreo." *Scripta ethnologica* 20 (1998): 31–72.

———. "Cosmología y sistema de representaciones entre los Ayoreo del Chaco Boreal." *Scripta ethnologica* 19 (1995).

———. "El daño mediante la palabra entre los Ayoreo del Chaco Boreal." Göteborgs Etnografiska Museum Arstryck, 1980.

———. "La muerte como cambio ontológico en los Ayoreo del Chaco Boreal." *Mitológicas* 5 (1990): 35–40.

———. "La significación de la mítica de Susmaningái (el coraje) en la cosmovisión Ayoreo." *Mitológicas* 13 (1989).

———. "Mito, temporalidad y estados alterados de conciencia: El surgimiento de relatos y las experiencias shamánicas entre los Ayoreo del Chaco Boreal." *Acta Americana: Journal of the Swedish Americanist Society* 5, no. 1 (1997): 61–76.

ILO. "The Second Session of the ILO Committee of Experts on Indigenous Labour." *International Labor Review* 70 (1954): 418–42.

Inciardi, James, and Hilary Surratt. "Children in the Streets of Brazil: Drug Use, Crime, Violence and HIV Risks." *Substance Use and Misuse* 33 (1998): 1461–80.

Infantas, Anna. "Bolivia's Ayoreo Indians, Devoured by the City." *Inter Press News Service.* December 18, 2012. Accessed January 12, 2013. http://www.ipsnews.net/2012/12/bolivias-ayoreo-indians-devoured-by-the-city/.

Informe Oficial. "Conclusiones." Unpublished document, 1987.

Iñigo Carrera, Nicolas. *Violence as an Economic Force: The Process of Proletarianization among the Indigenous People of the Argentinean Chaco, 1884–1930.* Copenhagen: International Work Group for Indigenous Affairs, 1982.

Ingold, Tim. *The Perception of the Environment: Essays on Livelihood, Dwelling and Skill.* London: Routledge, 2000.

International Labor Organization (ILO). "The Second Session of the ILO Committee of Experts on Indigenous Labor." *International Labor Review* 70 (1956): 422–23.

International Work Group for Indigenous Affairs. *The Case of the Ayoreo, Report 4.* Copenhagen: International Work Group for Indigenous Affairs, 2010.

Ivy, Marilyn. *Discourses of the Vanishing: Modernity, Phantasm, Japan.* Chicago: University of Chicago Press, 1995.

Jackson, Jean. "Culture, Genuine or Spurious? The Politics of Indianness in the Vaupes, Colombia." *American Ethnologist* 22, no. 1 (1995): 3–27.

———. "Rights to Indigenous Culture in Colombia." In *The Practice of Human Rights*, edited by Mark Goodale and Sally Merry, 204–37. Cambridge: Cambridge University Press, 2007.

Jackson, Jean, and Kay Warren. "Indigenous Movements in Latin America, 1992–2004: Controversies, Ironies and New Directions." *Annual Review of Anthropology* 34 (2005): 549–73.

Janzen, A. E. *The Moro's Spear.* Hillboro, Kansas: Board Missions of the Mennonite Brethren Church, 1962.

Johnson, Jean Dye. *God Planted Five Seeds.* New York: Harper and Row, 1966.

Johnston, Ken. *The Story of the New Tribes Mission.* Sanford, FL: New Tribes Mission Press, 1985.

Kaul, Inge, Isabelle Grunberg, and Marc Stern. "Introduction." In *Global Public Goods*, xix–1. Oxford: Oxford University Press, 1999.

Keane, Webb. "Sincerity, "Modernity" and the Protestants." *Cultural Anthropology* 17, no. 1 (2002): 65–92.

Keck, Margaret, and Kathryn Sikkink. *Activists Beyond Borders: Advocacy Networks in International Politics.* Ithaca, NY: Cornell University Press, 1998.

Keenan, Thomas. "The Point Is to (Ex)Change It: Reading Capital, Rhetorically." In *Fetishism as Cultural Discourse*, edited by Emily S. Apter and William Pietz, 152–85. Ithaca, NY: Cornell University Press, 1993.

Kelm, Heinz. "Das Jahrfest der Ayore (Ostbolivien)." *Baessler Archiv* 19 (1971).

———. "Die Zamuco (Ostbolivien)." *Zeitschrift für Ethnologie* 88 (1963): 66–85.

Kenrick, Justin, and Jerome Lewis. "Indigenous Peoples' Rights and the Politics of the Term 'Indigenous.'" *Anthropology Today* 20, no. 2 (2004): 4–9.

Kermode, Frank. *The Sense of an Ending: Studies in the Theory of Fiction.* Oxford: Oxford University Press, 1967.

Kidd, Stephen. "Los Totobiegosode reivindican su territorio tradicional." *Asuntos indígenas* 4. Copenhagen: International Work Group for Indigenous Affairs, 1993.

———. *The Working Conditions of Indigenous Peoples of the Gran Chaco.* Copenhagen: International Work Group for Indigenous Affairs, 1997.

Killeen, Timothy, et al. "Total Historical Land-Use Change in Eastern Bolivia: Who, Where, When and How Much?" *Ecology and Society* 13, no. 1 (2008): 36.

Kirmayer, Laurence, and Gail Valaskakis, eds. *Healing Traditions: Mental Health among Aboriginal Peoples in Canada.* Vancouver: University of British Columbia Press, 2009.

Klassen, Peter P. *The Mennonites in Paraguay.* Vol. 1, *Kingdom of God and Kingdom of This World.* Translated by Gunther H. Schmitt. Weierhof, Germany: Mennonitischer Buchversand, 2004.

———. *The Mennonites in Paraguay.* Vol. 2, *Encounter with Indians and Paraguayans.* Winnipeg: Mennonite Books, 2002.

Kleinman, Arthur, Veena Das, and Margaret Lock, eds. *Social Suffering.* Berkeley: University of California Press, 1997.

Knight, E. F. *The Cruise of the Falcon: A Voyage to South America and Up the River Plate in a Thirty Ton Yacht.* London: Samson Low Marsten Searle and Rivington, 1884.

Kroeber, Theodora. *Ishi in Two Worlds: A Biography of the Last Wild Indian in North America.* Berkeley: University of California Press, 1961.

Kunreuther, Laura. "Technologies of the Voice: FM Radio, Telephone, and the Nepali Diaspora in Kathmandu." *Cultural Anthropology* 21, no. 3 (2006): 323–53.

Kymlicka, Will. *Multicultural Citizenship.* Oxford: Clarendon, 1996.

La Prensa. "Ayoreo Sata Jaqinaka," February 16, 2006.

Latour, Bruno. "Perspectivism: Type or Bomb?" *Anthropology Today* 25, no. 2 (2009): 1–2.

———. *Reassembling the Social: An Introduction to Actor-Network Theory*. Oxford: Oxford University Press, 2005.

———. *We Have Never Been Modern*. Translated by Catherine Porter. Paris: La Découverte, 1993.

Lattas, Andrew. "Memory, Forgetting and the New Tribes Mission in West New Britain." *Oceania* 66 (1996).

Levi, Primo. *The Drowned and the Saved*. New York: Vintage, 1989.

———. *The Reawakening*. New York: Touchstone Books, 1995.

Lévi-Strauss, Claude. *Tristes tropiques*. Paris: Plon, 1955.

Lewis, Norman. *Eastern Bolivia: The White Promised Land*. Copenhagen: International Work Group for Indigenous Affairs, 1975.

Li, Tania. "Articulating Indigenous Identity in Indonesia: Resource Politics and the Tribal Slot." *Comparative Studies in Social History* 42, no. 1 (2000): 149–79.

———. "Indigeneity, Capitalism and the Management of Dispossession." *Current Anthropology* 51 (2010): 385–414.

Lind, Ulf. "Die Medizin der Ayoré-Indianer." Department of Anthropology, University of Bonn, unpublished dissertation, 1974.

———. "Los Ayoreos, un pueblo guerrero." *Mennoblatt* 4 (1971): 4–8.

Lozano, Pedro. *Descripción corográfica del Gran Chaco Gualamba*. Tucumán, Argentina: National University of Tucumán Press, 1989 [1733].

Lucero, José Antonio. "Locating the Indian Problem: Community, Nationality and Contradiciton in Ecuadorian Indigenous Politics." *Latin American Perspectives* 30 (2003): 23–48.

Lukács, George. *History and Class Consciousness: Studies in Marxist Dialectics*. Translated by Rodney Livingstone. London: Merlin Press, 1971.

Lussagnet, Susanne. "Présentation de arte de la lengua Zamuca, por el Padre Ignace Chomé." *Journal de la Société des Américanistes* 47 (1958): 121–23.

———. "Vocabulaires Samuku, Morotoko, Poturero et Guarañoka, précédés d'une étude historique sur les anciens Samuku du Chaco bolivien et leurs voisins." *Journal de la Société des Américanistes* 50 (1961): 185–243.

Lutz, Catherine. *Unnatural Emotions: Everyday Sentiments on a Micronesian Atoll and Their Challenges to Western Theory*. Chicago: University of Chicago Press, 1988.

MacCannell, Dean. *Empty Meeting Grounds: The Tourist Papers*. London: Routlege, 1992.

Maffi, Luisa. "Linguistic, Cultural and Biological Diversity." *Annual Review of Anthropology* 34 (2005): 599–617.

Martin, Emily. "Violence, Language and Everyday Life." *American Ethnologist* 34, no. 4 (2007): 741–45.

Marx, Karl. *Capital*. Vol. 1, *A Critique of Political Economy*. Translated by Ben Fowkes. London: Penguin, 1990 [1867].

Mashnshnek, Celia. "Expressions of Alterity among Some Ayoreo Mythical Beings." *Anthropos* 88, nos. 1–3 (1993): 201–5.

———. "La fenomenología como propuesta metodológica para la comprensión
de tres culturas etnográficas del Gran Chaco: Mataco, Pilagá y Ayoreo."
Scripta ethnologica 17 (1995): 29–35.

———. "La percepción del banco en la cosmovisión de los Ayoreo (Chaco Boreal)." *Scripta ethnologica supplemento* 11 (1991).

———. "Las nociones míticas en la economía de producción de los Ayoreos del
Chaco Boreal." *Scripta ethnologica* 8 (1990): 119–40.

Mauss, Marcel. "A Category of the Human Mind: The Notion of Person; The
Notion of Self." In *The Category of the Person: Anthropology, Philosophy, His-
tory*, edited by Michael Carrithers, Steven Collins, and Steven Lukes, 1–25.
Cambridge: Cambridge University Press, 1985.

Maybury-Lewis, David, and James Howe. *The Indian Peoples of Paraguay: Their
Plight and Their Prospects.* Cambridge: Cultural Survival, 1980.

Mbembe, Achille. "Necropolitics." *Public Culture* 15, no. 1 (2003): 11–40.

———. *On the Postcolony.* Berkeley: University of California Press, 2001.

McFalls, Laurence. "Benevolent Dictatorship: the Formal Logic of Humanitarian
Government." In *Contemporary States of Emergency*, edited by Didier Fassin
and Mariella Pandolfi, 317–34. New York: Zone, 2010.

Mendizabal, Miguel. "Influencia de la sal en la distribución geográfica de los
grupos indígenas de México." México DF: Imprenta del Museo nacional de
arqueología, historia y etnografía, 1928.

Metraux, Alfred. "Ethnography of the Gran Chaco." In *Handbook of South American Indians*, vol. 1, edited by Julian H. Steward, 197–370. Washington, DC:
Bureau of American Ethnology, 1946.

Mignolo, Walter. "Epistemic Disobedience, Independent Thought and Decolo-
nial Freedom." *Theory, Culture and Society* 26, nos. 7–8 (2009): 1–23.

Miller, Bruce. *Invisible Indigenes: The Politics of Non-Recognition.* Lincoln: University of Nebraska Press, 2003.

Miller, William Ian. *The Anatomy of Disgust.* Cambridge, MA: Harvard University
Press, 1997.

Minkowski, Eugene. *Lived Time: Phenomenological and Psychopathological Studies.*
Evanston, IL: Northwestern University Press, 1970.

Miyazaki, Hirokazu. "Economy of Dreams: Hope in Global Capitalism and Its
Critiques." *Cultural Anthropology* 21, no. 2 (2006): 147–72.

Muehlmann, Shaylih. "How Do Real Indians Fish? Neoliberal Multiculturalism
and Contested Indigeneities in the Colorado Delta." *American Anthropologist*
111, no. 4 (2009): 468–79.

Munn, Nancy. *The Fame of Gawa: A Symbolic Study of Value Transformation in a
Massim Society (Papua New Guinea).* Cambridge: Cambridge University Press,
1986.

Munzel, Mark. *The Ache Indians: Genocide in Paraguay.* Copenhagen: International Work Group for Indigenous Affairs, 1973.

Myers, Fred. "Locating Ethnographic Practice: Romance, Reality and Politics in the Outback." *American Ethnologist* 15, no. 4 (November 1988): 609–24.

———. "The Logic and Meaning of Anger among the Pintupi Aborigines." *Man*, New Series 23, no. 4 (1988): 589–61.

———. "The Meaning and Logic of Anger among Pintupi Aborigines." *Man* (1979): 589–61.

———. "Ontologies of the Image and Economies of Exchange." *American Ethnologist* 31, no. 1 (2004): 1–16.

———. *Painting Culture: The Making of an Aboriginal High Art.* Durham, NC: Duke University Press, 2002.

———. *Pintupi Country, Pintupi Self: Sentiment, Place, and Politics among Western Desert Aborigines.* Berkeley: University of California Press, 1991.

Nancy, Jean-Luc. *The Sense of the World.* Minneapolis: University of Minnesota Press, 1997.

Napolitano, Dora, and Aliya Ryan. "The Dilemma of Contact: Voluntary Isolation and the Impacts of Gas Exploitation on Health and Rights in the Kugapakori Nahua Reserve, Peruvian Amazon." *Environmental Research Letters* 2, no. 4 (2007): 1–12.

New Tribes Mission. "Ayoreos Emerge From Jungle." March 11, 2004. www.ntm.org.

———. "Economic Development at El Faro Moro: What Is Feasible?" Unpublished report, 1977.

———. "New Believers Face Problems." November 16, 2006, www.ntm.org.

———. "Pig People Clan Hearing God's Word." December 6, 2005, www.ntm.org.

———. "Truth Comes." February 3, 2005, www.ntm.org.

Nickson, Andrew, and Peter Lambert. "State Reform and the 'Privatized State' in Paraguay." *Public Administration and Development* 22, no. 2 (2002): 163–74.

Nietschmann, Bernard. "The Fourth World: Nations Versus States." In *Reordering the World: Geopolitical Perspectives on the Twenty-First Century*, edited by George J. Demko and William Wood, 225–42. Boulder, CO: Westview Press, 1994.

Niezin, Ronald. *The Origins of Indigenism: Human Rights and the Politics of Identity.* Berkeley: University of California Press, 2003.

Nostas, Mercedes, and Carmen Sanabria. *Detrás del cristal con que se mira: Mujres ayoreas- ayoredie, órdenes normativos e interlegalidad.* Santa Cruz de la Sierra/ La Paz, Bolivia: Coordinadora de la Mujer/ Editorial Presencia, 2009.

Nuñez, Carmen. "Asai, un personaje del horizonte mítico de los Ayoreo." *Scripta ethnologica* 5 (1979).

Oefner, Luis. "Apuntes sobre una tribu salvaje que existe en el Oriente de Bolivia." *Anthropos* 35/36 (1940): 100–108.

Pagden, Anthony. *The Fall of Natural Man: The American Indian and the Origins of Comparative Ethnology.* Cambridge: Cambridge University Press, 1982.

Pagels, Elaine. *The Origin of Satan: How Christians Demonize Jews, Pagans and Heretics.* New York: Random House, 1999.

Pages Larraya, Fernando. "El complejo cultural de la locura en los Moro-Ayoreo." *Acta psiquiátrica y psicológica de América Latina* 19, no. 4 (1973): 253–64.

Paine, Robert. "The Claim of the Fourth World." In *Native Power: The Quest for Autonomy and Nationhood of Indigenous Peoples.* Oxford: Oxford University Press, 1985.

Parellada, Alejandro, ed. *Pueblos indígenas en aislamiento voluntario y contacto inicial en la Amazonia y el Gran Chaco.* Copenhagen: International Work Group for Indigenous Affairs, 2007.

Pastore, Carlos. *El Gran Chaco en la formación territorial del Paraguay: Etapas de su incorporación.* Asunción, Paraguay: Criterio, 1989.

Pelleschi, Giovanni. *Eight Months on the Gran Chaco of the Argentine Republic.* London: Sampson Low, Marston, Searle and Rivington, 1886.

Perasso, José. *Cronicas de cacerias humanas: La tragedia Ayoreo.* Asunción, Paraguay: El Lector, 1987.

Perez-Diaz, A. A., and Francisco Salzano. "Evolutionary Implications of the Ethnography and Demography of Ayoreo Indians." *Journal of Human Evolution* 7 (1978).

Petryna, Adriana. *Life Exposed: Biological Citizens after Chernobyl.* Princeton, NJ: Princeton University Press, 2002.

Picanerai, Aquino. "De los dirigentes de la Union de Nativos Ayoreo de Paraguay," in *Pueblos indígenas en aislamiento voluntario,* edited by Parrellada, 234–36. Copenhagen: International Work Group for Indigenous Affairs, 2007.

POJOAJU (Asociación de ONGs del Paraguay). "Caso de allanamiento de la Oficina de la ONG Iniciativa Amotocodie." Press release, December 3, 2010.

"La policia marca diciembra como un mes de extremo cuidado." *El día.* December 5, 2011.

Portes, Alejandro. "Rationality in the Slum: An Essay on Interpretive Sociology." *Comparative Studies in Society and History* 14 (1972).

Possuelos, Sydney, and Vincent Brackelaire. "Insumos para una estrategia regional de protección." In *Pueblos indígenas,* edited by Parellada, 192–203.

Postero, Nancy. *Now We Are Citizens: Indigenous Politics in Post-Multicultural Bolivia.* Stanford, CA: Stanford University Press, 2007.

Povinelli, Beth. *The Cunning of Recognition: Indigenous Alterities and the Making of Australian Multiculturalism.* Durham, NC: Duke University Press, 2002.

———. "The State of Shame: Australian Multiculturalism and the Crisis of Indigenous Citizenship." *Critical Inquiry* 24, no. 2 (Winter 1998): 575–610.

Pratt, Mary. *Imperial Eyes: Travel Writing and Transculturation.* New York: Routledge, 1992.

Prieto, Esther. "Indigenous Peoples in Paraguay." In *Indigenous Peoples and Democracy in Latin America,* edited by D. L. Van Cott, 235–50. Hampshire, UK: MacMillan Press, 1994.

Prins, Harald. *The Mik'maq: Resistance, Accommodation and Cultural Survival*. Fort Worth, TX: Cengage, 1996.

———. "Visual Media and the Primitivist Perplex: Colonial Fantasies and Indigenous Imagination in North America." In *Media Worlds: Anthropology on New Terrain*, edited by F. Ginsburg, L. Abu-Lughod, and B. Larkin, 58–74. Berkeley: University of California Press, 2002.

Raboy, Marc. "Radio as an Emancipatory Cultural Practice." *Semiotexte* 6, no. 1 (1993): 129–36.

Ramos, Alcida. *Indigenism: Ethnic Politics in Brazil*. Madison: University of Wisconsin Press, 1998.

Read, Kenneth. *The High Valley*. New York: Columbia University Press, 1965.

Regehr, Walter. "Movimientos mesiánicos entre los grupos étnicos del Chaco Paraguayo." *Suplemento antropológico* 16, no. 2 (1981): 105–17.

Renshaw, Jonathan. "The Effectiveness of Symbols Revisited: Ayoreo Curing Songs." *Tipiti* 4, nos. 1–2 (2006).

———. *The Indians of the Paraguayan Chaco: Identity and Economy*. Lincoln: University of Nebraska Press, 2002.

———. "La propiedad, los recursos naturales y el concepto de la igualdad entre los indígenas del Chaco paraguayo." *Suplemento antropológico* 23, no. 1 (1988).

Riester, Jurgen. *En busca de la loma santa*. Santa Cruz de la Sierra: Los Amigos del Libro, 1976.

Riester, Jurgen, and Jutta Weber. *Nomadas de la llanura, nomadas del Asfalto: Una autobiografia del pueblo Ayoreode*. Santa Cruz de la Sierra: Los Amigos del Libro, 1998.

Rivas, Alex. "Lost pueblos indígenas en aislamiento: Emergenica, vulnerabilidad y necesidad de protección." *Cultura y representaciones sociales* 1, no. 2 (2007): 73–90.

Robbins, Joel. *Becoming Sinners: Christianity and Moral Torment in a Papua New Guinea Society*. Berkeley: University of California Press, 2004.

———. "Continuity Thinking and the Problem of Christian Culture: Belief, Time and the Anthropology of Christianity." *Current Anthropology* 48, no. 1 (2007): 5–3.

Roca Ortiz, Irene. "Del Chaco Boreal a la periferia urbana: Etnicidad Ayoreode en la ciudad de Santa Cruz de la Sierra." *Villa libre: Cuaderno de estudios sociales urbanos* 3 (2008): 73–102.

Roca Ortiz, Irene, Tania Cutamiño Dosape, and Rocío Picaneré Chiqueno. "Aspectos generales de la situación del derecho a la salud del pueblo Ayoreode." In *Pigasipiedie iji yoquijoningai*, edited by Irene Roca, 101–278. Santa Cruz de la Sierra: APCOB, 2012.

Rodriguéz-Piñero, Luis. *Indigenous Peoples, Postcolonialism and International Law: The ILO Regime (1919–1989)*. Oxford: Oxford University Press, 2005.

Rojas, Raquel. *Ayoreas. Vida sexual y reproductiva*. Asunción, Paraguay: Servilibro, 2004.

Rosaldo, Michelle. "Toward an Anthropology of Self and Feeling." In *Culture Theory: Essays on Mind, Self and Emotion*, edited by Richard A. Shweder and Robert A. LeVine, 137–57. Cambridge: Cambridge University Press, 1984.

Rosaldo, Renato. *Culture and Truth: The Remaking of Social Analysis*. Boston: Beacon, 1989.

Rose, Nikolas. *The Politics of Life Itself: Biomedicine, Power and Subjectivity in the 21st Century*. Princeton, NJ: Princeton University Press, 2007.

Sahlins, Marshall. "Goodbye Tristes Tropes: Ethnography in the Context of the Modern World System." *Journal of Modern History* 65 (1993): 1–25.

Salzano, Francisco, et al. "Unusual Blood Genetic Characteristics among the Ayoreo Indians of Bolivia and Paraguay." *Human Biology* 50, no. 2 (1978): 121–36.

Santos, Ricardo Ventura. "Indigenous Peoples, Postcolonial Contexts and Genomic Research in the Late 20th Century: A View from Amazonia (1960–2000)." *Critique of Anthropology* 22, no. 1 (2002): 81–104.

Santos, Ricardo Ventura, and Marcos Maio. "Race, Genomics, Identities and Politics in Contemporary Brazil." *Critique of Anthropology* 24, no. 4 (2004): 347–78.

Schieffelin, Bambi. "Marking Time: The Dichotomizing Discourse of Multiple Temporalities." *Current Anthropology* 43 (2002): 5–17.

Schieffelin, Edward. "Anger and Shame in the Tropical Forest: On Affect as a Cultural System in Papua New Guinea." *Ethos* 11 (1983).

Schmeda-Hirschmann, Guillerm. "Magic and Medicinal Plants of the Ayoreos of the Chaco Boreal." *Journal of Ethnopharmacology* 39, no. 2 (1993): 105–11.

———. "Tree Ash as an Ayoreo Salt Source in the Paraguayan Chaco." *Economic Botany* 48, no. 2 (1994): 159–62.

Scott, Heidi. "Paradise in the New World: An Iberian Vision of Tropicality." *Cultural Geographies* 17 (2010): 77–101.

Sebag, Lucien. "Le chamanisme Ayoreo (I)." *L'Homme* 5, no. 1 (1965): 5–32.

———. "Le chamanisme Ayoreo (II)." *L'Homme* 5, no. 2 (1965): 92–122.

Serageldin, Ismail. "Cultural Heritage as Public Good: Economic Analysis Applied to Historic Cities." In *Global Public Goods*, 240–63. Oxford: Oxford University Press, 1999.

Sieder, Rachel. "Introduction." In *Multiculturalism in Latin America: Indigenous Rights, Diversity and Democracy*, edited by Rachel Sieder. New York: Palgrave Macmillan, 2002.

Sikkink, Lynn, and Choque M. Braulio. "Landscape, Gender, and Community: Andean Mountain Stories." *Anthropological Quarterly* 72, no. 4 (1999): 167–82.

Stahl, Uwe, and Hedwig Stahl. *Reconciliation*. 60-minute video. Vancouver: Cine-Aesthetics Canada, 1988.

Stavenhagen, Rodolfo. "Indigenous Peoples and the State in Latin America: An Ongoing Debate." In *Multiculturalism in Latin America*, edited by Rachel Sieder, 24–44. New York: Palgrave Macmillan, 2002.

———. "Preface." In *Pueblos indígenas*, edited by Parellada, 11–13.

Stearman, Allyn Maclean. *Yuquí: Forest Nomads in a Changing World*. Chicago: Holt, Rinehart and Winston, 1989.

Steininger, Michael, et al."Clearance and Fragmentation of Tropical Deciduous Forest in the Tierras Bajas, Santa Cruz, Bolivia." *Conservation Biology* 15, no. 4 (2001): 856–66.

Stewart, Kathleen. *Ordinary Affects*. Durham, NC: Duke University Press, 2007.

———. *A Space on the Side of the Road: Cultural Poetics in an "Other" America*. Princeton, NJ: Princeton University Press, 1996.

Stewart, Kathleen, and Susan Harding. "Bad Endings: American Apocalypsis." *Annual Review of Anthropology* 28 (1999): 285–310.

Stoler, Ann. "Affective States." In *A Companion to the Anthropology of Politics*, edited by David Nugent and Joan Vincent, 4–20. New York: Blackwell, 2004.

———. "Imperial Debris: Reflections on Ruins and Ruination." *Cultural Anthropology* 23, no. 2 (2008): 191–219.

———. "On Degrees of Imperial Sovereignty." *Public Culture* 18, no. 1 (2006): 125–46.

———. *Race and the Education of Desire: Foucault's History of Sexuality and the Colonial Order of Things*. Durham, NC: Duke University Press, 1995.

———. "Rethinking Colonial Categories: European Communities and the Boundaries of Rule." *Comparative Studies in Society and History* 31 (1989): 134–61.

Suaznabar, Berta. "Género e identidad étnica, situación y posición de género de la mujer ayoré en un contexto de cambios socioculturales." Cochabamba, Bolivia: Universidad Mayor de San Simón, unpublished MA thesis, 1995.

Survival International. "57000 Sign Petition to Save Uncontacted Ayoreo-Totobiegosode." August 8, 2007. http://www.survivalinternational.org /news/2486.

———. "Ayoreo Indian Dies after First Contact." May 7, 2008. http://www .survivalinternational.org/news/3287.

———. "International Ad Campaign Launched to Protect Uncontacted Tribe." April 12, 2010. http://www.survivalinternational.org/news/5789

———. "Congress rejects bill to protect Isolated Indians," April 8, 2005. http:// www.survivalfrance.org/news/351.

———. "Last Uncontacted Indians South of the Amazon Make Contact." March 9, 2004. http://www.survivalfrance.org/news/122.

———. "Study Reveals World's Highest Deforestation Rate on Uncontacted Tribe's Land," January 21, 2014. http://www.survivalinternational.org /news/9911.

———. "Uncontacted Tribe's Forest Bulldozed for Beef." November 9, 2009. http://www.survivalinternational.org/news/5212.

———. "Unprecedented Fine for Ranchers in Uncontacted Tribe 'Cover-Up.' " May, 2010. http://www.survivalinternational.org/news/5918.

Susnik, Branislava. "La lengua de los ayoweos-moros: Nociones generales." *Boletín de la Sociedad Científica del Paraguay y del Museo Etnográfico* 8 (1963).

————. *Los Aborigenes del Paraguay*. Vol. 1. Asunción, Paraguay: Museo Etnografico "Andres Barbero," 1963.

Swain, Mike. "The Lost Tribe of Green Hell: Cultures Clash in a Battle for the Future of the Planet." *Daily Mirror*, December 31, 2010.

Taliman, Valerie. "Borders and Native Peoples: Divided, But Not Conquered." *Native Americas* 18, no. 1 (2001): 10–16.

Taussig, Michael. *The Devil and Commodity Fetishism in South America*. Chapel Hill: University of North Carolina Press, 1980.

————. *Mimesis and Alterity: A Particular History of the Senses*. New York: Routledge, 1993.

————. *My Cocaine Museum*. Chicago: University of Chicago Press, 2004.

————. *The Nervous System*. New York: Francis and Taylor, 1992.

————. *Shamanism, Colonialism and the Wild Man: A Study in Terror and Healing*. Chicago: University of Chicago Press, 1987.

Tax, Sol. "Action Anthropology." *Current Anthropology* 16, no. 4 (1975): 171–77.

Terceros Cuéllar, Elba. "Proceso de saneamiento de las tierras comunitarias de origen Ayoreo." In *Atlas de territorios indígenas en Bolivia: Situación de las tierras comunitarias de origen (TCOs) y proceso de titulación*, edited by José A. Martínez, 191–96. Santa Cruz de la Sierra: Centro de Planificación Territorial Indígena de la CIDOB, 2000.

Throop, Jason. *Suffering and Sentiment: Exploring the Vicissitudes of Experience and Pain in Yap*. Berkeley: University of California Press, 2010.

Trouillot, Michel Rolf. "Anthropology and the Savage Slot: The Poetics and Politics of Otherness." In *Recapturing Anthropology: Working in the Present*, edited by Richard Fox. Santa Fe, NM: School of American Research Press, 1991.

————. *Silencing the Past: Power and the Production of History*. Boston: Beacon, 1995.

Tsing, Anna. *Friction: An Ethnography of Global Connection*. Princeton, NJ: Princeton University Press, 2005.

Turner, Terence. "Anthropology and Multiculturalism: What Is Anthropology that Multiculturalists Should be Mindful of It?" *Cultural Anthropology* 8, no. 4 (1993): 411–29.

————. "The Crisis of Late Structuralism: Perspectivism and Animism: Rethinking Culture, Nature, Spirit and Bodiliness" *Tipiti* 7, no. 1 (2009): 3–42.

————. "Indigenous and Culturalist Movements in the Contemporary Global Conjuncture." In *Las identidades y las tensiones culturales de la modernidad*, edited by Francisco Del Riego et al. Santiago de Compostela, Spain: Federación de asociaciones de antropología del estado español, 1999.

————. "Representing, Resistance, Rethinking: Historical Transformation of Kayapó Culture and Anthropological Consciousness." In *Colonial Situations: Essays on the Contextualization of Ethnographic Knowledge; History of Anthropology*, vol. 7, edited by George Stocking, 285–313. Madison: University of Wisconsin Press, 1991.

Ultima Hora. "ONG reclama al gobierno de Paraguay proteger a los Ayoreos." August 9, 2007.

———. Untitled poem. July 11, 2009.

United Nations Development Programme. "Regional Human Development Report for Latin America and the Caribbean." New York: United Nations Development Programme, 2010.

United Nations Human Rights Council. *Draft Guidelines on the Protection of Indigenous Peoples in Voluntary Isolation and in Initial Contact of the Amazon Basin and El Chaco.* Geneva: United Nations document, 2009.

Urban, Greg. *Metaculture: How Culture Moves through the World.* Minneapolis: University of Minnesota Press, 2001.

Urban, Greg, and Joel Sherzer. *Nation-States and Indians in Latin America.* Austin: University of Texas Press, 1991.

Van Cott, Dorothy. *Indigenous Peoples and Democracy in Latin America.* New York: St. Martin's Press, 1994.

Vaudry, Jean. "Relación histórica sobre la reducción de San Ignacio de Zamucos." *Boletín de la Sociedad Geográfica de Sucre* 30 (1933): 324–26.

Vellard, Jehan. "Indiens Ayoreo du Grand Chaco." *Revue de l'Académie des Sciences* Série D, 274 (1972): 3245–47.

Vidal, John. "Natural History Museum Expedition Poses Genocide Threat to Paraguay Tribes." *Guardian*, November 8, 2010.

———. "Natural History Museum's Expedition to Arid Chaco Halted." *Guardian*, November 15, 2010.

———. "Paraguayan Government Raids Offices of NGO Opposing Chaco Research Trip." *Guardian*, December 15, 2010.

Vilaça, Aparecida. "Cultural Changes, Body Metamorphosis." In *Time and Memory in Indigenous Amazonia*, edited by Carlos Fausto and Michal Heckenberger, 169–93. Gainesville: University Press of Florida, 2007.

———. *Strange Enemies: Indigenous Agencies and Scenes of Encounter in Amazonia.* Durham, NC: Duke University Press, 2010.

Viveiros de Castro, Eduardo. "Cosmological Deixis and Amerindian Perspectivism." *Journal of the Royal Anthropological Institute* 4 (1998): 469–88.

———. *Cosmological Perspectivism in Amazonia and Elsewhere.* Manchester, UK: HAU Network of Ethnographic Theory, 2012.

———. "Perspectival Anthropology and the Method of Controlled Equivocation." *Tipiti* 2, no. 1 (2004): 3–22.

———. "The Untimely, Again." Introduction to *Archeology of Violence*, by Pierre Clastres, 9–52. Los Angeles: Semiotext(e), 2010.

Vysokolan, Oleg. "Los Totoviegosode Silvicolas: Entre el ganado vacuno y la biodiversidad." In *Pueblos indígenas*, edited by Parellada, 238–51.

Wacquant, Loïc. "Class, Race and Hyperincarceration in Revanchist America." *Daedalus* 139, no. 3 (2010): 74–90.

———. "The Rise of Advanced Marginality: Notes on Its Nature and Implications." *Acta Sociologica* 39 (1996): 121–39.

———. "Three Steps to a Historical Anthropology of Actually Existing Neoliberalism." *Social Anthropology* 20, no. 1 (2012): 66–79.

————. "Urban Desolation and Symbolic Denigration in the Hyperghetto." *Social Psychology Quarterly* 73, no. 3 (2010): 215–19.

————. *Urban Outcasts: A Comparative Sociology of Advanced Marginality.* Cambridge: Polity Press, 2008.

Wagner, Chuck. *Defeat of the Bird God, The Story of Missionary Bill Pencille, Who Risked His Life to Reach the Ayorés of Bolivia.* Grand Rapids, MI: Zondervan, 1967.

Warren, Kay. *Indigenous Movements and Their Critics: Pan Maya Activism in Guatemala.* Princeton, NJ: Princeton University Press, 1998.

Warren, Kay, and Jean Jackson, eds. *Indigenous Movements, Self-Representation and the State in Latin America.* Austin: University of Texas Press, 2002.

White, Hayden. "Forms of Wildness: Archaeology of an Idea." In *The Wild Man Within,* 3–38. Pittsburgh: University of Pittsburgh Press, 1972.

Wilbert, Johannes, and Karin Simoneau, eds. *Folk Literature of the Ayoreo Indians.* UCLA Latin American Studies, vol. 70. Los Angeles, 1989.

Williams, Raymond. *Marxism and Literature.* Oxford: Oxford University Press, 1977.

Wilmsen, Edwin. *Land Filled with Flies: A Political Economy of the Kalahari.* Chicago: University of Chicago Press, 1989.

Wolf, Eric. *Europe and the People without History.* Berkeley: University of California Press, 1982.

World Council of Churches. *Declaration of Barbados: For the Liberation of the Indians.* Document 1. Copenhagen: International Work Group for Indigenous Affairs, 1971.

Yashar, Deborah. *Contesting Citizenship in Latin America: The Rise of Indigenous Movements and the Postliberal Challenge.* Cambridge: Cambridge University Press, 2005.

Zak, Marcelo R., Marcelo Cabido, and John G. Hodgson. "Do Subtropical Seasonal Forests in the Gran Chaco, Argentina, Have a Future?," *Biological Conservation* 120, no. 4 (2004): 589–98.

Zanardini, José, ed. *Cultura del pueblo Ayoreo: Manual para docentes.* Asunción, Paraguay: CEADUC, 2003.

Zarzar, Alonso. *Tras las huellas de un antiguo presente: La problemática de los pueblos indígenas amazónicos en aislamiento y en contacto inicial.* Lima: Defensoría del pueblo, 2000.

Zolezzi, Graciela, and Carmina Sanabria. *Mujeres Ayoreas.* Unpublished text, n.d.

Zook, David H., Jr. *The Conduct of the Chaco War, 1932–1935.* New York: Bookman Associates, 1960.

Index

Aasi, 40, 84, 89–90, 104, 105, 134, 141, 159, 164, 221, 249n27
Abercrombie, Tom, 128, 146, 224–25, 228
Abujá (pl. *Abujádie*): anthropologists as, 22–29, 35, 39–41, 45–48, 270n48; applied to author, 22–29, 46, 52–53, 92; meaning of term, 22–23; mimetic excess of, 244n46; as term, 237n1; as trickster figure, 24, 45–46, 152, 163
Abujei, 156
Aché people, 85, 104–5
Achinguirai, 6, 126, 162, 163
adode myths, 24, 219; anthropologists' interest in, 23; Christian redefining of word, 45; post-contact Ayoreo views of, 44, 152; taboos recounted in, 175; as transformation narratives, 35–36, 37, 48
Adorno, Theodor, 124, 251n40
Agamben, Giorgio, 157
Aguilar, Jaime, 63, 247n15
AIDESEP, 265n12
ajengome (shame), 149–50, 152, 156–57, 163, 170
Ajidababia, 133
ajingaque (righteous anger), 149
Ajnisidai, 74–75
Ajua'nate, 163
Amnesty International, 207–8
Amomegosode Ayoreo, 55, 77
Amo'nate, 41–44

anthropologists: biolegitimacy and, 177–78, 189, 226–27; cultural project of, 39–41, 45–48, 140, 213–14; on Devil imagery of indigenous peoples, 25–32, 238n4; payment by for cultural performances, 25, 28–32, 48; shame and, 171; Totobiegosode perceptions of, 22–29, 45–48
anthropology: of affect, 258n1; binary societal schema in, 6–7; of emotion, 259n2; of indigeneity, 86; moral, 227; ontological, 19–20, 36–39; perspectivist, 19, 36, 40; political, 7, 11, 86, 107–8, 215, 228–29; primitivist narratives in, 7–8, 14–15, 19; salvage ethnography, 27–32, 46–47; tradition as ethnographic fetish, 32–41
apocalyptic futurism: appeal of, 144–46; bulldozers as sign and vehicle of, 128–34; in Christian belief, 136, 137; concept of, 256nn3–4; memory negated by, 139–42; miracle searches and, 142–43; negation and, 191, 228; New Tribes Mission philosophy and, 6, 15, 94–95; as ontological necessity for Ayoreo, 128–29, 136, 143–46; radical rupture as fulfilled prophecies, 16–17, 224; solar eclipse event and, 130–31; sympathetic magic's failure and, 135–36; temporal notions,